Lecture Notes in Economics and Mathematical Systems 613

Ahmad K. Naimzada · Silvana Stefani ·
Anna Torriero (Eds.)

Networks, Topology and Dynamics

Theory and Applications to Economics and Social Systems

Springer

Prof. Ahmad K. Naimzada
Università degli Studi di
Milano-Bicocca
Dipto. Economia Politica
Piazza dell' Ateneo Nuovo, 1
20126 Milano
Italy
ahmad.naimzada@unimib.it

Prof. Anna Torriero
Università Cattolica del Sarcro
Cuore
Largo A. Gemelli, 1
20123 Milano
Italy
anna.torriero@unicatt.it

Prof. Silvana Stefani
Università degli Studi di
Milano-Bicocca
Dipto. Economia Politica
Piazza dell' Ateneo Nuovo, 1
20126 Milano
Italy
silvana.stefani@unimib.it

ISBN 978-3-540-68407-7 ISBN 978-3-540-68409-1 (eBook)

DOI 10.1007/978-3-540-68409-1

Lecture Notes in Economics and Mathematical Systems ISSN 0075-8442

Library of Congress Control Number: 2008928718

The volume has been published with the financial contribution of the Quantitative Methods Department of the University of Milano-Bicocca, Italy.

Cover design: WMXDesign GmbH, Heidelberg

Printed on acid-free paper

9 8 7 6 5 4 3 2 1

springer.com

Preface

There is convergent consensus among scientists that many social, economic and financial phenomena can be described by a network of agents and their interactions. Surprisingly, even though the application fields are quite different, those networks often show a common behaviour. Thus, their topological properties can give useful insights on how the network is structured, which are the most "important" nodes/agents, how the network reacts to new arrivals. Moreover the network, once included into a dynamic context, helps to model many phenomena. Among the topics in which topology and dynamics are the essential tools, we will focus on the diffusion of technologies and fads, the rise of industrial districts, the evolution of financial markets, cooperation and competition, information flows, centrality and prestige.

The volume, including recent contributions to the field of network modelling, is based on the communications presented at NET 2006 (Verbania, Italy) and NET 2007 (Urbino, Italy); offers a wide range of recent advances, both theoretical and methodological, that will interest academics as well as practitioners.

Theory and applications are nicely integrated: theoretical papers deal with graph theory, game theory, coalitions, dynamics, consumer behavior, segregation models and new contributions to the above mentioned area. The applications cover a wide range: airline transportation, financial markets, work team organization, labour and credit market.

The volume can be used as a reference book for graduate and postgraduate courses on Network Theory and Complex Systems in Faculties of Economics, Mathematics, Engineering and Social Sciences. In Part I, the invited tutorials introduce Graph Theory from the theoretical point of view (Marusic) and the possible applications to economics (Battiston). In Part II, the contributions cover local and global interaction, complex behavior, network games, while in Part III they refer to Markov chains and topology. The applications are all placed in Part IV.

Fifteen papers have been selected among roughly thirty submitted extended abstracts; each paper has been reviewed by two referees. Space limitations are the main reason why no more papers have been accepted, although many of them were really interesting.

We are grateful to the scholars who have made NET 2006, NET 2007 and this book possible, to the members of the scientific committees of the two conferences and the referees:

- Gian-Italo BISCHI – University of Urbino, Italy
- Sergio CURARINI – University of Venice, Italy
- Domenico DELLI GATTI – Catholic University of Milan, Italy
- Andrea DE MONTIS – University of Sassari, Italy
- Mauro GALLEGATI – University of Ancona, Italy
- Laura GARDINI – University of Urbino, Italy
- Rosanna GRASSI – University of Milan-Bicocca, Italy
- Josef LAURI – University of Malta, Malta
- Raimondo MANCA – University of Rome "La Sapienza", Italy
- Marco A. MARINI – University of Urbino, Italy
- Dragan MARUSIC – University of Primorska and Liublijana, Slovenia
- Fausto MIGNANEGO – Catholic University, Italy
- Antonio PALESTRINI – University of Teramo, Italy
- Arturo PATARNELLO – University of Milan-Bicocca, Italy
- Aura REGGIANI – University of Bologna, Italy
- Massimo RICOTTILLI – University of Bologna, Italy
- Giulia RIVELLINI – Catholic University of Milan, Italy
- Pierluigi SACCO – IUAV Venezia, Italy
- Raffaele SCAPELLATO – Polytechnic of Milan, Italy
- Christos H. SKIADAS – Technical University of Crete, Greece
- Fabio TRAMONTANA – Marche Polytechnic University, Italy
- Giovanni ZAMBRUNO – University of Milan-Bicocca, Italy
- Luca ZARRI – University of Verona, Italy

Finally, we would like to acknowledge the support, hospitality and encouragement of the following institutions:

- Dipartimento di Metodi Quantitativi per le Scienze Economiche e Aziendali, Università degli Studi di Milano-Bicocca,
- Dipartimento di Discipline Matematiche, Finanza Matematica ed Econometria, Università Cattolica di Milano
- Istituto di Scienze Economiche, Università degli Studi di Urbino.

March 2008
Milan

Ahmad K. Naimzada
Silvana Stefani
Anna Torriero

Contents

Part I
Tutorials

Some Topics in Graph Theory

Klavdija Kutnar and Dragan Marušič

Abstract In this short introductory course to graph theory, possibly one of the most propulsive areas of contemporary mathematics, some of the basic graph-theoretic concepts together with some open problems in this scientific field are presented.

1 Some Basic Concepts

A *simple graph* X is an ordered pair of sets $X = (V, E)$. Elements of V are called *vertices* of X and elements of E are called *edges* of X. An edge joins two vertices, called its endvertices. Formally, we can think of the elements of E as subsets of V of size 2. A simple graph is thus an undirected graph with no loops or multiple edges.

If $u \neq v$ are vertices of a simple graph X and $\{u, v\}$ (sometimes shortened to uv) is an edge of X, then this edge is said to be *incident to u and v*. Equivalently, u and v are said to be *adjacent* or *neighbors*, and we write $u \sim v$. Phrases like, *"an edge joins u and v"* and *"the edge between u and v"* are also commonly used.

Graphs can be nicely represented with diagrams consisting of dots standing for vertices and lines standing for edges (see Fig. 1).

In a simple graph there is at most one edge joining a pair of vertices. In a *multigraph*, multiple edges are permitted between pairs of vertices. There may also be edges, called *loops*, that connect a vertex to itself (see Fig. 2).

As opposed to a simple graph where edges are undirected, a *directed graph* (in short, *digraph*) is an ordered pair of sets (V, E) where V is a set of vertices and E is

K. Kutnar
University of Primorska, FAMNIT, Glagoljaška 8, 6000 Koper, Slovenia
klavdija.kutnar@upr.si

D. Marušič
University of Primorska, FAMNIT, Glagoljaška 8, 6000 Koper, Slovenia and University of Ljubljana, IMFM, Jadranska 19, 1000 Ljubljana, Slovenia
dragan.marusic@upr.si

A.K. Naimzada et al. (eds.), *Networks, Topology and Dynamics*,
Lecture Notes in Economics and Mathematical Systems 613,

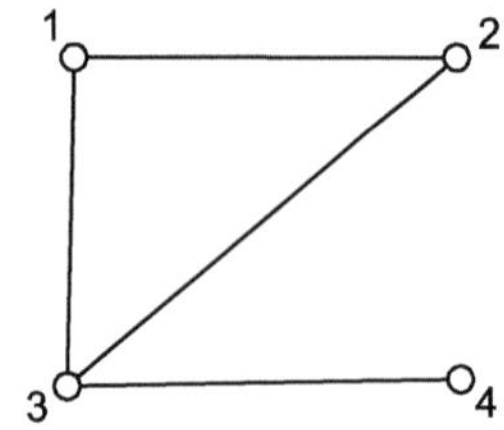

Fig. 1 A diagram of a graph $X = (V, E)$ where $V = \{1,2,3,4\}$ and $E = \{\{1,2\},\{1,3\},\{2,3\},\{3,4\}\}$

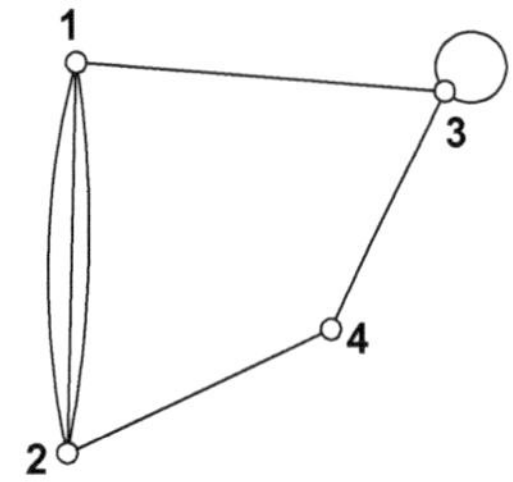

Fig. 2 An example of a multigraph with three edges between vertices 1 and 2 and loop on vertex 3

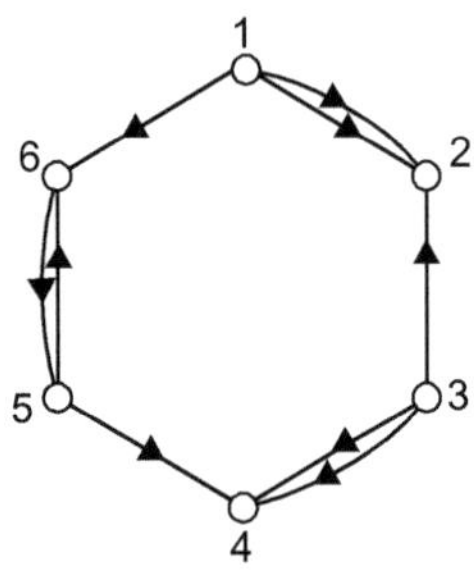

Fig. 3 A directed graph

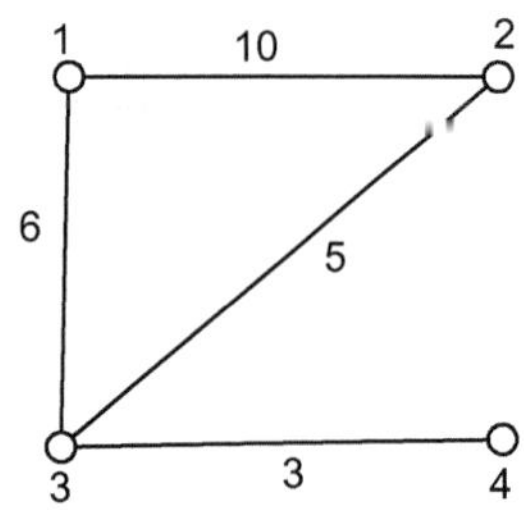

Fig. 4 A weighted graph

a subset of ordered pairs of vertices from V. Now the edges may be thought of as arrows going from a *tail (vertex)* to a *head (vertex)* (see Fig. 3).

Sometimes it is useful to associate a number, often called its *weight*, with each edge in a graph. Such graphs are called *edge-weighted* or simply *weighted graphs*; they may be simple, directed, etc. (see Fig. 4).

From now on by a graph we shall mean a simple and, unless otherwise specified, finite, undirected and connected graph. Let $X = (V, E)$ be a graph. The *degree* $d(v)$ of a vertex $v \in V$ is the number of edges with which it is incident. The set of

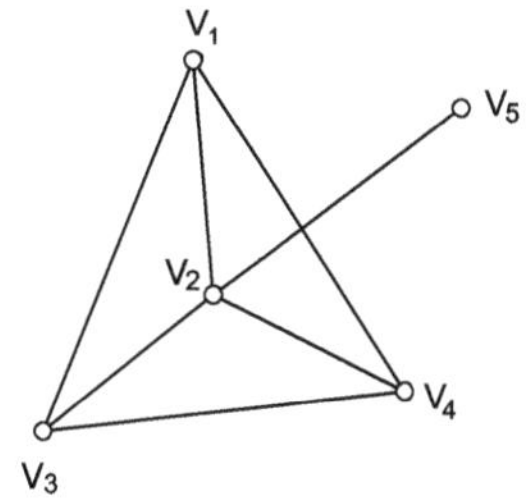

Fig. 5 $|N(v_1)| = |N(v_3)| = |N(v_4)| = 3$, $|N(v_2)| = 4$ and $|N(v_5)| = 1$

neighbors $N(v)$ of a vertex v is the set of vertices adjacent to v. Hence, $d(v) = |N(v)|$ (see Fig. 5). In the case of a directed graph $X = (V, E)$ the *indegree* and the *outdegree* of a vertex $v \in V$ is the number of edges having v as a head vertex and tail vertex, respectively. For example the vertex 5 in the graph shown in Fig. 3 has indegree 1 and outdegree 2.

When people at a party shake hands the total number of hands shaken is equal to twice the number of handshakes. Representing the party by a graph (with each person represented by a vertex and a handshake between two people represented by an edge between the corresponding vertices), the above fact, translated into graph-theoretic terminology, reads as follows.

Proposition 1 (Handshaking lemma). *The sum of degrees of all vertices in a simple graph equals twice the number of edges.*

Proof. Every edge contributes two to the sum of the degrees, one for each of its endvertices. □

As a corollary we have the following result.

Corollary 1. *In every graph, there is an even number of vertices of odd degree.*

Proof. Partitioning the vertices into those of even degree and those of odd degree, we know

$$\sum_{v \in V} d(v) = \sum_{d(v)\,\text{is odd}} d(v) + \sum_{d(v)\,\text{is even}} d(v)$$

The value of the left-handside of this equation is even, and the second summand on the right-handside is even since it is entirely a sum of even values. So the first summand on the right-handside must also be even. But since it is entirely a sum of odd values, it must contain an even number of terms. In short, there must be an even number of vertices with odd degree. □

A graph X is *d-regular* if all vertices have the same degree d. A 3-regular simple graph is usually called a *cubic graph.*

Corollary 2. *Every cubic graph has an even number of vertices.*

A *subgraph* of a graph X is a graph having all of its vertices and edges in X (see Fig. 6). A *spanning subgraph* of X is a subgraph containing all vertices of X.

The *complement* $\overline{X}$ of a graph X has $V(X)$ as its vertex set, and two vertices are adjacent in $\overline{X}$ if and only if they are not adjacent in X (see Fig. 7).

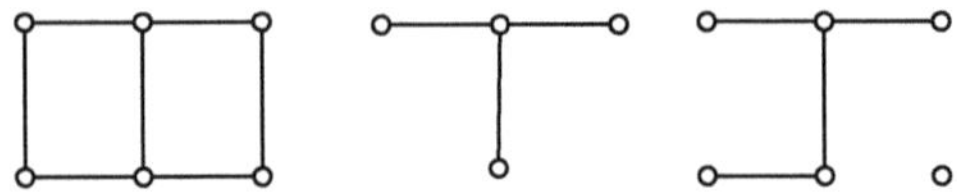

Fig. 6 A graph and two of its subgraphs

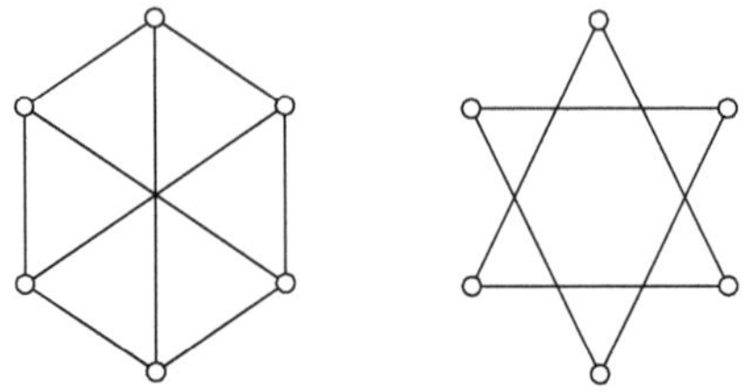

Fig. 7 A graph and its complement

Exercise 1. (a) Rewrite the proof of Corollary 1 more carefully as an inductive proof on the number of edges in a simple graph. (b) Extend Corollary 1 to multigraphs. (c) Extend Corollary 1 to digraphs.

Exercise 2. Show that a connected graph with n vertices and degree d, where $2 \leq d \leq (n-1)/2$, contains an induced 3-path. *(An induced subgraph is a subset of the vertices of a graph together with any edges whose endpoints are both in this subset. The m-path is a connected graph with two vertices of degree* 1*, and the other* $m-2$ *vertices of degree* 2.*)*

2 Traversability

A *walk* (or v_0v_k-*walk*) of a graph X is a sequence of vertices

$$v_0, v_1, v_2, \ldots, v_k$$

such that v_i is adjacent to v_{i+1} for every $i \in \{0,1,2,\ldots,k-1\}$. It is *closed* if $v_0 = v_k$, and it is *open* otherwise. It is a *path* if all the vertices are distinct. It is a *trail* if all the edges are distinct. A graph is *connected* if there is a path connecting any two distinct vertices, and it is *disconnected* otherwise. A *cycle* (also called a *circuit*) in a graph is a closed path $uv_1,\ldots,v_{k-1}u$ in which all u, v_i, $i \in \{1,\ldots,k-1\}$, are distinct. A *tree* is a connected graph with no cycles.

Let us give a few examples of some well known graphs. A *cycle* C_n is a connected graph with n vertices in which each vertex is of degree 2. A *complete graph* K_n is a graph with n vertices where any two vertices are adjacent. A *bipartite graph* is a graph whose vertex set V can be partitioned into two subsets Y and Y' in such a way that each edge has one end in Y and the other in Y'. Such a partition (Y,Y') is called a *bipartition* of the graph. A *complete bipartite graph* is a simple bipartite graph with bipartition (Y,Y') in which each vertex of Y is joined to each vertex of Y'. If $|Y| = m$ and $|Y'| = n$, such a graph is denoted by $K_{m,n}$ (see Fig. 8).

Fig. 8 Examples

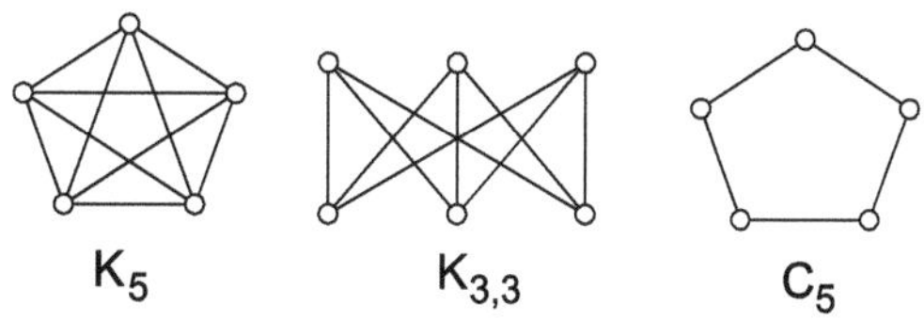

Proposition 2. *For a simple graph $X = (V,E)$, the following are equivalent:*

(1) X is connected and $|E| = |V| - 1$.
(2) X is connected and acyclic (X is a tree).
(3) X is connected, but removing any edge from X leaves a disconnected graph.
(4) There is a unique simple path between any two distinct vertices of X.

2.1 Graphs and Matrices

Let X be a graph. The *adjacency matrix* of X relative to the vertex labeling $v_1, v_2, \dots, v_n$ is the $n \times n$ matrix $A(X)$ whose entries a_{ij} are given by the rule

$$a_{ij} = \begin{cases} 1, & \text{if } v_i \text{ and } v_j \text{ are adjacent} \\ 0, & \text{otherwise} \end{cases}.$$

For example, the graph K_4 has adjacency matrix

$$A(K_4) = \begin{bmatrix} 0 & 1 & 1 & 1 \\ 1 & 0 & 1 & 1 \\ 1 & 1 & 0 & 1 \\ 1 & 1 & 1 & 0 \end{bmatrix}.$$

Note that the adjacency matrix of an undirected graph is symmetric. The following result gives one important use of powers of the adjacency matrix of a graph (see for example [12, 40]).

Proposition 3. *If A is the adjacency matrix of a graph X of order n relative to the vertex labeling $v_1, v_2, \dots, v_n$, the (i,j)-entry of A^r represents the number of distinct r-walks from vertex v_i to vertex v_j in the graph.*

Taking the square of the matrix $A(K_4)$ above gives

$$A^2 = A(K_4)^2 = \left(\begin{bmatrix} 0 & 1 & 1 & 1 \\ 1 & 0 & 1 & 1 \\ 1 & 1 & 0 & 1 \\ 1 & 1 & 1 & 0 \end{bmatrix} \right)^2 = \begin{bmatrix} 3 & 2 & 2 & 2 \\ 2 & 3 & 2 & 2 \\ 2 & 2 & 3 & 2 \\ 2 & 2 & 2 & 3 \end{bmatrix}.$$

The resulting matrix gives us the number of different paths using two edges between the vertices of the graph K_4.

We end this subsection with the following simple exercise.

Exercise 3. How many different paths using three edges between two vertices of the graph K_5 exist?

2.2 Eulerian Graphs

A bit of history: Koenigsberg was a city in Prussia situated on the Pregel River, which served as the residence of the dukes of Prussia in the sixteenth century. Today, the city is named Kaliningrad, and is a major industrial and commercial center of western Russia. The river Pregel flowed through the town, dividing it into four regions. In the eighteenth century, seven bridges connected the four regions. Koenigsberg people used to take long walks through town on Sundays. They wondered whether it was possible to start at one location in the town, travel across all the bridges without crossing any bridge twice and return to the starting point. This problem was first solved by the prolific Swiss mathematician Leonhard Euler, who, as a consequence of his solution invented the branch of mathematics now known as graph theory. Euler's solution consisted of representing the problem by a graph with the four regions represented by four vertices and the seven bridges by seven edges as shown in Fig. 9. Stated as a general graph theory problem, the problem is to construct a closed walk of the graph that traverses every edge exactly once. By Proposition 4 it follows that the trail of Koenigsberg bridges is not possible.

An *Eulerian trail* of an undirected graph X is a closed walk that traverses every edge of X exactly once.

Proposition 4. *If an undirected graph X has Eulerian trail, then X is connected and every vertex has even degree.*

Proof. The Eulerian trail naturally orients every edge. We enter a vertex as many times as we leave it. So every vertex must have indegree equal to outdegree. In particular, the degrees are all even. □

Proposition 5. *If X is a connected graph and every vertex has even degree, then X has an Eulerian trail.*

Proof. If X has only one vertex then it has an empty (1-vertex) Eulerian trail. So assume that in X each vertex has (even) degree at least 2. The result is clearly true if all vertices are of degree 2, as then X is a cycle. By induction, assume the result is true for all graphs with "average" degree smaller than that of X. As X is not a tree we can find a cycle C in X. Then by induction, $X - C$ has an Eulerian trail which together with C gives us the desired Eulerian trail in X. □

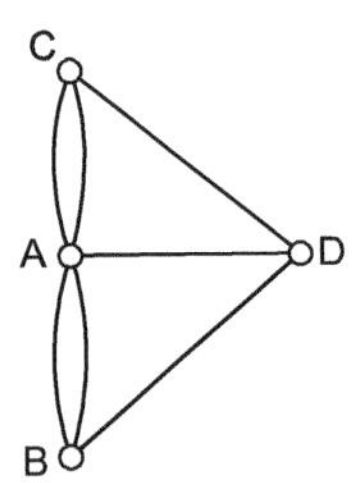

Fig. 9 Koenigsberg bridges in a graph

Exercise 4. Convince yourself that the following graphs are Eulerian: the cycle C_n for every n; the complete graph K_n if and only if n is odd; the bipartite graph $K_{n,m}$ if and only if both n and m are even.

2.3 Hamiltonian Graphs

A simple cycle that traverses every vertex exactly once is called a *Hamiltonian cycle* (*Hamiltonian circuit*). Similarly, a simple path that traverses every vertex exactly once is a *Hamiltonian path*. A *Hamiltonian graph* is a graph that possesses a Hamiltonian cycle.

Proposition 6 (Dirac theorem [31]). *Any graph with n vertices, $n \geq 3$, in which the minimum degree of each vertex is at least $n/2$, has a Hamiltonian cycle.*

Proof. Suppose by contradiction, that some non-Hamiltonian graph has n vertices, $n \geq 3$, and that the minimum degree of each vertex is at least $n/2$. If we add edges to this graph one at a time, we eventually end up with a complete graph, which does have a Hamiltonian cycle. Somewhere along this process we get a graph, call it X, that does not have a Hamiltonian cycle, but adding an edge uv yields a Hamiltonian graph, call it X'. We now get a contradiction for the graph X.

Since X' is Hamiltonian, X must have a Hamiltonian path $u = u_1, \dots, u_n = v$ joining u and v. By definition this path includes all the vertices of X. Now let us play with this path and turn it into a cycle, and thus get a contradiction. We use the fact that u and v each have degree at least $n/2$ to produce two edges to replace an edge (u_i, u_{i+1}) on this path.

Let us count: of the $n-2$ intermediate vertices on the path $u_2, \dots, u_{n-1}$ we know that at least $n/2$ are neighbors of u, and at least $n/2$ are neighbors of v. Consequently, by the Pigeon hole principle, there are two adjacent vertices, u_i and u_{i+1}, where u_i is a neighbor of v and u_{i+1} is a neighbor of u. Namely, let $S = \{i \mid u_{i+1} \sim u\}$ and $T = \{i \mid u_i \sim v\}$. Then each of S and T has at least $n/2$ elements. Since there are only $n-2$ possible values of i, some i must be in both sets. That is, u_i is a neighbor of v and u_{i+1} is a neighbor of u. To produce a Hamiltonian cycle in X is now an easy exercise, giving us the desired contradiction. □

To wrap up this subsection we give Hamilton-flavored exercises.

Exercise 5. Prove that the Petersen graph (the graph shown in Fig. 10) has no Hamiltonian cycle.

Exercise 6. Given any two vertices u and v of the dodecahedron, the graph shown in Fig. 11. Is there a Hamiltonian path between these two vertices? *(This problem is known as Hamilton's Icosian game proposed by W. R. Hamilton in T. P. Kirkman in 1857.)*

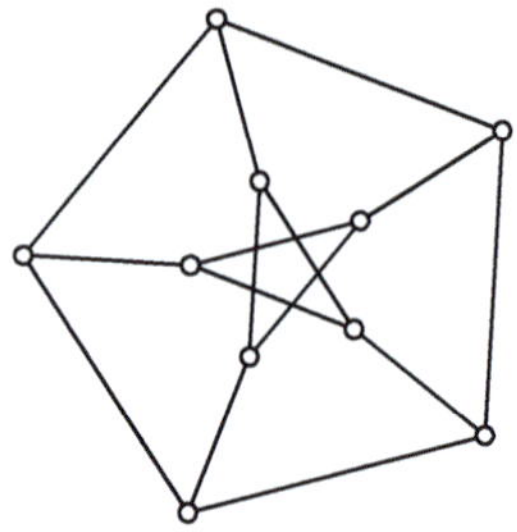

Fig. 10 The Petersen graph

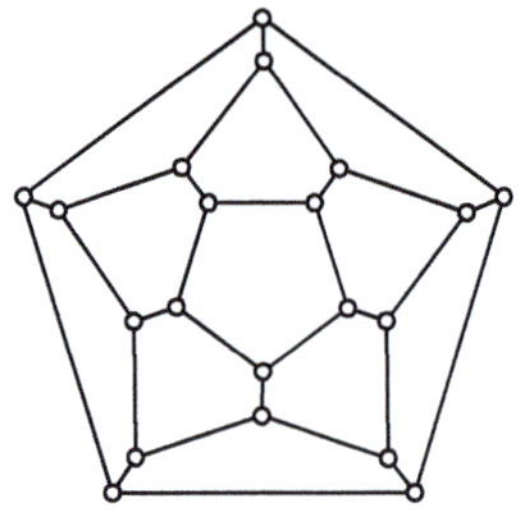

Fig. 11 The dodecahedron

Exercise 7. Prove that a graph with n vertices ($n > 3$) has a Hamiltonian cycle if for each pair of non-adjacent vertices, the sum of their degrees is n or greater. *(This problem is known as Ore theorem [62].)*

3 Factorizations, Colorings and Tournaments

A *factor* of a graph X is a spanning subgraph of X which is not totally disconnected. (A *totally disconnected* graph of order n is the complement $nK_1 = \bar{K}_n$ of the complete graph K_n.) We say that X is the *sum* of factors X_i if it is their edge-disjoint union, and such a union is called a *factorization* of X. An *n-factor* is a factor which is an n-regular graph. If X is the sum of n-factors, their union is called an *n-factorization* and X itself is *n-factorable*.

When $X = (V, E)$ has a 1-factor it is clear that $|V|$ is even and the edges of this 1-factor are vertex disjoint. In particular, K_{2n+1} cannot have a 1-factor, but K_{2n} certainly can as is shown in the proposition below.

Proposition 7. *The complete graph K_{2n} is 1-factorable.*

Proof. We need only display a partition of the set E of edges of K_{2n} into $(2n-1)$ 1-factors. Denote the vertices of K_{2n} by $v_1, v_2, \ldots, v_{2n}$. Define, for $i \in \{1, 2, \ldots, n-1\}$, the sets of edges

$$E_i = \{v_i v_{2n}\} \cup \{v_{i-j} v_{i+j} \mid j \in \{1, 2, \ldots, n-1\}\},$$

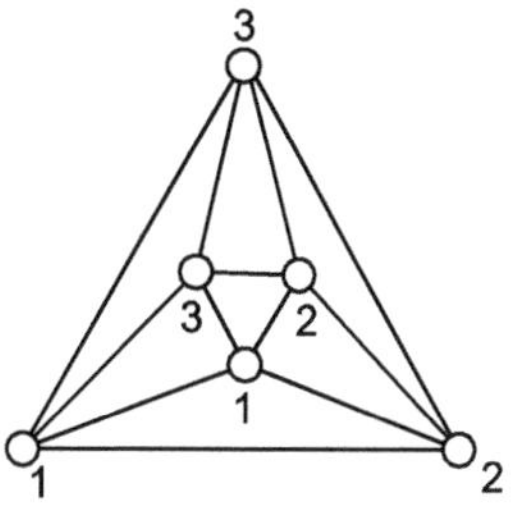

Fig. 12 Not a good scheduling

where each of the subscripts $i-j$ and $i+j$ is expressed as one of the numbers $1,2,\ldots,(2n-1)$ modulo $(2n-1)$. The collection $\{E_i\}$ is easily seen to give an appropriate partition of E, and the sum of the subgraphs X_i induced by E_i is a 1-factorization of K_{2n}. □

An example where the vertex colorings may be used is the exam scheduling problem. The Schedules Office needs to assign a time slot for each final exam. This is not easy, because some students are taking several classes with finals, and a student can take only one test during a particular time slot. The Schedules Office wants to avoid all conflicts, but wants to make the exam period as short as possible. Let each vertex represent a course for which final exams are taken. Put an edge between two vertices if there is some student taking both courses. Identify each possible time slot with a color. For example, Monday 9–12 is color 1, Monday 1–4 is color 2, Tuesday 9–12 is color 3, etc. If there is an edge between two vertices with the same color, then a conflict exam will have to be scheduled because there is a student who has to take exams for the courses represented by the vertices, but the exams are scheduled at the same time (see Fig. 12). Everyone wants to avoid conflict exams. So the registrar would like to color each vertex of the graph so that no adjacent vertices have the same color. To keep exam period as short as possible, the registrar would like to use the minimum possible number of colors.

The minimum number of colors needed to color the vertices of a graph X so that no two adjacent vertices are of the same color is called the *chromatic number* and is denoted by $\chi(X)$. For example: $\chi(C_n)=3$ if n is odd and $\chi(C_n)=2$ if n is even.

The following propositions gives the upper bound on the chromatic number.

Proposition 8. *For any graph X, $\chi(X)\leq 1+\Delta$ where Δ is the maximum degree of X.*

Proof. We use induction on the order n of the graph X. The only graph of order $n=1$ is the complete graph K_1 in which case $\Delta=1$ and $\chi(K_1)=1$. Now suppose that the result holds for all graphs of order less than or equal to $n-1$, and let X be a graph of order n and maximal degree Δ. Let Y be the graph obtained from X by deleting a vertex $v\in V(X)$ and all the edges having v as an endvertex. Since the order of Y is less than n and its maximal degree is less than or equal to Δ, by induction, $\chi(Y)\leq 1+\Delta$. Now a vertex coloring of X is obtained by coloring the vertex v with the color that is not the color of the neighboring vertices of v. Since v has at most Δ neighbors X is colored with at most $1+\Delta$ colors. □

Proposition 9 (Brook theorem [17]). *For a connected graph X different from a complete graph or an odd cycle,* $\chi(X) \leq \Delta$ *where* Δ *is the maximum degree of* X.

Now consider the task of coloring a political map. What is the minimum number of colors needed, with the obvious restriction that neighboring countries should have different colors? This is related to the so-called edge colorings of a graph.

The minimum number of colors needed to color the edges of a graph X in such a way that no two incident edges are of the same color is called the *edge chromatic number* $\bar{\chi}(X)$. By the famous Vizing theorem, given below, there are only two possibilities for the edge chromatic number of a graph.

Proposition 10 (Vizing theorem [71]). *For any graph X,* $\Delta \leq \bar{\chi}(X) \leq 1+\Delta$ *where* Δ *is the maximum degree of* X.

By Proposition 10 the edge chromatic number in a cubic graph is equal either to 3 or to 4. A *snark* is a bridgeless cubic graph with edge chromatic number 4. (A *bridge* is an edge that disconnects a graph.) The smallest snark is the Petersen graph shown in Fig. 10. A search for new snarks is an active topic of research (see [16, 41, 47, 61, 66]).

A *tournament* is an oriented complete graph. Tournaments are named so because an n-vertex tournament corresponds to a tournament in which each member of a group of n players plays all other $n-1$ players. The players are represented by vertices. For each pair of vertices an arc is drawn from the winner to the loser. The score sequence for a given tournament is obtained from the set of outdegrees sorted in nondecreasing order (see Fig. 13).

The following result holds.

Proposition 11 (Landau [42]). *The sequence* $d_1 \leq d_2 \leq \cdots \leq d_n$ *of positive integers is the score sequence of a tournament if and only if*

$$d_1 + d_2 + \cdots + d_t \geq \binom{t}{2}$$

for all $1 \leq t \leq n-1$, *and* $d_1 + d_2 + \cdots + d_n = \binom{n}{2}$.

Exercise 8. Give all possible tournaments on five vertices.

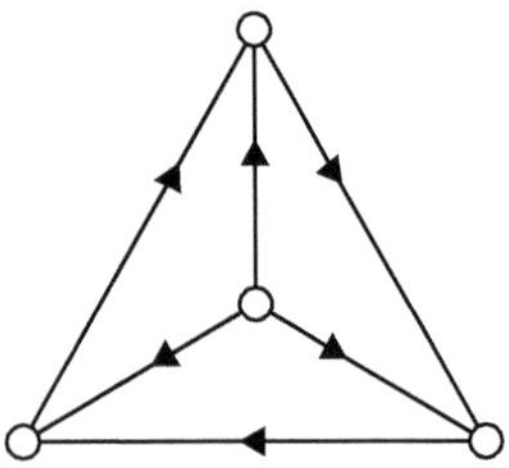

Fig. 13 An 4-vertex tournament

4 Graphs and Configurations

Suppose that in a presidential election with eight candidates a TV station wishes to organize eight TV debates with three candidates in each debate. Can candidates be arranged into eight groups of three such that no two meet more than once? This is typical problem that can be solved with the use of configurations.

An n_k *configuration* is a finite incidence structure with n points and n lines such that each line has k points and each point is on k lines. Also, two different lines intersect each other at most once and two different points are connected by at most one line. With each n_k configuration we can associate the so-called *Levi graph.* An *incidence graph* or *Levi graph* of an n_k configuration C is a bipartite k-regular graph of order $2n$, with n vertices representing the points of C and n vertices representing the lines of C, and with an edge joining two vertices if and only if the corresponding point and line are incident in C. In Figs. 14 and 15 the Moebius–Kantor configuration, the configuration solving the above election problem, and its Levi graph are shown. Since the Moebius–Kantor configuration is the only 8_3 configuration, this particular election problem has a unique solution.

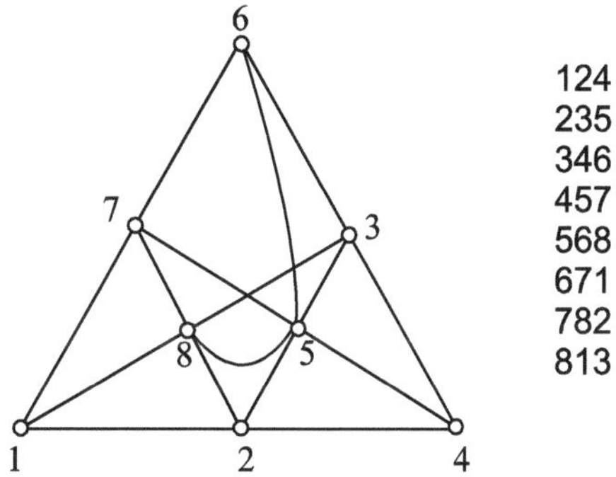

Fig. 14 The Moebius–Kantor configuration

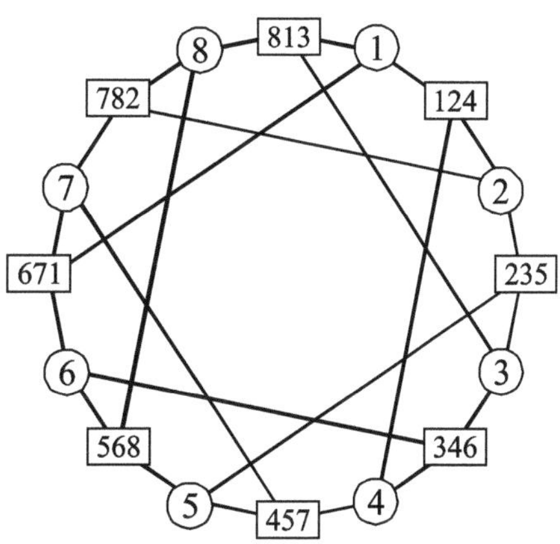

Fig. 15 The Levi graph of the Moebius–Kantor configuration

The property that two different points of a configuration are contained in at most one line implies that Levi graphs have girth (the length of the shortest cycle) at least 6. Conversely, each bipartite k-regular graph with girth at least 6 and with a chosen black-white coloring of the vertex set determines precisely one n_k configuration. If the coloring is not chosen in advance, such graphs determine a pair of dual configurations, that is, configurations with the role of their points and lines interchanged. Therefore the following result holds.

Proposition 12. *A k-regular graph of order $2n$ is a Levi graph of an n_k configuration if and only if it is bipartite and contains no 4-cycles.*

A slight modification of the above election problem may be posed: Can nine candidates be arranged into nine groups of size 3 such that no two meet more than once? The answer is again yes. But now there are three possible solutions (see Fig. 16). The upper 9_3 configuration on Fig. 16 allows no additional line, the lower allows one additional line and the middle allows three additional lines. This middle 9_3 configuration is known as the Pappus configuration. Among others this configuration solves the classical Orchard Planting Problem for $n = 9$, $k = 3$ and $r(n,k) = 10$ which is as follows: Can n trees be planted so as to produce $r(n,k)$ straight rows with k trees in each row.

More information on graphs and configurations can be found in [11, 29, 39].

Exercise 9. Show that the 8_3 configuration given in Fig. 15 is the only 8_3 configuration.

Exercise 10. Show that the three 9_3 configurations given in Fig. 16 are the only 9_3 configurations.

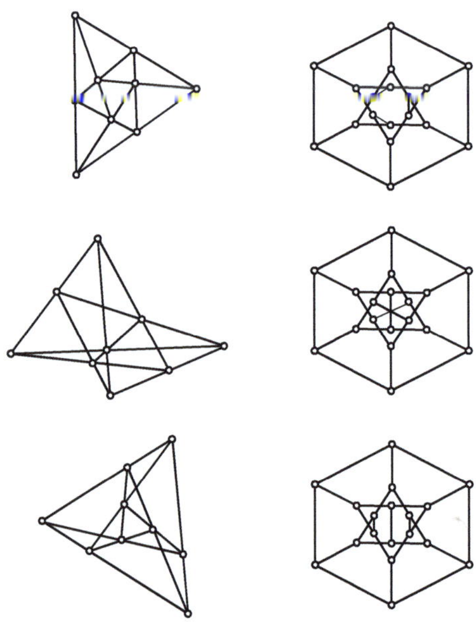

Fig. 16 The 9_3 configurations on the left handside pictures and the corresponding Levi graphs on the right handside pictures

5 Symmetries in Graphs

Digraphs $X_1 = (V_1, E_1)$ and $X_2 = (V_2, E_2)$ are *isomorphic* if and only if there is a bijection $f : V_1 \to V_2$ such that for all $u, v \in V_1$

$$(u,v) \in E_1 \Leftrightarrow (f(u), f(v)) \in E_2.$$

The bijection f is called an *isomorphism* between the graphs.

Example 1. Let X_1 be a simple graph whose vertices are integers $1,\dots,2n$ with an edge between two vertices if and only if they are of the same parity. Let X_2 be a simple graph whose vertices are integers $-1,\dots,-2n$ with an edge between two vertices if and only if either both vertices are less than $-n$ or both are more than or equal to $-n$. Then the function $f : \{-1,\dots,-2n\} \to \{1,\dots,2n\}$, where $f(k) = -2k$ if $-n \le k \le -1$ and $f(k) = 2(2n+k)+1$ if $-2n \le k < -n$, is an isomorphism between X_1 and X_2.

A nonempty set G together with a binary operation $\cdot$ is a *group* if the following four axioms are satisfied: (1)*(Closure)* For all $g_1, g_2 \in G$, $g_1 \cdot g_2 \in G$. (2) *(Associativity)* For all $g_1, g_2, g_3 \in G$, $g_1 \cdot (g_2 \cdot g_3) = (g_1 \cdot g_2) \cdot g_3$. (3) *(Identity)* There is an element e in G such that $e \cdot g = g \cdot e = g$ for all $g \in G$. (4) *(Inversion)* For each $g \in G$, there is an element denoted g^{-1} such that $g^{-1} \cdot g = g \cdot g^{-1} = e$.

An *automorphism* of a graph X is an isomorphism of X with itself. Thus each automorphism α of X is a permutation of the vertex set V which preserves adjacency. The set of automorphisms of a graph X is a group, called the *automorphism group* of X and denoted by $\mathrm{Aut}(X)$. For example, the automorphism group of the Petersen graph is isomorphic to the symmetric group $\mathrm{Aut}(GP(5,2)) = S_5$

Proposition 13. *A graph and its complement have the same automorphism group* $\mathrm{Aut}(X) = \mathrm{Aut}(\overline{X})$.

If G is a group and Ω is a set, then a *(right) group action* of G on Ω is a binary function $\Omega \times G \to \Omega$ with notation $(\omega, g) \mapsto \omega^g$ which satisfies the following two axioms:

(1) $\omega^1 = \omega$ for every $\omega \in \Omega$ (where 1 denotes the identity element of G).
(2) $(\omega^g)^h = \omega^{gh}$ for all $g, h \in G$ and $\omega \in \Omega$.

In a similar way we can define *(left) group action*.

For a group G acting on the set Ω the set

$$Orb_G(\omega) = \omega^G = \{\omega^g \mid g \in G\},$$

where $\omega \in \Omega$, is called a *G-orbit* (in short an *orbit* if the group G is clear from the contest) of the element ω with respect to the action of G. If the group orbit $Orb_G(\omega)$ is equal to the entire set Ω for some element ω in Ω, then G is *transitive*. For $\omega \in \Omega$ the set $G_\omega = \{g \in G \mid \omega^g = \omega\}$, the *stabilizer of the element* ω, is a subgroup of G. If $|G_\omega| = 1$ for every element $\omega \in \Omega$ then we say that G acts *semiregularly*. If

G acts on Ω transitively and $|G\omega| = 1$ for every element $\omega \in \Omega$ we say that G acts *regularly* (G is *regular*). For a more detail discussion on group actions we refer the reader to [64, 72].

A graph is said to be *vertex-transitive*, *edge-transitive*, and *arc-transitive* (also called *symmetric*) if its automorphism group acts vertex-transitively, edge-transitively and arc-transitively, respectively. Given a group G and a subset S of $G \setminus \{1\}$, the *Cayley graph* $X = Cay(G,S)$ has vertex set G and edges of the form $\{g, gs\}$ for all $g \in G$ and $s \in S$. Every Cayley graph is vertex-transitive but there exist vertex-transitive graphs that are not Cayley. Sabidussi [65] characterized Cayley graph in the following way: A graph is a Cayley graph of a group G if and only if its automorphism group contains a regular subgroup isomorphic to G.

Example 2. The graph shown in Fig. 5 is neither vertex-transitive nor edge-transitive. Graphs shown in Fig. 8 are all vertex-transitive, arc-transitive and Cayley. The Petersen graph and the dodecahedron (graphs shown in Figs. 10 and 11) are vertex-transitive and arc-transitive but they are not Cayley graphs.

Example 3. Let $n \geq 3$ be a positive integer, and let $k \in \{1, \ldots, n-1\} \setminus \{n/2\}$. The *generalized Petersen graph* $GP(n,k)$ is defined to have the following vertex set and edge set:

$$V(GP(n,k)) = \{u_i \mid i \in \mathbb{Z}_n\} \cup \{v_i \mid i \in \mathbb{Z}_n\},$$
$$E(GP(n,k)) = \{u_i u_{i+1} \mid i \in \mathbb{Z}_n\} \cup \{v_i v_{i+k} \mid i \in \mathbb{Z}_n\} \cup \{u_i v_i \mid i \in \mathbb{Z}_n\}.$$

There are infinitely many vertex-transitive but only 7 arc-transitive generalized Petersen graphs: $GP(4,1)$, $GP(5,2)$, $GP(8,3)$, $GP(10,2)$, $GP(10,3)$, $GP(12,5)$, $GP(24,5)$.

Clearly, a graph that is arc-transitive is also vertex-transitive and edge-transitive. But the converse is not true in general. In 1966 Tutte proved that the converse is true for graphs of odd valency.

Proposition 14 (Tutte [70]). *A vertex-transitive and edge-transitive graph of valence k, where k is odd, is also arc-transitive.*

Corollary 3. *Every cubic vertex-transitive and edge-transitive graph is arc-transitive.*

There exist graphs that are vertex-transitive and edge-transitive but not arc-transitive. These graphs are called *half-arc-transitive graphs*. By Proposition 14 the valency of such a graph is necessarily even. Moreover, in 1970 Bouwer [14] proved the existence of half-arc-transitive graphs of any given even valency. The smallest half-arc-transitive graph is the Doyle–Holt graph, the graph of order 27 shown in Fig. 17. The proof of the next proposition is straightforward.

Proposition 15. *Every graph that is edge-transitive but not vertex-transitive is bipartite.*

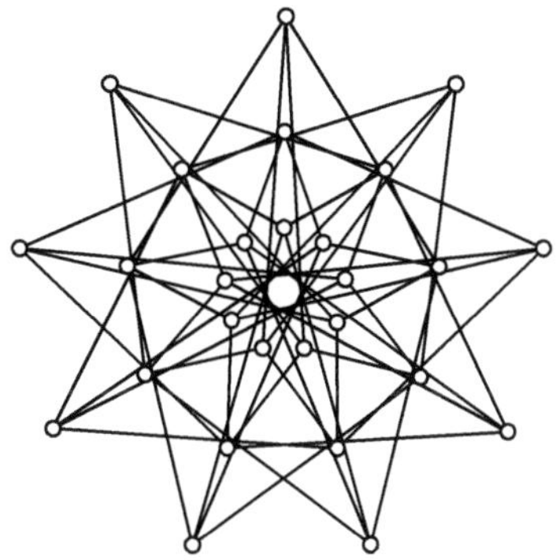

Fig. 17 The Doyle–Holt graph

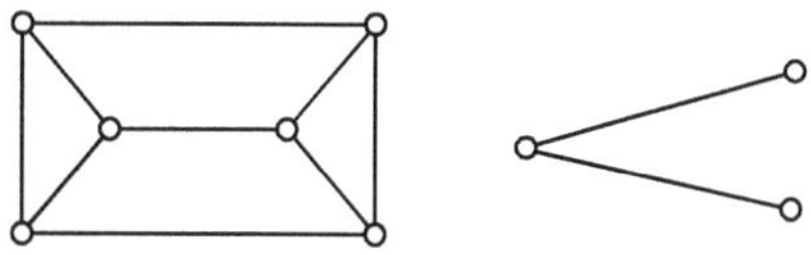

Fig. 18 The smallest graph that is vertex-transitive but not edge-transitive (the 3-prism $K_3 \square K_2$) and the smallest graph that is edge-transitive but not vertex-transitive (the star $K_{1,2}$)

Fig. 19 The Folkman graph

Examples of edge-transitive but not vertex-transitive graphs are non-regular graphs $K_{m,n}$, $n \neq m$ (see also Fig. 18). A regular edge-transitive but not vertex-transitive graph is called *semisymmetric*. Research on these graphs has initiated and the foundations of the theory laid out by Jon Folkman [36]. The smallest such a graph has 20 vertices and is now known as the Folkman graph (see Fig. 19). The smallest cubic semisymmetric graph is the Gray graph shown in Fig. 20 (see also [24]).

Additional information on symmetries in graphs can be found in [38]. To wrap up this section we give the following two exercises.

Exercise 11. Prove that the Petersen graph is the smallest vertex-transitive graph that is not a Cayley graph.

Exercise 12. Prove Proposition 15.

Fig. 20 The Gray graph

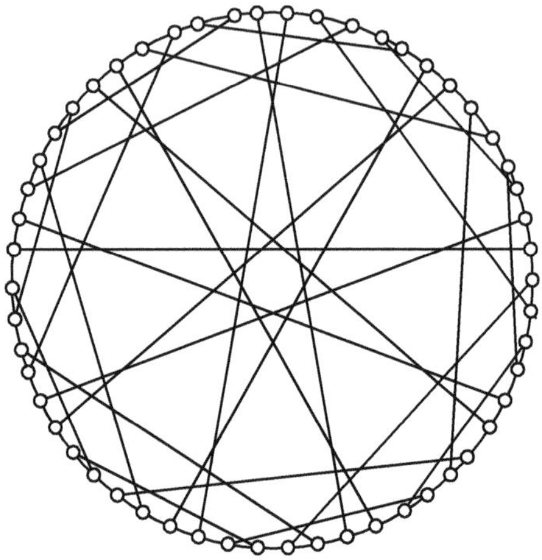

6 Some Well-Known Open Problems

The following unsolved problem proposed by Lovász can be thought of as the main motivation for the study of vertex-transitive graphs.

Open problem 1 (Lovász [46]) *Does every connected vertex-transitive graph have a Hamiltonian path?*

With the exception of K_2, only four connected vertex-transitive graphs that do not have a Hamiltonian cycle are known to exist. These four graphs are the Petersen graph, the Coxeter graph and the two graphs obtained from them by replacing each vertex by a triangle. The fact that none of these four graphs is a Cayley graph has led to a folklore conjecture that every Cayley graph is hamiltonian (see [7, 9, 10, 30, 37, 52] for the current status of this conjecture). Coming back to vertex-transitive graphs, it was shown in [32] that, with the exception of the Petersen graph, a connected vertex-transitive graph whose automorphism group contains a transitive subgroup with a cyclic commutator subgroup of prime-power order, is hamiltonian. Furthermore, it has been shown that connected vertex-transitive graphs of orders p, $2p$ (except for the Petersen graph), $3p$, $4p$, p^2, p^3, p^4 and $2p^2$, where p is a prime, are hamiltonian (see [1, 19, 44, 53–55, 68]). On the other hand, connected vertex-transitive graphs of orders $5p$ and $6p$ are only known to have Hamiltonian paths (see [45, 59]). Some other partial results related to this problem are known (see [2–6, 8]).

The following open problem was posed in [18, 51].

Open problem 2 ([18, 51]) *Does every vertex-transitive graph have a semiregular automorphism?*

An element of a permutation group is *semiregular*, more precisely (m,n) – semiregular, if it has m orbits of size n. It is known that every cubic vertex-transitive graph has a semiregular automorphism [60]. Recently this result have been extended to quartic vertex-transitive graphs [33], but the problem of existence of semiregular automorphisms in vertex-transitive graphs [18, 51] is still open for larger valencies. However, some results with restriction to certain orders of the graphs in question are also known (see [51]). As for arc-transitive graphs of valency pq, where p and q are primes, recently Xu [73] proved existence of semiregular automorphisms in the case when their automorphism groups have a nonabelian minimal normal subgroup with at least three orbits on the vertex set of the graph.

Open problem 3 *Classification and structural results for cubic symmetric graphs of different transitivity degrees ($s = 1, 2, 3, 4$ or 5).*

In 1947 Tutte [69] proved that in a cubic symmetric graph the order of a vertex stabilizer is 3×2^s where $s \leq 4$. All cubic symmetric graphs on up to 2,048 vertices and some partial results related to the above problem are known (see [15, 20–22, 26–28]).

Open problem 4 *Classification and structural results for quartic half-arc-transitive graphs.*

The list of all quartic half-arc-transitive graphs on up to 500 vertices is close to being completed. Although quartic half-arc-transitive graphs are an active topic of research these days the above problem is still open (see [25, 34, 43, 48, 49, 56, 58, 67]).

Open problem 5 *Classification and structural results for semisymmetric graphs.*

The list of all cubic semisymmetric graphs on up to 768 vertices in given in [24]. There exist 43 cubic semisymmetric graphs on up to 768 vertices, 21 with solvable automorphism group and 22 with nonsolvable automorphism group. Partial results on the above problem were proven in [13, 23, 35, 50, 57, 63].

References

1. Alspach B (1979) Hamiltonian cycles in vertex-transitive graphs of order $2p$. In: Proceedings of the tenth southeastern conference on combinatorics, graph theory and computing, Florida Atlantic University, Boca Raton, FL, 1979, pp 131–139. Congress Numer XXIII–XX, Utilitas Math, Winnipeg, MB
2. Alspach B (1983) The classification of hamiltonian generalized Petersen graphs. J Combin Theory B 4:293–312
3. Alspach B (1989a) Hamilton cycles in metacirculant graphs with prime power cardinal blocks. In: Graph theory in memory of G. A. Dirac (Sandbjerg, 1985). North-Holland, Amsterdam, pp 7–16. Ann Discrete Math 41
4. Alspach B (1989b) Lifting Hamilton cycles of quotient graphs. Discrete Math 78:25–36
5. Alspach B, Chen CC, McAvaney K (1996) On a class of Hamiltonian laceable 3-regular graphs. Discrete Math 151:19–38
6. Alspach B, Durnberger E, Parsons TD (1985) Hamilton cycles in metacirculant graphs with prime cardinality blocks. In: Cycles in graphs (Burnaby, B.C., 1982). North-Holland, Amsterdam, pp 27–34. Ann Discrete Math 27
7. Alspach B, Locke S, Witte D (1990) The Hamilton spaces of Cayley graphs on abelian groups. Discrete Math 82:113–126

8. Alspach B, Parsons TD (1982) On Hamiltonian cycles in metacirculant graphs, Algebraic and geometric combinatorics. Ann Discrete Math 15:1–7
9. Alspach B, Qin YS (2001) Hamilton-connected Cayley graphs on Hamiltonian groups. European J Combin 22:777–787
10. Alspach B, Zhang CQ (1989) Hamilton cycles in cubic Cayley graphs on dihedral groups. Ars Combin 28:101–108
11. Betten A, Brinkmann G, Pisanski T (2000) Counting symmetric configurations v_3. Discrete Appl Math 99:331–338
12. Biggs N (1993) Algebraic graph theory, 2nd edn. Cambridge University Press, Cambridge
13. Bouwer IZ (1968) An edge but not vertex transitive cubic graph. Bull Canad Math Soc 11:533–535
14. Bouwer IZ (1970) Vertex and edge-transitive but not 1-transitive graphs. Canad Math Bull 13:231–237
15. Bouwer IZ (ed) (1988) The Foster census. Charles Babbage Research Centre, Winnipeg
16. Brinkmann G, Steffen E (1998) Snarks and reducibility. Ars Combin 50:292–296
17. Brooks RL (1941) On coloring the nodes of a network. Proc Cambridge Philos Soc 37:194–197
18. Cameron PJ (ed) (1997) Problems from the fifteenth British combinatorial conference. Discrete Math 167/168:605–615
19. Chen YQ (1998) On Hamiltonicity of vertex-transitive graphs and digraphs of order p^4. J Combin Theory B 72:110–121
20. Conder MDE (2006) Trivalent symmetric graphs on up to 2048 vertices. http://www.math.auckland.ac.nz/conder/symmcubic2048list.txt
21. Conder MDE, Dobcsanyi P (2002) Trivalent symmetric graphs on up to 768 vertices. J Combin Math Combin Comp 40:41–63
22. Conder MDE, Lorimer PJ (1989) Automorphism groups of symmetric graphs of valency 3. J Combin Theory B 47:60–72
23. Conder MDE, Malnič A, Marušič D, Pisanski T, Potočnik P (2005) The edge-transitive but not vertex-transitive cubic graph on 112 vertices. J Graph Theory 50:25–42
24. Conder MDE, Malnič A, Marušič D, Potočnik P (2006) A census of semisymmetric cubic graphs on up to 768 vertices. J Algebraic Combin 23:255–294
25. Conder MDE, Marušič D (2003) A tetravalent half-arc-transitive graph with nonabelian vertex stabilizer. J Combin Theory B 88:67–76
26. Conder MDE, Morton M (1995) Classification of trivalent symmetric graphs of small order. Australas J Combin 11:139–149
27. Conder MDE, Nedela R (2007) Symmetric cubic graphs of small girth. J Combin Theory B 97:757–768
28. Conder MDE, Nedela R (2008) A more detailed classification of symmetric cubic graphs (preprint)
29. Coxeter HSM (1950) Self-dual configurations and regular graphs. Bull Am Math Soc 56:413–455
30. Curran S, Gallian JA (1996) Hamiltonian cycles and paths in Cayley graphs and digraphs – a survey. Discrete Math 156:1–18
31. Dirac GA (1952) Some theorems on abstract graphs. Proc London Math Soc (3) 2:69–81
32. Dobson E, Gavlas H, Morris J, Witte D (1998) Automorphism groups with cyclic commutator subgroup and Hamilton cycles. Discrete Math 189:69–78
33. Dobson E, Malnič A, Marušič D (2007) Semiregular automorphisms of vertex-transitive graphs of certain valencies. J Combin Theory B 97:371–380
34. Du SF, Xu MY (1998) Vertex-primitive $1/2$-arc-transitive graphs of smallest order. Comm Algebra 27:163–171
35. Du SF, Xu MY (2000) A classification of semisymmetric graphs of order $2pq$. Comm Algebra 28:2685–2715
36. Folkman J (1967) Regular line-symmetric graphs. J Combin Theory 3:215–232

37. Glover HH, Marušič D (2007) Hamiltonicity of cubic Cayley graph. J Eur Math Soc 9:775–787
38. Godsil C, Royle G (2001) Algebraic graph theory. Springer, New York
39. Gropp H (1993) Configurations and graphs. Discrete Math 111:269–276
40. Harary F (1969) Graph theory. Addison, Reading, MA
41. Kochol M (1996) A cyclically 6-edge-connected snark of order 118. Discrete Math 161:297–300
42. Landau HG (1953) On dominance relations and the structure of animal societies: III The condition for a score structure. Bull Math Biol 15:143–148
43. Li CH, Sims HS (2001) On half-transitive metacirculant graphs of prime-power order. J Combin Theory B 81:45–57
44. Kutnar K, Marušič D (2008) Hamiltonicity of vertex-transitive graphs of order $4p$. Eur J Combin 29(2):423–438. doi:10.1016/j.ejc.2007.02.002
45. Kutnar K, Šparl P (2008) Hamilton cycles and paths in vertex-transitive graphs of order $6p$. Discrete Math (submitted)
46. Lovász L (1970) Combinatorial structures and their applications. In: Proc. Calgary Internat. Conf., Calgary, Alberta, 1969, pp 243–246. Problem 11. Gordon and Breach, New York
47. Máčajová E, Škoviera M (2006) Irreducible snarks of given order and cyclic connectivity. Discrete Math 306:779–791
48. Malnič A, Marušič D (1999) Constructing 4-valent $1/2$-transitive graphs with a nonsolvable group. J Combin Theory B 75:46–55
49. Malnič A, Marušič D (2002) Constructing $1/2$-transitive graphs of valency 4 and vertex stabilizer $\mathbb{Z}_2 \times \mathbb{Z}_2$. Discrete Math 245:203–216
50. Malnič A, Marušič D, Wang C (2004) Cubic edge-transitive graphs of order $2p^3$. Discrete Math 274:187–198
51. Marušič D (1981) On vertex symmetric digraphs. Discrete Math 36:69–81
52. Marušič D (1983) Hamilonian circuits in Cayley graphs. Discrete Math 46:49–54
53. Marušič D (1985) Vertex transitive graphs and digraphs of order p^k. In: Cycles in graphs (Burnaby BC, 1982). North-Holland, Amsterdam, pp 115–128. Ann Discrete Math 27
54. Marušič D (1987) Hamiltonian cycles in vertex symmetric graphs of order $2p^2$. Discrete Math 66:169–174
55. Marušič D (1988) On vertex-transitive graphs of order qp. J Combin Math Combin Comput 4:97–114
56. Marušič D (1998) Half-transitive group actions on finite graphs of valency 4. J Combin Theory B 73:41–76
57. Marušič D (2000) Constructing cubic edge- but not vertex-transitive graphs. J Graph Theory 35:152–160
58. Marušič D (2005) Quartic half-arc-transitive graphs with large vertex stabilizers. Discrete Math 299:180–193
59. Marušič D, Parsons TD (1982) Hamiltonian paths in vertex-symmetric graphs of order $5p$. Discrete Math 42:227–242
60. Marušič D, Scapellato R (1998) Permutation groups, vertex-transitive digraphs and semiregular automorphisms. Eur J Combin 19:707–712
61. Nedela R, Škoviera M (1996) Decompositions and Reductions of Snarks. J Graph Theory 22:253–279
62. Ore O (1960) Note on Hamilton circuits. Am Math Monthly 67:55
63. Parker CW (2007) Semisymmetric cubic graphs of twice odd order. Eur J Combin 28:572–591
64. Passman D (1968) Permutation groups. Benjamin, New York
65. Sabidussi G (1958) On a class of fixed-point-free graphs. Proc Am Math Soc 9:800–804
66. Steffen E (1998) Classifications and characterizations of snarks. Discrete Math 188:183–203
67. Šparl P (2008) A classification of tightly attached half-arc-transitive graphs of valency 4. J Combin Theory B, doi:10.1016/j.jctb.2008.01.001, in press

68. Turner J (1967) Point-symmetric graphs with a prime number of points. J Combin Theory 3:136–145
69. Tutte WT (1947) A family of cubical graphs. Proc Cambridge Philos Soc 43:459–474
70. Tutte WT (1966) Connectivity in graphs. University of Toronto Press, Toronto
71. Vizing VG (1964) On an estimate of the chromatic class of a p-graph. Metody Diskret Analiz 3:25–30
72. Wielandt H (1964) Finite permutation groups. Academic, New York
73. Xu J (2008) Semiregular automorphisms of arc-transitive graphs with valency pq. Eur J Combin 29(3):622–629. doi:10.1016/j.ejc.2007.04.008

From Graph Theory to Models of Economic Networks. A Tutorial

Michael D. König and Stefano Battiston

Abstract Networks play an important role in a wide range of economic phenomena. Despite this fact, standard economic theory rarely considers economic networks explicitly in its analysis. However, a major innovation in economic theory has been the use of methods stemming from graph theory to describe and study relations between economic agents in networks. This recent development has lead to a fast increase in theoretical research on economic networks. In this tutorial, we introduce the reader to some basic concepts used in a wide range of models of economic networks.

1 Introduction

Networks are ubiquitous in social and economic phenomena. The use of methods from graph theory has allowed economic network theory to improve our understanding of those economic phenomena in which the embeddedness of individuals in their social inter-relations cannot be neglected. In this tutorial will give a brief overview of network models, starting from simple network constructions to more complex models that allow for the strategic formation of links.

When discussing these models we try to introduce the reader to the most important concepts of economic networks. However, the literature that is discussed in this tutorial is far from being exhaustive. For a more detailed introduction to economic network theory we recommend the books Vega-Redondo [84], Jackson [52] and Goyal [42] as well as the lecture notes by Calvó-Armengol [15], Zenou [93]. A more mathematical treatment of complex networks can be found in Chung and Lu [20] and Durrett [30]. Standard references for graph theory are Bollobas [8], Diestel [28], West [91].

M.D. König and S. Battiston
Chair of Systems Design, ETH Zurich, Kreuzplatz 5, 8032 Zurich, Switzerland
mkoenig@ethz.ch

A.K. Naimzada et al. (eds.), *Networks, Topology and Dynamics*,
Lecture Notes in Economics and Mathematical Systems 613,

This tutorial is organized as follows. First, we will argue in Section 2 that standard economic theory is in the need of incorporating networks in its analysis. In Section 3 we will mention several applications of economic network theory and we will introduce the basic terminology used to describe networks in Section 4. We will proceed by discussing several prominent network models with an increasing degree of complexity ranging from Poisson random networks in Section 5.1, its generalization in Section 5.2, growing random networks in Section 6 to models of strategic network formation in Section 7.

2 Why Networks in Economics?

Gallegati and Kirman [36], Kirman [60] propose that the aggregate behavior of an economy cannot be investigated in terms of the behavior of isolated individuals, as it is usually done in standard economic theory. Firms interact only with a few other firms, out of all firms present in the economy. Moreover, there are different ways in which firms interact, and they may learn over time to adapt their interactions, meaning that they strengthen profitable ones while they cut costly ones. All this is based on their previous experience. We may then view the economy as an evolving network.

Viewing the economy as an evolving network is different from what a standard neoclassical model[1] of the economy would look like. In such a model it is assumed that anonymous and autonomous individuals take decisions independently and interact only through the price system which they cannot influence at all. This situation refers to a market with perfect competition. However, competition easily becomes imperfect because, if agents have only minimal market power, they will anticipate the consequences of their actions and anticipate the actions of others. In order to overcome this deficiency, game theorists have tried to integrate strategically interacting firms into a general equilibrium [2] framework. But still they leave two questions unanswered. First, it is assumed that the behavior is fully optimizing considering all possible actions as well as all possible actions of others. This leads to agents with extremely sophisticated information processing capabilities. Such ability of passing these enormous amounts of information in short times cannot be found in any realistic setting of human interaction. Advances in weakening that assumption are referred to as "bounded rationality" [40]. Second, the problem of coordination of activities is not addressed in the standard equilibrium model of the economy. Instead

[1] A standard neoclassical model includes the following assumptions [35]: (1) perfect competition, (2) perfect information, (3) rational behavior, (4) all prices are flexible (all markets are in equilibrium). The resulting market equilibrium (allocation of goods) is then efficient. See [49] for a discussion of these assumptions.

[2] The individual decision making process is represented as maximizing a utility function. A utility function is a way of assigning a number to every possible choice such that more-preferred choices have a higher number than less-preferred ones [82]. The gradients of the utility function are imagined to be like forces driving people to trade, and from which economic equilibria emerge as a kind of force balance [32].

it is assumed that every agent can interact and trade with every other agent, which becomes quite unrealistic for large systems. One has to specify the framework within the individual agents take price decisions and thus limit the environment within which they operate and reason. An obvious way is to view the economy as a network in which agents interact only with their neighbors. In the case of technological innovation, neighbors might be similar firms within the same industry, but these firms will then be linked either through customers or suppliers with firms in other industries. Through these connections innovations will diffuse throughout the network. The rate and extent of this diffusion then depends on the structure and connectivity of the network.

Finally, the evolution of the network itself should be made endogenous. In this case the evolution of the link structure is dependent on the agents' experience from using the links (respectively contacts) available to them. Individuals learn and adapt their behavior and this in turn leads to an evolution of the network structure which then feeds back into the incentives of agents to form or sever links. We will briefly discuss this coupled dynamic interaction between individuals' incentives and the network dynamics in Section 4.4.

3 Examples of Networks in Economics

In this section we point to several applications of network models in economics. We have restricted ourselves to a few applications but this list could of course be greatly expanded [see e.g. also 84, p. 10].

Corporate Ownership and Boars of Directors

Ownership relations between firms, as well as members in common in the boards of directors, give raise to intricate networks. On one hand ownership relations are instruments to exert corporate control and several works have studied indirect ownership relations [13] and patterns such as the so-called pyramids and cross-shareholdings [19], as well as business groups [33, 45]. Other works have also studied the financial architecture of corporations in national or global economies [5, 21, 37, 61].

On the other hand, interlocked directors among firms are known to convey information and power [4, 27]. The spread of corporate practices through the director network and the role of inter-organizational imitation of managers has been studied by Davis and Greve [26]. Moreover, it has been shown that the structure of the intrlock network has implications for the decision making process [3].

Labor markets

A wide range of empirical studies of labor markets have shown that a significant fraction of all jobs are found through social networks. The role of informal social networks in labor markets has been emphasized first by Granovetter [44]. He found that over 50% of jobs were found through personal contacts. In a recent paper, Jackson and Calvo-Armengol [53] introduce a network model of job information transmission. The model reproduces the empirically stylized fact that the employment situation of individuals that are connected, either directly or indirectly, is correlated. Further, they show that the topology of the network influences the length and correlation of unemployment among individuals. Finally, with this model the authors can explain the pervasive inequalities in wages, employment and drop-out rates.

Diffusion in Networks

In economics diffusion is usually related to the spread of a technology through a society or industry. A new technology or idea might be generated by an innovator and then be subsequently adopted by others over time. The literature on technological diffusion focuses on alternative explanations of the dominant stylized fact: that the usage of new technologies over time typically follows an S-curve. Geroski [39] gives an excellent survey on models of technological diffusion [see also 24, 79, 85]. Most models assume that there are no restrictions on the interactions between agents and the path along which knowledge can flow. This assumption is clearly not supported by the restrictions and limited contacts firms realistically maintain [47, 74]. In particular, if knowledge diffuses through social contacts or personal interrelations then the diffusion of a technology critically depends on the underlying network structure. Thus, a proper understanding of the diffusion of innovations needs to be grounded in economic network theory.

Formal and Informal Organizations

The central question in the theory of organizations is how a complex decision problem can be efficiently decomposed into distinct tasks, distributed among the different units of an organization. A network can represent the paths along which these tasks are distributed in an organization [see 80, 81, for a general discussion of network forms of organization].

One can distinguish between formal or informal networks in an organization. Formal network usually refer to the hierarchical structure of an organizational chart. On the other hand, informal organizational networks are usually referred to "communities of practice". They can serve as a complement to the formal organizational structure [11, 14, 65]. Beyond the formal working relationships institutionalized in the organizational chart, informal working relationships may coexists or may even play a predominant role [16]. In principle. a hierarchical formal organization assumes that a central coordinator can distribute tasks efficiently among the members of an organization. However, central coordination may not be feasible when the number of agents in the organization is large, the problems the organization has to solve are highly complex and their nature varies considerably such that they cannot be decomposed and distributed.

The existing literature has mainly focussed on the formal organizational structure whereas recent works try to incorporate both the formal as well as the informal communication networks [51] among individuals in an organization. A recent example is the work by Dodds et al. [29]. The authors find a particular organizational network structure that enhances the robustness of the organization and reduces the possibility of a communication overload among its members.

R&D Collaborations

There exist many theoretical works in the literature on industrial organization trying to explain the effects and incentives of R&D collaborations between competing firms [see e.g. 59] and Veugelers [86, for a review]. However, these works do not address the heterogeneity of inter-firm collaborations that have been observed in empirical studies [e.g. 72]. In recent works by Goyal and Moraga-Gonzalez [43] and Vega-Redondo [83] R&D collaborations are investigated in a network setup in which these collaborations are not exogeneously given but the endogeneous outcome of the incentives of firms to collaborate. In this way, heterogeneous interaction profiles are possible. Their equilibrium analysis, however, leads to simple network structures. These simple networks are in contradiction to the empirical literature that shows that R&D networks can have complex network topologies, in general characterized by high clustering, sparseness and a heterogeneous degree distribution. A recent example of a model that tries to incorporate these empirical stylized facts can be found in König et al. [62], König et al. [63] and we will give a brief overview of this model in Section 7.3.

4 Characterization of Networks

If the links in a network do not change over time (we have a static network) we can associate a state variable to the nodes based on their position in the network. If, for example, an agent has many neighbors which in turn have many neighbors, for instance, she may have much better opportunities to gather information from others compared to an agent that maintains only a few connections to other loosely connected agents. We can assign this agent a high centrality in the network (see Section 4.3.3). But a highly central agent may also be much more frequently exposed to any threat propagating through the network, e.g. viruses or avalanches of insolvencies.

In this section, we start with some concepts of graph theory that deals with the properties of static networks. We then review some of the measures that are used to characterize networks, discussing briefly their meaning in economic systems and point to some relevant literature. Finally, we introduce a possible classification scheme for models of networks in economics.

4.1 Elements of Graph Theory

In this section we follow closely West [91] to which we refer the reader for further details. In the following, we will use the terms graph and network as synonyms. The same holds for nodes and nodes as well as links and edges.

A *graph* G is a pair, $G = (V,E)$, consisting of a node set $V(G)$ and an edge set $E(G)$. The edge set $E(G)$ induces a symmetric binary relation on $V(G)$ that is called the adjacency relation of G. Nodes i and j are adjacent if $e_{ij} \in E(G)$.

The *degree*, d_i, of a node i is the number of edges incident to it. A graph can either be *undirected* or *directed*, where in the latter case one has to distinguish between *indegree*, d_i^-, and *outdegree*, d_i^+, of node i. In the case of an undirected graph, the (first-order) neighborhood of a node i in G is $N_i = \{w \in V(G) : e_{wi} \in E(G)\}$. The degree of a node i is then $d_i = |N_i|$. The second-order neighborhood of node i is $\bigcup_{u \in N_i} N_u \backslash \{i \cup N_i\}$. Similarly, higher order neighborhoods can be defined (as well as neighborhoods for directed graphs). A graph G is *regular* if all nodes have the same degree. A graph G is $k -$ *regular* if every node has degree k.

The *adjacency matrix*, $\mathbf{A}(G)$, of G, is the $n \times n$ matrix in which the entry a_{ij} is 1 if the edge $e_{ij} \in E(G)$, otherwise a_{ij} is 0. For an undirected graph $\mathbf{A}$ is symmetric, i.e. $a_{ij} = a_{ji}\ \forall i, j \in V(G)$. An example of a simple directed graph on four nodes and its associated adjacency matrix $\mathbf{A}$ is given in Figure (1).

The *eigenvalues* of the adjacency matrix $\mathbf{A}$ are the numbers λ such that $\mathbf{Ax} = \lambda\mathbf{x}$ has a nonzero solution vector, which is an *eigenvector* associated with λ. The term λ_{PF} denotes the *largest real eigenvalue* of $\mathbf{A}$ [the *Perron-Frobenius eigenvalue*, cf. 50, 75], i.e. all eigenvalues λ of $\mathbf{A}(\mathbf{G})$ satisfy $|\lambda| \leq \lambda_{\mathrm{PF}}$ and there exists an associated nonnegative eigenvector $\mathbf{v} \geq 0$ such that $\mathbf{Av} = \lambda_{\mathrm{PF}}\mathbf{v}$. For a connected graph G the adjacency matrix $\mathbf{A}(G)$ has a unique largest real eigenvalue λ_{PF} and a positive associated eigenvector $\mathbf{v} > 0$.

A *walk* is an alternating list, $\{v_0, e_{01}, v_1, \ldots, v_{k-1}, e_{k-1k}, v_k\}$, of nodes and edges. A *trail* is a walk with no repeated edge. A *path* is a walk with no repeated node. The shortest path between two nodes is also known as the *geodesic distance*. If the endpoints of a trail are the same (a closed trail) then we refer to it as a *circuit*. A circuit with no repeated node is called a *cycle*. In particular, C_n denotes the *cycle* on

$$A = \begin{pmatrix} 0 & 1 & 0 & 0 \\ 0 & 0 & 1 & 1 \\ 0 & 0 & 0 & 1 \\ 1 & 0 & 0 & 0 \end{pmatrix}$$

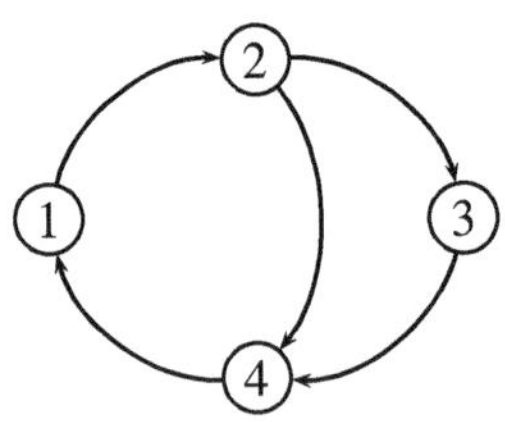

Fig. 1 (Right) a directed graph consisting of 4 nodes and 5 edges. (Left) the corresponding adjacency matrix A. For example, in the first row in $\mathbf{A}$ with elements, $a_{11} = 0, a_{12} = 1, a_{13} = 0, a_{14} = 0$, the element $a_{12} = 1$ indicates that there exist an edge from node 1 to node 2 while node 1 has not other outgoing links

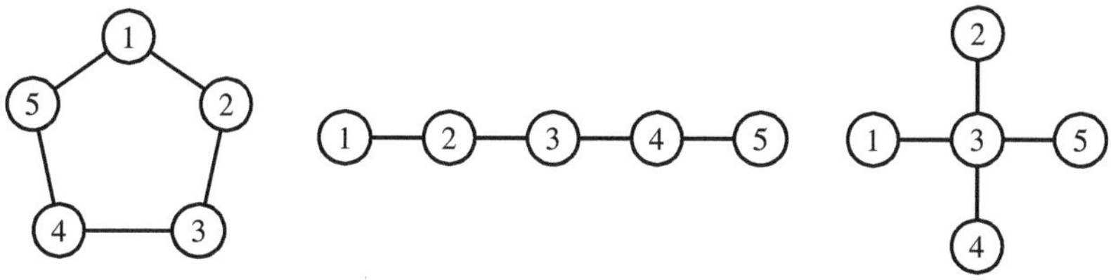

Fig. 2 A cycle C_5 (left), a path P_5 (middle) and the star $K_{1,4}$ (right). All graphs are undirected and contain 5 nodes

n nodes. Note that a cycle is also a circuit but a circuit is not necessarily a cycle. Examples of simple graphs are shown in Figure (2).

The kth power of the adjacency matrix is related to walks of length k in the graph. In particular, $(\mathbf{A}^k)_{ij}$ gives the number of walks of length k from node i to node j [41].

A *subgraph*, G', of G is the graph of subsets of the nodes, $V(G') \subseteq V(G)$, and edges, $E(G') \subseteq E(G)$. A graph G is *connected*, if there is a path connecting every pair of nodes. Otherwise G is disconnected. The *components* of a graph G are the maximal connected subgraphs. A graph is said to be *complete* if every node is connected to every other node. K_n denotes the *complete graph* on n nodes.

4.2 Graphs and Matrices

We will state some useful facts about matrices and graphs in this section. The study of irreducible and primitive graphs is important in linear dynamic network models. We will present the theory here and discuss a particular application in Section 7.3. Next, we introduce bipartite graphs and show how they can be applied to study networks between members of boards of different companies.

Irreducible and Primitive Graphs

If a graph G is not connected then its adjacency matrix $\mathbf{A}(G)$ can be decomposed in blocks, each block correspond to a connected component. An $n \times n$ matrix $\mathbf{A}$ is said to be a reducible matrix if and only if for some permutation matrix $\mathbf{P}$, the matrix $\mathbf{P}^T\mathbf{AP}$ is block upper triangular. If a square matrix is not reducible, it is said to be an irreducible matrix. If a graph is connected then there exists a path from every node to every other node in the graph. The adjacency matrix of a connected graph is irreducible [50] and in particular it cannot be decomposed in blocks. Irreducible matrices can be primitive or cyclic (imprimitive) [75]. This distinction is relevant for several results on the convergence of linear systems Boyd [12], Horn and Johnson [50] and we will apply it in Section 7.3.

A non-negative matrix $\mathbf{A}$ is primitive if $\mathbf{A}^k > 0$ for some positive integer $k \leq (n-1)n^n$. This means that, $\mathbf{A}$ is primitive if, for some k, there is a walk of length

k from every node to every other node. Notice that this definition is a much more restrictive than the one of irreducible (or connected) graphs in which it is required that there exits a walk from every node to every other node, but not necessarily of the same length. A graph G is said to be primitive if its associated adjacency matrix $\mathbf{A}(G)$ is primitive.

It is useful to look at an alternative but equivalent way to characterize a primitive graph. A graph G is primitive if and only if it is connected and the greatest common divisor of the set of length of all cycles in G is 1 [50, 92]. This means for instance that the connected graph consisting of two connected nodes is not primitive as the only cycle has length 2 (since the link is undirected a walk can go forward and backward along the link). Similarly, a chain or a tree is also not primitive, since all cycles have only even length. However, if we add one link in order to form a triangle, the graph becomes primitive. The same is true, if we add links in order to form any cycle of odd length. In general, if the graph of interaction between agents is connected, the presence of one cycle of odd length is a sufficient condition for the primitivity of the graph.

Bipartite Graphs

In a *bipartite* graph G, $V(G)$ is the union of two disjoint independent sets V_1 and V_2. In a bipartite graph, if $e_{12} \in E(G)$ then $v_1 \in V_1$ and $v_2 \in V_2$. In other words, the two endpoints of any edge must be in different sets. The *complete bipartite graph* with partitions of size $|V_1| = n_1$ and $|V_2| = n_2$ is denoted K_{n_1,n_2}. A special case is the *star* which is a complete bipartite graph with one partition having size $n_1 = 1$ and $n_2 = n - 1$, denoted as $K_{1,n-1}$ in Figure (2).

A bipartite graph can be 'projected' into two one-mode networks. For sake of clarity let us take the following example. Assume that in Figure (3) each node denoted with a number represents the board of directors of a company, while each node denoted with a letter represents a person. A link, say, between person B and board 1 represents the fact that person B serves in board 1. Notice that B serves also in board 2. The one-mode projection on the directors is a new graph in which there is a link between two persons if they serve together in one or more boards. In doing this projection some information is lost: consider for instance three directors connected in a triangle (not shown). The links do not specify whether each pair of directors sit in a different board or whether the three directors sit all in the same board. Denote $\mathbf{C}$ the adjacency matrix of our network of boards and persons,

$$C_{\alpha i} = \begin{cases} 1 \text{ if } \alpha \text{ sits in board } i \\ 0 \text{ otherwise.} \end{cases} \tag{1}$$

$\mathbf{C}$ is an $M \times N$ matrix, M being the number of persons, and N being the number of boards. This is a binary matrix, and in general it is neither square, nor symmetric. For the one-mode projection relative to the boards, we should take into account that the number of directors sitting in boards i and j, is equivalent to the number of paths of length 2 connecting i and j in the bipartite graph. Therefore, this number can be

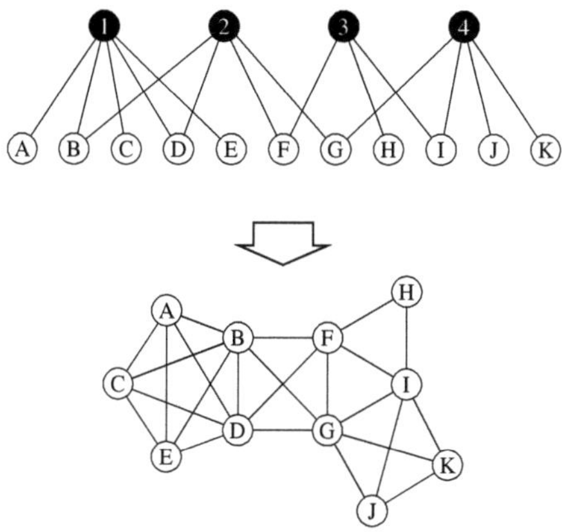

Fig. 3 Example of bipartite network (top). There are two classes of nodes and links are assigned only between nodes that do not belong to the same class. A one-mode projection is a new graph consisting only of nodes of one class in which a link between two nodes implies that, in the original bipartite graph, the two nodes where connected to a same third node

assigned as the weight of the connection between i and j, and result in a natural way from the follwong operation on the adjacency matrix. If we define the adjacency matrix of the board network as

$$B_{ij} = \begin{cases} w_{ij} \text{ if } i \text{ and } j \text{ are connected with weight } w_{ij} \\ 0 \text{ if } i \text{ and } j \text{ are not connected.} \end{cases} \quad (2)$$

then it holds that

$$B_{ij} = \sum \alpha C_{\alpha i} C_{\alpha j}. \quad (3)$$

In terms of matrix product this means $\mathbf{B} = \mathbf{C}^T \mathbf{C}$. In analogous way, the adjacency matrix of the director network is related to the initial board-person network as follows

$$D_{\alpha\beta} = \sum_i C_{\alpha i} C_{\beta i}. \quad (4)$$

which is equivalent to $\mathbf{D} = \mathbf{C}\mathbf{C}^T$. While the off-diagonal entries correspond to the edge weights, the diagonal entries, are, respectively, the size B_{ii} of board i (the number of directors serving on it), and the number $D_{\alpha\alpha}$ of boards which director α serves on.

4.3 Network Measures

This section covers only a few network measurements. For a more extensive survey see Costa et al. [22] and also Newman [67] as well as Wasserman and Faust [87]. The following definitions assume undirected graphs.

4.3.1 Average Path Length

The *average path length* $\mathcal{L}$ is the mean geodesic (i.e. shortest) distance between node pairs in a graph

$$\mathcal{L} = \frac{1}{\frac{1}{2}n(n-1)} \sum_{i \geq j}^{n} d_{ij}$$

where d_{ij} is the geodesic distance from node i to node j. The average path length is important for instance in networks in which agents benefit from the knowledge of the others (so called knowledge spillovers, see Section 7.2 and 7.3 for examples). The smaller is the average distance among agents the more intense is the knowledge exchange.

For Poisson random graphs (Section 5.1) we obtain $\mathcal{L} = \frac{\ln n}{\ln z}$ where n denotes the number of nodes in the graph and z the average degree. For a regular graph the average path length is $\mathcal{L} = \frac{n}{2z}$. For a complete graph K_n it is trivially $\mathcal{L} = 1$. For a cycle C_n it is half the length of the cycle $\mathcal{L} = \frac{n}{2}$ and for scale free networks (see Section 6.3) it is $\mathcal{L} = \frac{\ln n}{\ln \ln n}$. [1].

4.3.2 Clustering

For each node i, the *local clustering coefficient*, $\mathcal{C}_l(i)$, is simply defined as the fraction of pairs of neighbors of i that are themselves neighbors. The number of possible links between the neighbors of node i is simply $d_i(d_i - 1)/2$. Thus we get

$$\mathcal{C}_l(i) = \frac{|\{e_{jk} \in E(G) : e_{ij} \in E(G) \wedge e_{ik} \in E(G)\}|}{d_i(d_i - 1)/2}$$

The *global clustering coefficient* $\mathcal{C}_l$ is then given by $\mathcal{C}_l = \frac{1}{n} \sum_{i=1}^{n} \mathcal{C}_l(i)$.

A high clustering coefficient $\mathcal{C}_l$ means (in the language of social networks), that two of your friends are likely to be also friends of each other. It also indicates a high redundancy of the network. For a complete graph K_n it is trivially $\mathcal{C}_l = 1$. Let $\langle d \rangle$ denote the average degree then we get for a Poisson random graph $\mathcal{C}_l = \frac{\langle d \rangle}{n-1}$ and for a cycle $\mathcal{C}_l \sim \frac{3}{4}$ for large n [1].

4.3.3 Centrality

Centrality measures the importance of a node on the basis of its position in the network [9, 10, 34]. We can look at a simple example. Consider the star $K_{1,n-1}$ in Figure (2). the most central node is node 3 which has the highest centrality, and all other nodes have minimum centrality. Actually, the star is also the most centralized graph [87].

In the following paragraphs we will introduce different measures of centrality which incorporate different aspects of a nodes position in the network. Degree centrality counts the number of links incident to a node. Closeness centrality measures how many steps it takes to reach any other node in the network. Betweenness centrality measures how many paths between any pair of nodes pass through a node. Finally, eigenvector centrality measures the importance of a node as a function of the importance of its neighbors. The different measures of centrality capture different aspects of the position of an agent in a network and therefore the choice of the right measure depends on the particular application under consideration.

Degree Centrality

The degree centrality of node i is just the number of links d_i . We have that $d_i = \sum_{j=1}^{n} a_{ij} = \sum_{j=1}^{n} a_{ji}$ (since $\mathbf{A}$ is symmetric). If we consider the degree of an agent as a measure of centrality then her centrality depends on the size of the network (with maximum centrality given by $n-1$). In order to overcome this bias one can consider the normalized degree centrality that divides the degree by $n-1$, yielding a measure in $[0,1]$. There are several applications of degree centrality, for example the popularity in friendship networks, the diffusion of information and the spread of infections.

Closeness Centrality

The closeness $C_C(i)$ of i is the reciprocal of the sum of geodesic distances to all other nodes in the graph, that is

$$C_C(i) = \sum_{v \neq i} \frac{1}{d_{iv}}. \tag{5}$$

If an agent has high closeness centrality she can quickly interact with other agents and gather information from them since she has short communication paths to the others.

Betweenness Centrality

The betweenness centrality of i, denoted by $C_B(i)$ is defined as follows.

$$C_B(i) = \sum_{u,v \neq i} \frac{g_{uv}(i)}{g_{uv}}. \tag{6}$$

More precisely, if g_{uv} is the number of geodesic paths d_{uv} from u to v and $g_{uv}(i)$ is the number of paths from u to v that pass through i, then $\frac{g_{uv}(i)}{g_{uv}}$ is the fraction of geodesic paths from u to v that pass through i. Normalized betweenness divides simple betweenness by its maximum value. Agents who are not directly connected might depend on another agent if she lies on a path connecting them. If an agent lies on many such path connecting different components in a network then she has a high betweenness centrality.

Eigenvector Centrality

Eigenvector centrality measures the importance of a node from the importance of its neighbors. Even if a node is only connected to a few others (thus having a low degree centrality) its neighbors may be important, and therefore the node is important too, giving it a high eigenvector centrality. Let's assume that the importance of a node i is measured by x_i. Then the eigenvector centrality of node i is proportional to the sum of the eigenvector centralities of all nodes which are connected to i [68].

$$x_i = \frac{1}{\lambda} \sum_{j \in N_i} x_j = \frac{1}{\lambda} \sum_{j=1}^{n} a_{ij} x_j, \tag{7}$$

where N_i is the set of nodes that are connected to node i, n is the total number of nodes and λ is a constant. In matrix-vector notation we can write $\mathbf{A}\mathbf{x} = \lambda \mathbf{x}$, which is the eigenvector equation. If the proportionality factor λ is given by the largest eigenvalue λ_{PF} (Section 4.1) of the adjacency matrix $\mathbf{A}$ then all the elements in the eigenvector must be positive [50] and we get a proper measure of centrality.

4.4 Dynamics of State Variables and Network Evolution

In the following we introduce a classification of network models in four types. This classification has mainly a didactic value and it should help readers to find their way in the growing landscape of network models.

As mentioned in the beginning of this section, the agents $N = \{1,...,n\}$ in an economic network G can be associated with a state variable x_i, representing agent i's wealth, firm i's output or, in the case of R&D collaborations, knowledge. The links between the agents i and j can be indicated by the elements a_{ij} of an adjacency matrix $\mathbf{A}$. It is important to distinguish between (1) the dynamics taking place on the state variables $\mathbf{x}(t)$ and (2) the evolution of the network $\mathbf{A}(t)$. In the first, the state variables are changed as a result of the interaction among connected nodes. In the latter, nodes or edges are added to/removed from the network by a specific mechanism. For example, the value of the assets of a firm depends on the value of the firms it holds shares in. Even if the links do not change the asset value may change. On the other hand, the links may change in time, depending or not on the asset value. Consequently, there are four types of dynamics that can be investigated in models of economic networks, as illustrated in Figure (4).

In socio-economic systems dynamics and evolution are often coupled. The utility of agents depends on their links to the other agents and agent modify their links

		state variables	
		static	dynamic
network	static	$\frac{dx_i}{dt}=0, \frac{da_{ij}}{dt}=0$	$\frac{dx_i}{dt}\neq 0, \frac{da_{ij}}{dt}=0$
	dynamic	$\frac{dx_i}{dt}=0, \frac{da_{ij}}{dt}\neq 0$	$\frac{dx_i}{dt}\neq 0, \frac{da_{ij}}{dt}\neq 0$

Fig. 4 Possible combinations of static and dynamic state variables x_i associated with the nodes and fixed or changing links indicated by a_{ij} between the nodes for $i, j \in N$

over time depending on the utility they expect or they experience from a link. So, in principle, all systems should be studied with models in which the state variables and the network are dynamic, they co-evolve. However, evolution and dynamics do not necessarily have the same time scale.

Assume that agents have a certain inertia for creating new links and evaluating their existing ones. The rate at which links are formed is much slower than the rate at which the state variables change. In other words, there are two different time scales in our dynamical system: the fast dynamics of the state variables and the slow evolution of the network. The state variables immediately reach their quasi-equilibrium state, whereas the network remains unchanged during this short adaptation time. An illustration can be seen in Figure (4.4). One can say that the variables with the fast dynamics are "slaved" by the variables with the slow dynamics [48], [see also 46, for a review][3]. We will introduce such an approach in Section 7.3 when studying the evolution of R&D networks.

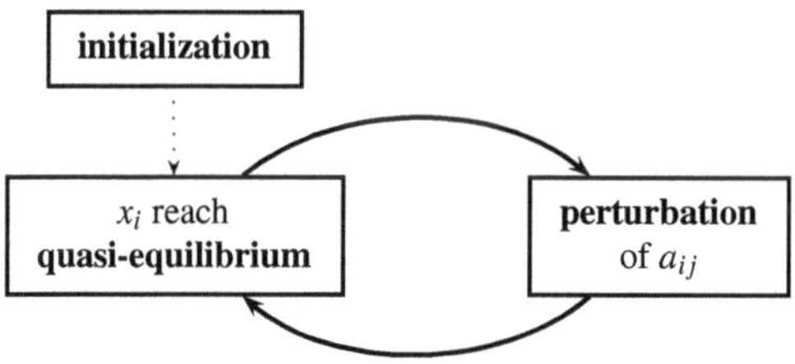

Another example for the coupling of a dynamic network with dynamic state variables are credit relations among firms. The links may represent credit relations among firms, established through contracts. Many financial variables (such as total asset value or solvency ration) of a firm are affected when financial variables change in the connected firms. Despite that, some relations maybe fixed until the expiration of the contract. Therefore, while links may be modified on a time scale of, say, several months, financial variables may vary on a time scale of days.

In the following sections we will discuss several models of networks. According to the classification we have introduced in this section, the models in Section 5 and Section 6 do not consider a state variable attached to the nodes. These models

[3] This principle has been used e.g by [57, 58] in the context of evolutionary biology and by [64] in order to explain the sustainability of informal knowledge exchange in innovation networks.

consider different ways how networks can be constructed in a stochastic network formation process. The process can be viewed as a network evolution. Since these models do not consider a dynamic state variable, they are easier to analyze and so we take them as a starting point before moving on to more complex network models. More complex models follow in the next sections. Both models in Section 7 introduce a state variable attached to the nodes. The nodes are interpreted as agents and the state variable is their utility. The model in Section 7.2 considers the case of a dynamic network but does not assume any dynamics on the state variables (even though the state variables depend on the network). Finally, in Section 7.3 we discuss a model that includes both a dynamic state variable and a dynamic network and it assumes a time-scale separation between the two.

5 Random Network Constructions

In this section we present some basic models of networks. In this discussion we follow Newman [67] as well as Vega-Redondo [84]. For a more detailed mathematical treatment see Chung and Lu [20] and Durrett [30]. The network construction algorithms introduced in this section can be simulated with the Java package "econnet" available upon request to the authors[4]. The algorithms used there serve for educational purposes only and we refer to Batagelj and Brandes [2] for an efficient implementation.

5.1 Poisson Random Graphs

We denote the Poisson random graph by $G(n,p)$ with n nodes and in which every edge is present with probability p. The expected degree is $z = \frac{2p}{n}\binom{n}{2} = p(n-1)$ where $\binom{n}{2}$ is the number of edges in the complete graph K_n. The degree distribution of $G(n,p)$ is given by

$$p_k = \binom{n-1}{k} p^k (1-p)^{n-1-k}, \tag{8}$$

where p_k is the probability that a randomly chosen node has degree k. We have that

$$\lim_{n\to\infty} p_k = \frac{z^k e^{-z}}{k!} = \text{Pois}(z;k). \tag{9}$$

Many results on the topological properties and phase transitions can be derived for Poisson random graphs. We refer to West [91], Chung and Lu [20], Durrett [30] and Bollobas [7] for the interested reader.

---

[4] Mail to mkoenig@ethz.ch.

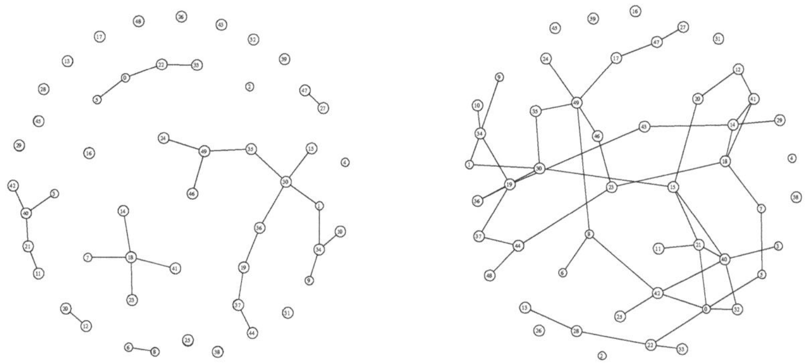

Fig. 5 Poisson random graph $G(n,p)$ 50% below the phase transition $p = \frac{1}{n-1}$ (left) and at the phase transition (right). The graph was generated with the Java package "econnet" and the ARF layout algorithm [38]

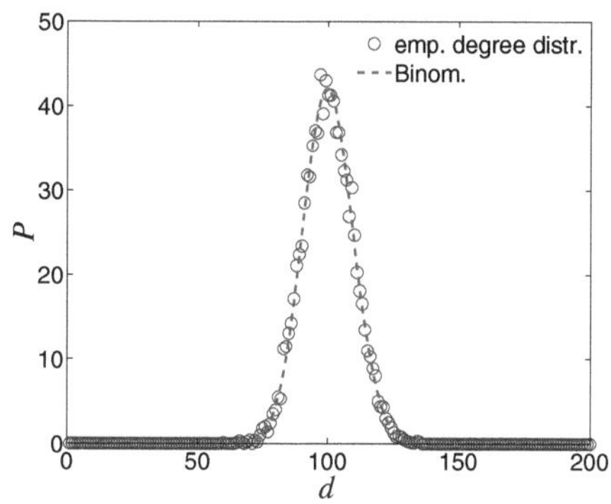

Fig. 6 Degree distribution of the Poisson random graph $G(n,p)$ with $p = 0.1$, $n = 1000$ and averaged over 10 realizations

5.2 *Generalized Random Graphs*

In the following we give a short introduction to random graphs with arbitrary degree distributions. For a detailed discussion (including all the material presented here) see Newman et al. [69, 70].

5.2.1 Random Graph Construction

Consider a set of nodes $N = \{1,...,n\}$. A degree sequence of a graph is a list of node degrees $d_1 \geq d_2 \geq ... \geq d_n$ with the property that $\sum_{k=1}^{n} d_k$ must be even. We construct the random graph G by creating d_i half-edges attached to node i, and then pair the half-edges at random. The resulting graph may have loops and multiple edges.

5.2.2 Neighborhood Size, Diameter, Phase Transition and Clustering

The probability of a randomly chosen node having degree k is given by

$$p_k = \frac{1}{n}|\{i \in N : d_i = k\}|. \tag{10}$$

Its generating function is defined by [31]

$$G_0(x) = \sum_{k=0}^{\infty} p_k x^k. \tag{11}$$

p_k is the probability that a randomly chosen node has degree k. The distribution p_k is assumed to be correctly normalized, so that

$$G_0(1) = \sum_{k=0}^{\infty} p_k = 1. \tag{12}$$

$G_0(x)$ is finite for all $|x| \leq 1$. If the distribution is Poisson, $p_k = z^k e^{-z}/k!$, then the generating function is

$$G_0(x) = \sum_k \frac{1}{k!} e^{-z} z^k x^k = e^{-z} \sum_k \frac{(zx)^k}{k!} = e^{z(x-1)}. \tag{13}$$

The probability p_k is given by the kth derivative of G_0 according to

$$p_k = \frac{1}{k!} \frac{d^k G_0}{dx^k}\bigg|_{x=0}. \tag{14}$$

Thus, the function $G_0(x)$ encapsulates all the information of the discrete probability distribution p_k.

The mean (first-order moment), e.g. the average degree z of a node, is given by

$$z = \langle k \rangle = \sum_{k=0}^{\infty} k p_k = G_0'(1). \tag{15}$$

Higher order moments of the distribution can be calculated from higher derivatives. In general we have

$$\langle k^n \rangle = \sum_{k=0}^{\infty} k^n p_k = \left(x \frac{d}{dx} \right)^n G_0(x)\bigg|_{x=1}. \tag{16}$$

For the first two moments of the Poisson distribution we obtain

$$x \frac{d}{dx} e^{z(x-1)} = z \tag{17}$$

$$\left(x \frac{d}{dx} \right)^2 e^{z(x-1)} = z(1+z). \tag{18}$$

If we select a node i then the number of neighbors has distribution p. However, the distribution of the first neighbors of a node is not the same as the degree distribution of nodes on the graph as a whole. Because high-degree nodes have more edges connected to it, there is a higher probability that a randomly chosen edge is incident to it, in proportion to the node degree. The number of nodes with degree k is np_k. The number of edges incident to nodes with degree k is given by knp_k. This is equal the number of possibilities to select an edge which is incident to a node with degree k. Thus, the probability that a node incident to a randomly chosen edge has degree k is proportional to kp_k and not just p_k. Through normalization we get that the probability distribution of the degree among neighbors of a randomly selected node i is given by [67]

$$q_k = \frac{kp_k}{\sum_s sp_s}. \tag{19}$$

The average degree of a neighboring node is then

$$\sum_k kq_k = \frac{\sum_k k^2 p_k}{\sum_s sp_s} = \frac{\langle k^2 \rangle}{\langle k \rangle}. \tag{20}$$

The corresponding generating function is

$$\begin{aligned}\sum_k q_k x^k &= \frac{\sum_k kp_k x^k}{\sum_s sp_s} \\ &= \frac{1}{\langle k \rangle} x \underbrace{\sum_k p_k k x^{k-1}}_{G'_p(x)} \\ &= \frac{xG'_p(x)}{G'_p(1)}.\end{aligned} \tag{21}$$

If we are interested in the (excess) distribution p^*_k of links of a node that can be reached along a randomly chosen edge, other then the one we arrived along, $p^* = q_{k+1} \propto (k+1)p_{k+1}$, then its generating function is

$$\begin{aligned}G_{p^*}(x) &= \sum_k \frac{(k+1)p_{k+1}}{\sum_s sp_s} x^k \\ &= \frac{1}{G'_p(1)} \sum_k kp_k x^{k-1} \\ &= \frac{G'_p(x)}{G'_p(1)}.\end{aligned} \tag{22}$$

In order to compute the expected number of second neighbors we have to exclude node i from the degree count of its neighboring node and obtain

$$q_{k-1} = \frac{kp_k}{\sum_s p_s}, \tag{23}$$

or equivalently

$$q_k = \frac{(k+1)p_{k+1}}{\sum_s s p_s}. \tag{24}$$

The average (excess) degree of such a node is then

$$\begin{aligned}\sum_k k q_k &= \frac{\sum_k (k+1)p_{k+1}}{\sum_s s p_s} \\ &= \frac{\sum_k (k-1)k p_k}{\sum_s s p_s} \\ &= \frac{\langle k^2 \rangle - \langle k \rangle}{\langle k \rangle}.\end{aligned} \tag{25}$$

The average total number of second neighbors of a node is given by the average degree of the node times the excess degree of the first neighbours:

$$z_2 = \langle k \rangle \frac{\langle k^2 \rangle - \langle k \rangle}{\langle k \rangle} = \langle k^2 \rangle - \langle k \rangle. \tag{26}$$

The average number of second neighbors is then equal to the difference between the second- and first-order moments of the degree distribution p. The expectation of the first neighbors is $z_1 = G'_p(1)$ and for the second neighbors one derives $z_2 = G''_p(1)$. Note that in general the number of rth neighbors is not simply the rth derivative of the generating function.

The average number of edges leaving from a second neighbor is given by Equation (25). This also holds for any distance m away from a randomly chosen node. Thus, the average number of neighbors at distance m is

$$\begin{aligned}z_m &= \frac{\langle k^2 \rangle - \langle k \rangle}{\langle k \rangle} z_{m-1} \\ &= \frac{z_2}{z_1} z_{m-1} \\ &= \left(\frac{z_2}{z_1}\right)^{m-1} z_1,\end{aligned} \tag{27}$$

where $z_1 = \langle k \rangle$ and z_2 is given by Equation (26). Depending on whether z_2 is greater than z_1 or not, this expression will either diverge or converge exponentially as m becomes large so that the average number of neighbors of a node is either finite or infinite for $n \to \infty$. We call this abrupt change a phase transition at $z_1 = z_2$. This condition can be written as $\langle k^2 \rangle - 2\langle k \rangle = 0$ or

$$\sum_k k(k-2)p_k = 0. \tag{28}$$

In the above sum isolated nodes and nodes with degree one do not contribute since they can be removed from a graph without changing its connectivity.

We assume that $z_2 \gg z_1$ so that there exists a giant component essentially including all the nodes and most of the nodes are far from each other, at around distance D, the diameter of the graph. This means that

$$n \sim z_D = \left(\frac{z_2}{z_1}\right)^{D-1} z_1, \tag{29}$$

which leads to

$$\begin{aligned} \ln\frac{n}{z_1} &\sim (D-1)\ln\frac{z_2}{z_1} \\ D &\sim \frac{\ln\frac{n}{z_1}}{\ln\frac{z_2}{z_1}} + 1. \end{aligned} \tag{30}$$

For the special case of a Poisson network with $z_1 = z$ and $z_2 = z^2$ we obtain for large n

$$D \sim \frac{\ln\frac{n}{z}}{\ln z} + 1 = \frac{\ln n}{\ln z} \tag{31}$$

In the following we study the clustering coefficient C_l of a random graph. For this, we consider a particular node i. The jth neighbor of i has k_j links emanating from it other than the edge e_{ij} and k_j is distributed according to the distribution q. The probability that node j is connected to another neighbor s is $\frac{k_j k_s}{nz}$, where k_s is distributed according to q. The average of this probability is precisely the clustering coefficient

$$\begin{aligned} C_l &= \frac{\langle k_j k_s \rangle}{nz} \\ &= \frac{1}{nz}\left(\sum_k k q_k\right)^2 \\ &= \frac{z}{n}\left(\frac{\langle k^2\rangle\langle k\rangle}{\langle k\rangle^2}\right)^2 \\ &= \frac{z}{n}\left(c_v^2 + 1 - \frac{1}{\langle k\rangle}\right)^2, \end{aligned} \tag{32}$$

where $c_v = \frac{\langle(k-\langle k\rangle)\rangle}{\langle k\rangle^2}$ is the coefficient of variation of the degree distribution - the ratio of the standard deviation to the mean. For Poisson networks we get $z_2 = \langle k^2\rangle - \langle k\rangle = \langle k\rangle^2 = z^2$ and the clustering coefficient is $C_l = \frac{z}{n}$. For arbitrary degree distributions we still have that $\lim_{n\to\infty} C_l = 0$ but the leading term in (32) may be higher.

5.2.3 Average Component Size Below the Phase Transition

With similar methods one can compute the average size of the connected component a node belongs to. Here we closely follow the discussion in Baumann and Stiller [6]. The computation is valid under following assumptions:

(i) The network contains no cycles. One can show that this assumption is a good approximation for big, sparse random networks.
(ii) For any edge e_{uv} of a node u the degree of v is distributed independently of u's neighbors and independently of the degree of u.

We then choose an edge e uniform at random among the edges in $E(G)$. We select one of the incident nodes of e at random, say v. Let p^0 denote the distribution of the size of the component of v in the graph of $E(G) \setminus e$. Further, let p^* be the distribution of the degree of v in $E(G) \setminus e$. Then $p^0(1) = p^*(0)$. If the degree of v is k then we denote the neighbors of v in $E(G) \setminus e$ as $n_1, ..., n_k$. We define the following probability: $P_k(s-1)$ is the probability that the size of the components of the k nodes $n_1, ..., n_k$ in $E(G) \setminus \{e, e_{vn_1}, ..., e_{vn_k}\}$ sum up to $s-1$. Then we can write

$$p_s^0 = \sum_k p_k^* P_k(s-1). \tag{33}$$

Now let S denote a random variable that is the sum of m independent random variables $X_1, ..., X_m$, that is

$$S = X_1 + ... + X_m, \tag{34}$$

then the generating function of S is given by

$$G_S(x) = G_{X_1}(x) G_{X_2}(x) \cdots G_{X_m}(x). \tag{35}$$

Consider the distribution of the sum of the degrees of two nodes when p_k is the distribution of a single node. Then the sum of the degrees has a generating function $G_0(x)^m$. For two nodes we get

$$\begin{aligned} G_0(x)^2 &= \left(\sum_k p_k x^k\right)^2 \\ &= \left(\sum_k p_k x^k\right)\left(\sum_j p_j x^j\right) \\ &= \sum_{j,k} p_j p_k x^{j+k} \\ &= p_0 p_0 x^0 + (p_0 p_1 + p_1 p_0) x^1 + (p_2 p_0 + p_1 p_1 + p_0 p_2) x^2 + \cdots . \end{aligned} \tag{36}$$

The coefficients of the powers of x^n are clearly the sum of all products $p_j p_k$ such that $j + k = n$ and hence it gives the probability that the sum of the degrees of the two nodes will be n. We can use a similar argument to prove that higher order powers of generating functions can be computed in the same way.

Following our assumptions, the edges e_{vn_i} are chose independently and uniform at random among all edges in $E(G)$. Therefore, P_k is distributed as the sum of k random variables, which are in turn distributed according to p^0. Using the powers of generating functions we have that $G_{P_k} = G_{p^0}(x)^k$. Moreover, the generating function of p^0 is

$$\begin{aligned} G_{p^0}(x) &= \sum_s p^0 x^s \\ &= \sum_s x^s \sum_k p_k^* P_k(s-1) \\ &= x \sum_k p_k^* \underbrace{\sum_s x^{s-1} P_k(s-1)}_{G_{P_k}(x) = G_{p^0}(x)^k} \\ &= x \sum_k p_k^* G_{p^0}(x)^k \\ &= x G_{p^*}\left(G_{p^0}(x)\right). \end{aligned} \tag{37}$$

The quantity we are actually interested in is the distribution of the size of the component a randomly chosen node belongs to. The number of edges emanating from such a node is distributed according to the degree distribution p_k. Each such edge leads to a component whose size is drawn from the distribution generated by the function $G_{p^0}(x)$. In a similar way to the derivation of Equation (37), one can show that the size of the component to which a randomly selected node belongs is generated by

$$\begin{aligned} G_{\tilde{p}}(x) &= x \sum_k p_k G_{p^0}(x)^k \\ &= x G_p\left(G_{p^0}(x)\right). \end{aligned} \tag{38}$$

The expected component size of a randomly selected node can be computed directly from above. The expectation of a distribution is the derivative of its generating function evaluated at point 1. Therefore the mean component size $\langle s \rangle$ is given by

$$\langle s \rangle = G'_{\tilde{p}}(1) = \underbrace{G_p\left(G_{p^0}(1)\right)}_{=1} + G'_p\left(G_{p^0}(1)\right) G'_{p^0}(1). \tag{39}$$

where we used the normalization of the generating function. From (37) we know that

$$\begin{aligned} G'_{p^0}(1) &= G_{p^*}\left(G_{p^0}(1)\right) + G'_{p^*}\left(G_{p^0}(1)\right) G'_{p^0}(1) \\ &= 1 + G'_{p^*}(1) G'_{p^0}(1), \end{aligned} \tag{40}$$

and thus $G'_{p^0}(1) = \frac{1}{1-G'_{p^*}(1)}$. Inserting this equation into Equation (39) yields

$$\langle s \rangle = 1 + \frac{G'_p(1)}{1 - G'_{p^*}(1)}. \tag{41}$$

We further have that

$$
\begin{aligned}
G'_p(1) &= \sum_k k p_k = \langle k \rangle = z_1 \\
G'_{p^*}(1) &= \frac{\sum_k k(k-1)p_k}{\sum_l l p_l} \\
&= \frac{\langle k^2 \rangle - \langle k \rangle}{\langle k \rangle} \\
&= \frac{z_2}{z_1}.
\end{aligned}
\tag{42}
$$

Therefore, the average component size below the transition is

$$
\langle s \rangle = 1 + \frac{z_1^2}{z_1 - z_2}. \tag{43}
$$

The above expression diverges for $z_1 = z_2$ which signifies the formation of the giant component. We can also write the condition for the phase transition as $G'_{p^*}(1) = 1$. We see that for $p = 0$ $\langle s \rangle = 1$ (an empty graph contains only isolated nodes). For the Poisson random graph $z_1 = z = p(n-1)$, $z_2 = z^2$ and thus we get $\langle s \rangle = 1 + \frac{p(n-1)}{1-p(n-1)}$.

5.3 *The Watts-Strogatz "Small-World" Model*

The model draws inspiration from social systems in which most people have friends among their immediate neighbors, but everybody has one or two friends who are a far away - people in other countries, old acquaintances, which are represented by the long-range edges obtained by rewiring. Empirically, in social networks the average distance turns out to be "small"; the fact that any two persons in the US are separated on average by only six acquaintances is the so called "Small-World" phenomenon discovered by Milgram [66]. Watts and Strogatz [89] introduced a "Small-World" network model which has triggered an avalanche of works in the field. Their model generates a one-parameter family of networks laying in between an ordered lattice and a random graph. We will explain how such a "Small-World" network can be constructed in the next section.

5.3.1 "Small-World" Network Construction

The initial network is a one-dimensional ring of n nodes (if each node has only two neighbors it is a cycle) as shown in Figure (7), with periodic boundary conditions, each node being connected to its z nearest neighbors. The nodes are then visited one after the other: each link connecting a node to one of its $\frac{z}{2}$ neighbors in the clockwise order is left in place with probability $1-p$, and with probability p is reconnected to a randomly chosen node. With varying p the system exhibits a transition between order ($p = 0$) and randomness ($p = 1$).

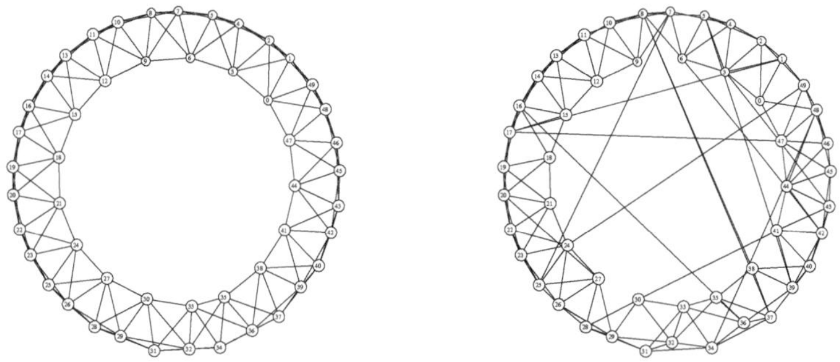

Fig. 7 Regular (lattice) graph with $n = 50$ nodes and neighborhood size $z = 6$ (left). Small World graph with $n = 50$ nodes, neighborhood size $z = 6$ of the underlying lattice and rewiring probability $p = 0.1$ (right). The graph was generated with the Java package "econnet" and the ARF layout algorithm [38]

5.3.2 Degree Distribution

For $p = 0$, each node has the same degree, z. On the other hand, a non-zero value of p introduces disorder in the network, in the form of a non-uniform degree distribution, while maintaining a fixed average degree $\langle X \rangle = z$. Let us denote $P(X = k)$ the probability of the degree of a node being equal k.

Since $\frac{z}{2}$ of the original z edges are not rewired by the above procedure, the degree of node i can be written as [1].

$$X = \frac{z}{2} + n_0 + n_+ \tag{44}$$

with $n_0 + n_+ \geq 0$. n_0 denotes the number of links that have been left in place during the rewiring procedure (with probability $1 - p$) and n_0 denotes the number of links that have been rewired to node i from other nodes (with probability $p/(n-1)$, since there are $n - 1$ other nodes). This sequence of independent events (the links left in place as well as the rewired links) is actually a Bernoulli process. Thus, the probabilities are given by Binomial distributions

$$P(n_0 = s) = \binom{\frac{z}{2}}{s} (1-p)^s p^{\frac{z}{2}-s}, \tag{45}$$

with $0 \leq s \leq \frac{z}{2}$ and

$$P(n^+ = s) = \binom{(n-1)\frac{z}{2}}{s} \left(\frac{p}{n-1}\right)^s \left(1 - \frac{p}{n-1}\right)^{(n-1)\frac{z}{2}-s}, \tag{46}$$

where $0 \leq s \leq (n-1)\frac{z}{2}$ and $n - 1$ is the number of other nodes, $\frac{z}{2}$ the maximum number of edges that can be rewired by other nodes. If we define

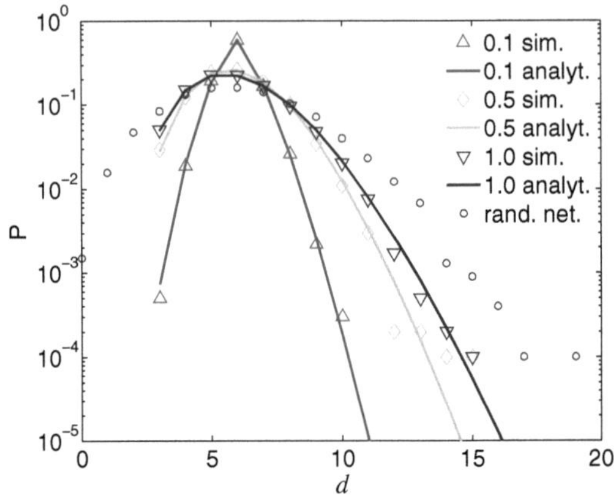

Fig. 8 Empirical and theoretical degree distributions of the "Small-World" network for $n = 500$, neighborhood size $z = 6$ and and different values of the rewiring probability $p \in \{0.1, 0.5, 1.0\}$

$$\begin{aligned} N &= (n-1)\tfrac{z}{2} \\ q &= \tfrac{p}{n-1} \\ \lambda &= Nq = \tfrac{z}{2}(n-1)\tfrac{p}{n-1} = nq, \end{aligned} \tag{47}$$

we get the standard form of the Binomial distribution

$$P(n^+ = s) = \binom{N}{s} q^s (1-q)^{N-s}. \tag{48}$$

For $N \to \infty$ respectively $n \to \infty$ we obtain the Poisson distribution

$$P(n^+ = s) = \frac{\lambda^s e^{-\lambda}}{s!} = \frac{\left(\frac{pz}{2}\right)^s e^{-\left(\frac{pz}{2}\right)}}{s!}. \tag{49}$$

Thus, we get for $k \geq \frac{z}{2}$ (k links remain unchanged by construction)

$$P(X = k) = \sum_{i=0}^{\min\{k-\frac{z}{2},\frac{z}{2}\}} \binom{\frac{z}{2}}{2} (1-p)^i p^{\frac{z}{2}-i} \frac{\left(\frac{pz}{2}\right)^{k-\frac{z}{2}-i}}{\left(k-\frac{z}{2}-i\right)!} e^{-\frac{pz}{2}}. \tag{50}$$

The upper bound in the sum above guarantees that $n_0 \leq \frac{z}{2}$. Since any degree $k > \frac{z}{2}$ must come from new edges. Figure (8) shows the degree distribution for different values of p.

5.3.3 Average Path Length and Clustering Coefficient

For a cycle ($p = 0$) we have a linear chain of nodes and we find for the average path length (defined in Section 4.3) for large n [1]

$$L(p=0) = \frac{n(n+z-2)}{2z(n-1)} \sim n/2z \gg 1. \tag{51}$$

Moreover, for $p = 0$ each node has z neighbors and the number of links between these neighbors is $\frac{3z(z/2-1)}{4}$ and it follows that for large n [1]

$$\mathcal{C}_l(p=0) = \frac{3(z/2-1)}{2(z-1)} \sim 3/4. \tag{52}$$

Thus, $\mathcal{L}$ scales linearly with the system size, and the clustering coefficient is large and independent of n. On the other hand, for $p \to 1$ the model converges to a random graph for which $\mathcal{L}(p=1) \sim \ln(n)/\ln(z)$ and $\mathcal{C}_l(p=1) \sim z/n$ when n is large, thus $\mathcal{L}$ scales logarithmically with n and the clustering coefficient decreases with n. Based on these scaling relationships, one could expect that a large (small) value of $\mathcal{C}_l$ is always associated with a large (small) value of $\mathcal{L}$. Unexpectedly, it turns out that there is a broad range of values of p in which $\mathcal{L}(p<1)$ is close to $\mathcal{L}(p=1)$ and yet $\mathcal{C}_l(p<1) \gg \mathcal{C}_l(p=1)$. The coexistence of small $\mathcal{L}$ and large $\mathcal{C}_l$ means that the network is a "Small-world" like a random graph *and* has high clustering like a lattice. Interestingly, this feature is found in many real networks. In a regular lattice ($p=0$) the clustering coefficient $\mathcal{C}_l$ does not depend on the system size but only on its topology. As the edges of the network are randomized, the clustering coefficient remains close to $\mathcal{C}_l(p=0)$ up to relatively large values of p, while the average path length $\mathcal{L}$ drops quite rapidly. This is the reason of the onset of the small world regime. We show examples for the clustering coefficient and the average path length in the "Small-World" network in Figure (9).

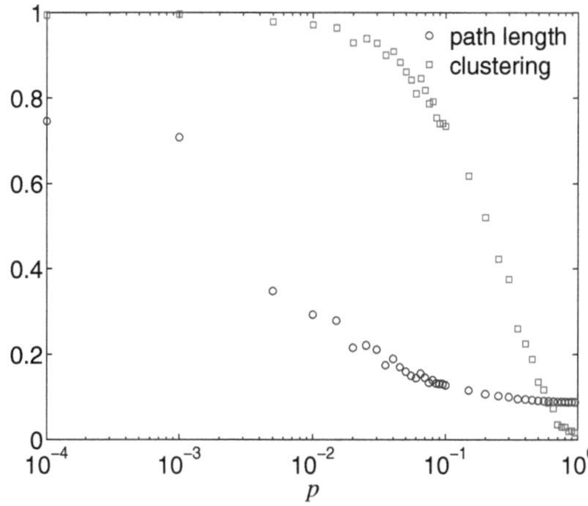

Fig. 9 Clustering coefficient $\mathcal{C}_l$ and average path length $\mathcal{L}$ of the "Small-World" network for with $n = 500$, neighborhood size $z = 6$ and and different values of the rewiring probability p. The average path length $\mathcal{L}$ is normalized to the corresponding value of the lattice. For $p = 1$ the normalized path length (proportional to $\ln n/n$) converges to zero for large n

6 Growing Random Networks

In the next sections we derive the degree distributions for two types of networks, the uniform and the preferential attachment network, illustrated in Figure (10) and their corresponding degree distributions in Figure (11). Both networks are generated by continuously adding nodes to the existing network. The difference is the following: in the uniform attachment network new nodes form links uniformly to the existing nodes and in the preferential attachment network new nodes form links more likely to existing nodes with higher degree. In the derivation of the degree distribution we follow closely Vega-Redondo [84].

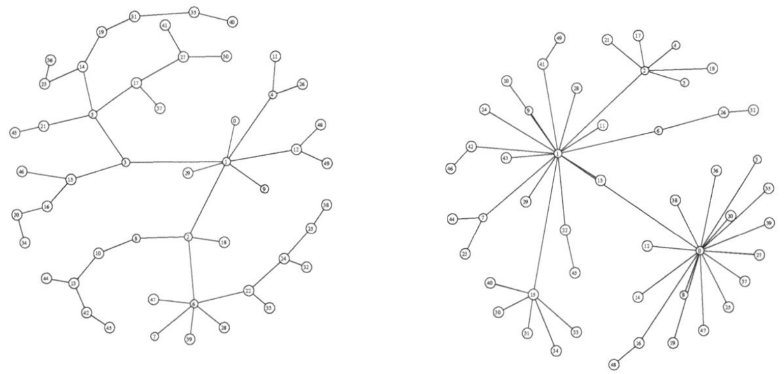

Fig. 10 Uniform attachment (left) and preferential attachment (right) networks with $n = 50$ nodes. The graph was generated with the Java package "econnet" and the ARF layout algorithm [38]

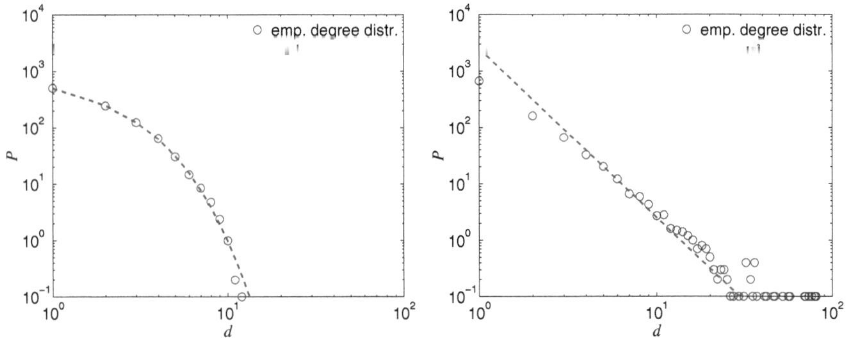

Fig. 11 Degree distribution of the uniform (left) and preferential attachment (right) networks for $n = 1000$ averaged over 10 realizations

6.1 Uniform Attachment Network Construction

The network is constructed as follows. Times is measured at countable dates $t \leq 0$. A node that enters the network at time t is attached the label t. We initialize nodes $1,2$ and the edge 12. Then, at every step $t > 2$ we add a new node t and create the edge e_{ts}, where node s is selected uniformly at random from the set $\{1,...,t-1\}$ of already existing nodes in the network.

6.2 Degree Distribution

In the following we derive the degree distribution if edges are attached to existing nodes with uniform probability. Denote by $q_t(s,k)$ the probability that a particular node s has degree k at time t where $s \leq t$. Any existing node s enjoys degree $k \geq 1$ at time $t+1$ if, and only if, one of the following events occurs: (i) Node s had degree $k-1$ at time t (with probability $q_t(s,k-1)$) and is chosen to be linked by the entering node at time t (with probability $\frac{1}{t+1}$), or (ii) node s already had degree k at time t (with probability $q_t(s,k)$) and is not chosen by the new node (with probability $1-\frac{1}{t+1}$).

Thus we get the following master equation [73, 90] and Vega-Redondo [84, p. 272]

$$q_{t+1}(s,k) = \frac{1}{t+1} q_t(s,k-1) + \left(1 - \frac{1}{t+1}\right) q_t(s,k), \tag{53}$$

with the boundary conditions[5]

$$\begin{aligned} q_1(0,k) &= q_1(1,k) = \delta_{k,1} \\ q_t(t,k) &= \delta_{k,1}. \end{aligned} \tag{54}$$

Denote $p_t(k)$ the probability that a randomly selected node has any given degree k at time t. $p_t(k)$ is the degree distribution at time t. Assuming that the selection of nodes is a sequence of stochastically independent events, it follows that

$$p_t(k) = \frac{1}{t+1} \sum_{s=0}^{t} q_t(s,k) \tag{55}$$

Summation over all nodes $s = 0,...,t$ in Equation (53) yields

$$\sum_{s=0}^{t} q_{t+1}(s,k) = \frac{1}{t+1} \sum_{s=0}^{t} q_t(s,k-1) + \left(1 - \frac{1}{t+1}\right) \sum_{s=0}^{t} q_t(s,k), \tag{56}$$

and further adding the term $q_{t+1}(t+1,k)$ on both sides gives

[5] The Kronecker-Delta is defined as $\delta_{ij} = 1$ if $i = j$ and $\delta_{ij} = 0$ if $i \neq j$.

$$\sum_{s=0}^{t+1} q_{t+1}(s,k) = \frac{1}{t+1}\sum_{s=0}^{t} q_t(s,k-1) + \left(1 - \frac{1}{t+1}\right)\sum_{s=0}^{t} q_t(s,k) + \delta_{k,1} \tag{57}$$
$$= p_t(k-1) + t p_t(k) + \delta_{k,1},$$

where we used the boundary condition $q_{t+1}(t+1,k) = \delta_{k,1}$. This reflects the fact that, in every period $t+1$, the entering node $t+1$ always represents a unit contribution to the set of nodes with degree 1 (and only these nodes). Then, with

$$(t+2)\frac{1}{t+2}\sum_{s=0}^{t+1} q_{t+1}(s,k) = (t+2)p_{t+1}(k), \tag{58}$$

we may write Equation (57) as follows

$$(t+2)p_{t+1}(k) - t p_t(k) = p_t(k-1) + \delta_{k,1}, \tag{59}$$

which is the law of motion of the degree distribution. In the limit $t \to \infty$, $p_t(k)$ attains its stationary distribution $p(k)$.

$$2p(k) = p(k-1) + \delta_{k,1} \tag{60}$$

We can solve the above equation for $k > 1$ ($\delta_{k,1} = 0$):

$$p(k) = 2^{-k}. \tag{61}$$

Since there are no disconnected nodes in the network we have that $p(0) = 0$. For $k = 1$ we thus find that Equation (61) also solves Equation (60) for any $k = 1, 2, \ldots$. This means that the long run stationary degree distribution is geometric.

6.3 Preferential Attachment Network Construction

The network is constructed in a similar way as in the uniform attachment network formation process. We initialize nodes $1,2$ and edge 12, setting $t = 3$. Let $k_t(s)$ denote the degree of node s at time t. Then, at every step t we add a node t and create the edge e_{ts} with probability $k_t(s)/\sum_{r=0}^{t-1} k_t(r)$.

6.4 Degree Distribution

The master equation for the probabilities $q_t(s,k)$ that any node s has degree $k \geq 1$ at time t, $s \leq t$ is given by

$$q_{t+1}(s,k) = \frac{k-1}{2t} q_t(s,k-1) + \left(1 - \frac{k}{2t}\right) q_t(s,k). \tag{62}$$

There are two exclusive events that may lead node s to have degree k in time step $t+1$: (i) Node s had degree $k-1$ at time t and the new node $t+1$ establishes a link to s, or (ii) node s had degree k at time t and the new node $t+1$ does not form a link to it.

The probability of event (i) is given by $q_t(s,k-1)$ multiplied by the ratio of the degree, $k-1$, to the sum of the degrees, that is $2t$. The probability of the event (ii) is the complement of the probability that the new node establishes a link to s with degree k, that is $1-\frac{k}{2t}$ times $q_t(s,k)$. Summing over all nodes $s \leq t+1$ in Equation (62) and adding the term $q_{t+1}(t+1,k)$ on both sides, we arrive at the law of motion for the degree distribution

$$\sum_{s=0}^{t+1} q_{t+1}(s,k) = \frac{k-1}{2t} \sum_{s=0}^{t} q_t(s,k-1) + \left(1 - \frac{k}{2t}\right) \sum_{s=0}^{t} q_t(s,k) + \delta_{k,1}. \tag{63}$$

We have that

$$\begin{aligned} \sum_{s=0}^{t+1} q_{t+1}(s,k) &= \frac{1}{2}\frac{t+1}{t} \left[(k-1)\frac{1}{t+1} \sum_{s=0}^{t} q_t(s,k-1) - k\frac{1}{t+1} \sum_{s=0}^{t} q_t(s,k) \right] \\ &\quad + (t+1)\frac{1}{t+1} \sum_{s=0}^{t} q_t(s,k) + \delta_{k,1} \\ &= \frac{1}{2}\frac{t+1}{t} ((k-1p_t(k-1) - kp_t(k))) + (t+1)p_t(k) + \delta_{k,1}. \end{aligned} \tag{64}$$

Using the fact that

$$\begin{aligned} \sum_{s=0}^{t+1} q_{t+1}(s,k) &= (t+2)\frac{1}{t+2} \sum_{s=0}^{t+1} q_{t+1}(s,k) \\ &= (t+2)p_{t+1}(k), \end{aligned} \tag{65}$$

we get

$$(t+2)p_t(k) = \frac{1}{2}\frac{t+1}{t} ((k-1)p_t(k-1) - kp_t(k)) + (t+1)p_t(k) + \delta_{k,1}. \tag{66}$$

In the limit, as $t \to \infty$, and each $p_t(k)$ converges to its stationary distribution $p(k)$,we obtain

$$p(k) = \frac{1}{2}((k-1)p(k-1) - kp(k)) + \delta_{k,1}, \tag{67}$$

since $p_{t+1}(k) = p_t(k)$ in the stationary state and for large t, $t+2 \sim t+1 \sim t$. The solution for $k > 1$ of Equation (67) is given by

$$p(k) = \frac{4}{k(k+1)(k+2)}. \tag{68}$$

One can write Equation (67) in the form

$$\begin{aligned} p(k) &= \frac{1}{2}\left(k\left[p(k-1)-p(k)\right]-p(k-1)\right)+\delta_{k,1} \\ &= -\frac{1}{2}\left(k\frac{p(k)-p(k-\Delta k)}{\Delta k}+p(k-\Delta k)\right)+\Delta_k, \end{aligned} \tag{69}$$

where $\Delta k = 1$. Taking the limit $\Delta k \to 0$ one obtains the continuous form of (67)

$$\begin{aligned} p(k) &= -\frac{1}{2}\left(k\frac{dp}{dk}+p(k)\right) \\ &= -\frac{1}{2}\frac{d}{dk}\left(kp(k)\right). \end{aligned} \tag{70}$$

The solution of this equation is given by

$$p(k) = 2k^{-3}, \tag{71}$$

where the factor 2 comes from the normalization condition $\int_1^\infty p(k)dk = 1$. We find, therefore, that the degree distribution satisfies a power law of the form $p(k) \propto k^{-\gamma}$. If the frequency of nodes with a degree k is proportional to $k^{-\gamma}$, then the distribution is scale-free.

7 Strategic Network Formation

In the preceding sections we have studied the formation of networks under different stochastic processes governing the way in which links are formed between nodes. However, in social and economic settings the choice of forming a link or not is governed by individual incentives and the potential benefits versus costs that arise from the establishment or withdrawal from a relationship. Strategic network formation[6] thus constitute strategic settings in which the payoffs of agents are interdependent and this interdependency is rooted in a network structure.

7.1 Efficiency and Pairwise Stability

If we want to model network formation based on individual incentives then we first need to introduce a utility function that describes the net benefits an agent enjoys from being part of the network. This can formally be done via a utility function $u_i : G \to \mathbb{R}$ that assigns each agent $i \in N = \{1,...,n\}$ a utility from the network G.

[6] We restrict our discussion in this tutorial to non-cooperative games on networks [see also 42, 55, 84, 93, for an excellent introduction]. Cooperative games on networks have been treated in [76]. For algorithmic issues we refer to Nisan et al. [71].

Based on a properly defined utility function we can address the question of how efficient or stable certain network structures are. We treat both of these issues in the next paragraphs.

A measure of the global performance of the network is introduced by its efficiency. The total utility of a network is defined by $U(G) = \sum_{i=1}^{n} u_i(G)$. A network is considered efficient if it maximizes the total utility of the network $U(G)$ among all possible networks, G with n nodes [56].

Definition 1. Denote the set of networks with n nodes by $\mathcal{G}(n)$. A network G is *efficient* if $U(G) = \sum_{i=1}^{n} u_i(G) \geq U(G') = \sum_{i=1}^{n} u_i(G')$ for all $G' \in \mathcal{G}(n)$.

The evolution of the network is the result of strategic interactions between agents when they decide to create or delete links. In the following we consider a particularly simple network formation process. At every time step a pair of agents is chosen at random and tries to establish a new link between them or delete an already existing one. If a link is added, then the two agents involved must both agree to its addition, with at least one of them strictly benefiting (in terms of a higher utility) from its formation. Similarly a deletion of a link can only take place in a mutual agreement. The subsequent addition and deletion of links creates a sequence of networks. If no new links are accepted nor old ones are deleted then the network reaches an equilibrium. An equilibrium under the above described network formation process leads us to the notion of pairwise stability, introduced by Jackson and Wolinsky [56].

Definition 2. A network G is *pairwise stable* if and only if

(i) for all $e_{ij} \in E(G)$, $u_i(G) \geq u_i(E \backslash e_{ij})$ and $u_j(G) \geq u_j(E \backslash e_{ij})$,
(ii) for all $e_{ij} \notin E(G)$, if $u_i(G) < u_i(E \cup e_{ij})$ then $u_i(G) > u_j(E \cup e_{ij})$.

A network is pairwise stable if and only if (i) removing any link does not increase the utility of any agent, and (ii) adding a link between any two agents, either does not increase the utility of any of the two agents, or if it does increase one of the two agents' utility then it decreases the other agent's utility.

The point here is that establishing a new link with an agent requires the consensus, that is, an increase in utility, of both of them. The notion of pairwise stability can be distinguished from the one of Nash equilibrium[7] which is appropriate when each agent can establish or remove unilaterally a connection with another agent.

In Section 7.2 and in Section 7.3 we will give specific examples for different utility functions. As we will show, the particular choice of the utility function significantly shapes individual incentives to form or severe links. As a result, different incentive structures translate into network outcomes that can vary considerably in terms of efficiency and stability.

---

[7] Considering two agents playing a game (e.g. trading of knowledge) and each adopting a certain strategy. A Nash equilibrium is characterized by a set of strategies where each strategy is the optimal response to all the others.

7.2 The Connections Model

In the Connections Model introduced in Jackson and Wolinsky [56] agents receive information from others to whom they are connected to. Through these links they also receive information from those agents that they are indirectly connected to, that is, trough the neighbors of their neighbors, their neighbors, and so on[8].

The utility, $u_i(G)$, agent i receives from network G with n agents is a function $u_i : G \to \mathbb{R}$ with

$$u_i(G) = \sum_{j=1}^{n} \delta^{d_{ij}} - \sum_{j \in N_i} c, \tag{72}$$

where d_{ij} is the number of edges in the shortest path between agent i and agent j. $d_{ij} = \infty$ if there is no path between i and j. $0 < \delta < 1$ is a parameter that takes into account the decrease of the utility as the path between agent i and agent j increases. $N(i)$ is the set of nodes in the neighborhood of agent i.

There exists a tension between stability and efficiency in the connections model. This will become clear, after we state the following two propositions.

Proposition 1. *The unique efficient network in the symmetric Connections Model is*

(i) the complete graph K_n *if* $c < \delta - \delta^2$,
(ii) a star encompassing everyone if $\delta - \delta^2 < c < \delta + \frac{n-2}{2}\delta^2$,
(iii) the empty graph (no links) if $\delta + \frac{n-2}{2}\delta^2 < c$.

Proof. (i) We assume that $\delta^2 < \delta - c$. Any pair of agents that is not directly connected can increase its utility (the net benefit for creating a link is $\delta - c - \delta^2 > 0$) and thus the total utility, by forming a link. Since every pair of agents has an incentive to form a link, we will end up in the complete graph K_n, where all possible links have been created and no additional links can be created any more.

(ii) Consider a component of the graph G containing m agents, say G'. The number of links in the component G' is denoted by k, where $k \geq m-1$, otherwise the component would not be connected. E.g. a path containing all agents would have $m-1$ links. The total utility of the direct links in the component is given by $k(s\delta - 2c)$. There are at most $\frac{m(m-1)}{2} - k$ left over links in the component, that are not created yet. The utility of each of these left over links is at most $2\delta^2$ (it has the highest utility if it is in the second order neighborhood). Therefor the total utility of the component is at most

$$k2(\delta - c) + \left(\frac{m(m-1)}{2} - k\right) 2\delta^2. \tag{73}$$

Consider a star $K_{1,m-1}$ with m agents. The star has $m-1$ agents which are not in the center of the star. An example of a star with 4 agents is given in Figure (12). The utility of any direct link is $2\delta - 2c$ and of any indirect link $(m-2)\delta^2$, since

[8] Here only the shortest paths are taken into account.

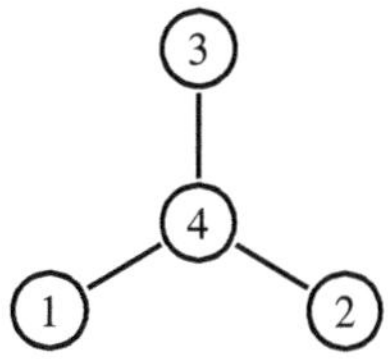

Fig. 12 A star encompassing 4 agents

any agent is 2 links away from any other agent (except the center of the star). Thus the total utility of the star is

$$\underbrace{(m-1)(2\delta-2c)}_{\text{direct connections}}+\underbrace{(m-1)(m-2)\delta^2}_{\text{indirect connections}}. \tag{74}$$

The difference in total utility of the (general) component and the star is just $2(k-(m-1))(\delta-c-\delta^2)$. This is at most 0, since $k \geq m-1$ and $c > \delta-\delta^2$, and less than 0 if $k > m-1$. Thus, the value of the component can equal the value of the star only if $k = m-1$. Any graph with $k = m-1$ edges, which is not a star, must have an indirect connection with a distance longer than 2, and getting a total utility less than $2\delta^2$. Therefore the total utility from indirect connections of the indirect links will be below $(m-1)(m-2)\delta^2$ (which is the total utility from indirect connections of the star). If $c < \delta-\delta^2$, then any component of a strongly efficient network must be a star.
Similarly it can be shown [56] that a single star of $m+n$ agents has a higher total utility than two separate stars with m and n agents. Accordingly, if an efficient network is non-empty, it must be a star.

(iii) A star encompassing every agent has a positive value only if $\delta + \frac{n-2}{2}\delta^2 > c$. This is an upper bound for the total achievable utility of any component of the network. Thus, if $\delta + \frac{n-2}{2}\delta^2 < c$ the empty graph is the unique strongly efficient network. □

Moreover, Jackson and Wolinsky [56] also determine the stable networks in the Connections Model.

Proposition 2. *Consider the Connections Model in which the utility of each agent is given by Equation (72).*

(i) A pairwise stable network has at most one (non-empty) component.
(ii) For $c < \delta-\delta^2$, the unique pairwise stable network is the complete graph K_n.
(iii) For $\delta-\delta^2 < c < \delta$ a star encompassing every agent is pairwise stable, but not necessarily the unique pairwise stable graph.
(iv) For $\delta < c$, any pairwise stable network that is non-empty is such that each agent has at least two links (and thus is inefficient).

Proof. (i) Lets assume, for the sake of contradiction, that G is pairwise stable and has more than one non-empty component. Let u^{ij} denote the utility of agent i

having a link with agent j. Then, $u^{ij} = u_i(G+e_{ij}) - u_i(G)$ if $e_{ij} \notin E(G)$ and $u^{ij} = u_i(G) - u_i(G-e_{ij})$ if $e_{ij} \in E(G)$. We consider now $e_{ij} \in E(G)$. Then $u^{ij} \geq 0$. Let e_{kl} belong to a different component. Since i is already in a component with j, but k is not, it follows that $u^{jk} > u^{ij} \geq 0$, because agent k will receive an additional utility of δ^2 from being indirectly connected to agent i. For similar reasons $u^{jk} > u^{lk} \geq 0$. This means that both agents in the separate component would have an incentive to form a link. This is a contradiction to the assumption of pairwise stability.

(ii) The net change in utility from creating a link is $\delta - \delta^2 - c$. Before creating the link, the geodesic distance between agent i and agent j is at least 2. When they create a link, they gain δ but they lose the previous utility from being indirectly connected by some path whose length is at least 2. So if $c < \delta - \delta^2$, the net gain from creating a link is always positive. Since any link creation is beneficial (increases the agents' utility), the only pairwise stable network is the complete graph, K_n.

(iii) We assume that $\delta - \delta^2 < c - \delta$ and show that the star is pairwise stable. The agent in the center of the star has a distance of 1 to all other agents and all other agents are separated by 2 links from each other. The center agent of the star cannot create a link, since she has already maximum degree. She has no incentive to delete a link either. If she deletes a link, the net gain is $c - \delta$, since there is no path leading to the then disconnected agent. By assumption, $\delta - \delta^2 < c < \delta$, $c - \delta < 0$ and the gain is negative, and the link will not be removed. We consider now an agent that is not the center of the star. She cannot create a link with the center, since they are both already connected. The net gain of creating a link to another agent is $\delta - \delta^2 - c$, which is strictly negative by assumption. So she will not create a link either. The star is pairwise stable.

Now consider the star encompassing all agents. Suppose an agent would not be connected to the star. If the center of the star would create a link to this isolated agent, the net gain would be $\delta - c > 0$ and the benefit of the isolated agent is again $\delta - c > 0$. So both will create the link.

The star is not the unique pairwise stable network. We will show that for 4 agents, the cycle, C_4 is also a pairwise stable network (see Figure (13)).

If agent 3 removes a link to agent 4, then her net gain is $c - \delta - \delta^3$. For the range of costs of $\delta - \delta^2 < c < \delta - \delta^3 < \delta$, she will never do it. If agent 3 adds a link to agent 1, Figure (13), the net gain is $\delta - \delta^2 < 0$. Thus, for $n = 4$ and

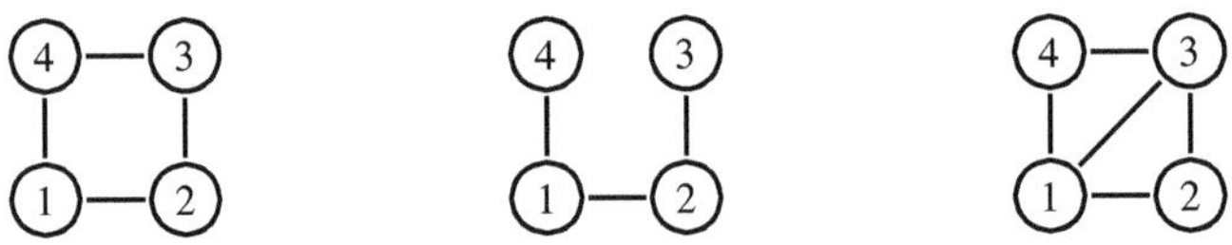

Fig. 13 A cycle of 4 agents (left) and the resulting graph after the deletion of a link from agent 3 to agent 4 (middle) and and the resulting graph after the creation of a link from agent 3 to agent 1 (right)

$\delta - \delta^2 < c < \delta - \delta^3$, then there are at least two pairwise stable networks: the star and the cycle.

(iv) For $\delta < c$ the star is not a pairwise stable network because the agent in the center of the star would gain $c - \delta$ from deleting a link. Moreover, it can be shown [56] that any connected agent has at least 2 links. □

One can see, from the two propositions described above, that a pairwise stable network is not necessarily efficient. For high cost $c > \delta$ there are non-empty pairwise stable networks but they are not efficient. Moreover, Watts [88] shows that if the benefit from maintaining an indirect link of length two is greater than the net benefit from maintaining a direct link ($\delta^2 > \delta - c > 0$) then the probability that the unique efficient network, the star $K_{1,n-1}$, is reached vanishes for large n.

The existence of inefficient equilibria is of interest because it indicates that the system, let alone to evolve, does not always reach an efficient configuration. In this respect, the result is important from the point of view of designing of policies that help the system to reach an efficient configuration.

Finally, we note that Jackson and Rogers [54] have proposed an extension of the Connections Model in which stable networks show the properties of a "Small-World" (see Section 5.3).

7.3 A Model of Dynamic Innovation Networks

We now briefly discuss a recent model of dynamic innovation networks in which agents compete for the most valuable knowledge for production, while knowledge can only be created through collaborations and knowledge exchange [62, 63]. By knowledge exchange we mean R&D partnerships (either informal or formal), characterized by bilateral interactions among agents. We characterize the emerging network topologies in terms of their efficiency (total knowledge growth) and in terms of the individual agents' knowledge growth.

Consider a set of agents, $N = \{1,...,n\}$, represented as nodes of an undirected graph G, with an associated variable x_i representing the knowledge of agent i. A link e_{ij}, represents the transfer of knowledge between agent i and agent j. Knowledge is shared among an individual's direct and indirect acquaintances. The level of knowledge an agent holds is proportional to the knowledge levels of its neighbors. We assume that knowledge $\mathbf{x} = (x_1,...,x_n)$ grows, starting from positive values, $x_i(0) > 0\ \forall i \in N$, according to the following linear ordinary differential equation

$$\dot{x}_i = \sum_{j=1}^{n} a_{ij} x_j, \tag{75}$$

where $a_{ij} \in \{0,1\}$ are the elements of the adjacency matrix $\mathbf{A}$ of the graph G. In vector notation we have $\dot{\mathbf{x}} = \mathbf{A}\mathbf{x}$. Similar to Carayol and Roux [17, 18] we assume that the gross return of agent i is proportional to her knowledge growth rate, with proportionality constant set to 1 for sake of simplicity[9]. We also assume that maintaining a link induces a constant marginal cost $c \geq 0$ for both agents connected via the link. Therefore the utility $u_i(t)$ of agent i is given by

$$u_i(t) = \frac{\dot{x}_i(t)}{x_i(t)} - cd_i, \tag{76}$$

where d_i denotes the degree of agent i. From Equation (75) one can show that [50] $\lim_{t\to\infty} \frac{\dot{x}_i(t)}{x_i(t)} = \lambda_{\mathrm{PF}}(G_i)$ where λ_{PF} is the largest real eigenvalue (Section 4.1) of the connected component G_i to which agent i belongs. Therefore, for large times t the utility function of agent i in a network G is given by

$$u_i = \lambda_{\mathrm{PF}}(G_i) - cd_i. \tag{77}$$

We can compare the utility functions of the Connections Model introduced in Equation (72) and the utility function from Equation (77). In both the utility depends on the position of an agent in the network. In Equation (77) and Equation (72) the cost term is identical. However, while the utility function in the Connections Model considers the length d_{ij} of the shortest path from node i to node j, the utility function in Equation (77) takes into account all paths of all lengths (in fact, λ_{PF} is proportional to the asymptotic growth rate of walks in a graph [25]). Indeed, it has been argued that knowledge gets transferred not only along the shortest path but also along all other paths in a network [78, 87]. Accordingly, all agents to which agent i is indirectly connected to along path of any length, contribute to the utility of agent i in this model.

Based on the definition of efficiency we can derive the efficient networks for certain values of the marginal cost. For a full characterization of efficient networks in this model we refer to König et al. [63].

Proposition 3. *The complete graph K_n is efficient for $c < \frac{1}{2}$. For costs $c \geq n$ the empty graph is efficient.*

Proof. Since for the complete graph it is $\lambda_{\mathrm{PF}} = n-1$ and $m = \frac{n(n-1)}{2}$, its aggregate utility is $U(K_n) = n(n-1) - 2\frac{n(n-1)}{2}c = n(n-1)(1-c)$.

On the other hand, the largest real eigenvalue λ_{PF} of a graph G with m edges is bounded from above so that $\lambda_{\mathrm{PF}} \leq \frac{1}{2}(\sqrt{8m+1}-1)$ [77]. For the aggregate utility of the network we then have

$$\begin{aligned} U(G) &= \sum_{i=1}^{n} \lambda_{\mathrm{PF}}(G_i) - 2mc \leq n \max_{1\leq i\leq n} \lambda_{\mathrm{PF}}(G_i) - 2mc \\ &\leq \frac{n}{2}(\sqrt{8m+1}-1) - 2cm := b(n,m,c), \end{aligned} \tag{78}$$

[9] The detailed derivation of the relation between an agents profits and her knowledge growth rate can be found in König et al. [63].

with $n \leq m \leq \binom{n}{2}$. For fixed cost c and number of nodes n, the number of edges maximizing Equation (78) is given by $m^* = \frac{n^2-c^2}{8c^2}$ if $\frac{n^2-c^2}{8c^2} < \binom{n}{2}$ and $m^* = \frac{n(n-1)}{2}$ if $\frac{n^2-c^2}{8c^2} > \binom{n}{2}$. The graph with the latter number of edges is the complete graph. Inserting m^* into Equation (78) yields

$$b(n,m^*,c) = \begin{cases} \frac{n}{2}(\sqrt{\frac{n^2-c^2}{c^2}+1}-1) - \frac{n^2-c^2}{4c} & c > \frac{n}{2n-1} \\ n(n-1)(1-c) = \Pi(K_n) & c < \frac{n}{2n-1}. \end{cases} \tag{79}$$

The bound for $c \leq \frac{n}{2n-1} \sim \frac{1}{2}$ coincides with the aggregate utility of the complete graph K_n for large n. K_n is therefore the efficient graph. If instead $c = n$ then $m^* = 0$. The efficient graph is the empty graph. This concludes the proof. □

In the following we make an important assumption. The network evolution process is assumed to be much slower than the knowledge growth of Equation (75), so that agents make decisions based on the asymptotic growth rate of knowledge and the utility function in Equation (77). Thus, we introduce a coupling of fast knowledge growth coupled with a slow network evolution, as illustrated in Figure (4.4) in Section 4.4.

Given the utility function in Equation (77) and the network evolution introduced in Section 7.1 one can show that the network evolution can lead to pairwise stable networks [63]. However, there exists a multiplicity of different equilibria, as the next proposition reveals (for the proof see König et al. [63]).

Proposition 4. *Consider costs c, $c' = \alpha c$, $\alpha \in [0,1]$ and the network G with n nodes such that*[10] $\lceil \frac{2}{c} \rceil \leq n \leq \lfloor \frac{1+c'^2(6+c'^2)}{4c'^2} \rfloor$. *If there exists an integer $k \leq n$,* $\mod(n,k) = 0$ *such that* $\lceil \frac{1+c(1-c)}{c} \rceil \leq k \leq \lfloor \frac{2-c'(1-c')}{c'} \rfloor$ *then G can be stable for at least two cases:*

(i) G consists of disconnected cliques $K_k^1, \ldots, K_k^d$, $n = kd$ or
(ii) G consists of a spanning star $K_{1,n-1}$.

There are at least two stable networks for the same level of marginal cost c (degenerate cost region).

The variety of the possible equilibria is not only restricted to cliques and stars of different sizes but it also includes networks with complex topologies which are characteristic of many real word networks. Examples of such equilibrium networks can be seen in Figure (14). Differently to the Connections Model this model is able to reproduce some stylized empirical facts of R&D networks, namely that networks are sparse, locally dense and show heterogeneous degree distributions [23, 72].

We have seen that in this model there exist multiple equilibrium networks, some of them being inefficient. Depending on the cost of interactions the system can get stuck in stable but inefficient structures. However, in König et al. [62] it is shown

[10] In the following, $\lceil x \rceil$, where x is a real valued number $x \in \mathbb{R}$, denotes the smallest integer larger or equal than x (the ceiling of x). Similarly, $\lfloor x \rfloor$ the largest integer smaller or equal than x (the floor of x).

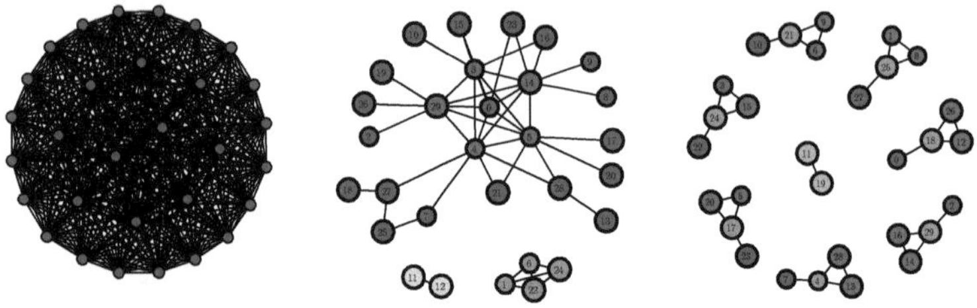

Fig. 14 Equilibrium networks for vanishing cost $c = 0$ (left), intermediate cost $c = 0.2$ (middle) and high cost $c = 0.5$ (right) for $n = 50$ agents (without link removal in the network formation process) adopted from König et al. [62]. Clearly, the higher the cost c, the sparser and the more clustered are the equilibrium networks

that if it is difficult to break up already existing collaborations and agents maintain R&D collaborations even when, in the short run, they may be unprofitable, then emerging networks are efficient.

7.4 Summary and Conclusion

In this tutorial we have given an introduction to models of economic networks and we have tried to show the wide applicability and importance of these models. Since the field of economic network theory is growing at an increasing pace, this tutorial is far from being exhaustive. However, the models presented here can serve as a starting point for interested students and prospective researchers in the field.

References

1. Barrat, A. and Weigt, M. (2000). On the properties of small-world network models. *The European Physical Journal B*, (13):547–560.
2. Batagelj, V. and Brandes, U. (2005). Efficient generation of large random networks. *Physcal Review E*, 71.
3. Battiston, S., Bonabeau, E., and Weisbuch, G. (2003). Decision making dynamics in corporate boards. *Physica A: Statistical Mechanics and its Applications*, 322:567–582.
4. Battiston, S. and Catanzaro, M. (2004). Statistical properties of corporate board and director networks. *The European Physical Journal B-Condensed Matter*, 38(2):345–352.
5. Battiston, S., Rodrigues, J. F., and Zeytinoglu, H. (2007). The network of inter-regional direct investment stocks across europe. *Advances in Complex Systems (ACS)*, 10:29–51.
6. Baumann, N. and Stiller, S. (2005). *Network Analysis: Methodological Foundations*, chapter Network Models. Springer.
7. Bollobas, B. (1985). *Random Graphs*. Cambridge University Press, 2nd edition.
8. Bollobas, B. (1998). *Modern Graph Theory*. Graduate Texts in Mathematics. Springer.
9. Borgatti, S. (2005). Centrality and network flow. *Social Networks*, 27(1):55–71.

10. Borgatti, S. and Everett, R. (2006). A graph-theoretic perspective on centrality. *Social Networks.*
11. Bowles, S. and Gintis, H. (November 2002). Social capital and community governance. *The Economic Journal*, 112:419–436(18).
12. Boyd, S. (2006). Linear dynamical systems. Lecture Notes, Stanford University.
13. Brioschi, F., Buzzacchi, L., and Colombo, M. (1989). Risk capital financing and the separation of ownership and control in business groups. *Journal of Banking and Finance*, 13(1):747 – 772.
14. Brown, J. S. and Duguid, P. (1991). Organizational learning and communities-of-practice: Toward a unified view of working, learning, and innovation. *Organization Science*, 2(1):40–57. Special Issue: Organizational Learning: Papers in Honor of (and by) James G. March.
15. Calvó-Armengol, A. (2007). Networks in economics. lecture notes.
16. Calvó-Armengol, A. and Martí, J. d. (2007). On optimal communication networks. Working paper, Universitat Autonoma de Barcelona.
17. Carayol, N. and Roux, P. (2003). Self-organizing innovation networks: When do small worlds emerge? Working Papers of GRES - Cahiers du GRES 2003-8, Groupement de Recherches Economiques et Sociales.
18. Carayol, N. and Roux, P. (2005). *Nonlinear Dynamics and Heterogeneous Interacting Agents*, chapter Collective Innovation in a Model of Network Formation with Preferential Meeting, pages 139–153. Springer.
19. Chapelle, A. (2005). Separation of ownership and control: Where do we stand? *Corporate Ownership and Control*, 15(2).
20. Chung, F. and Lu, L. (2007). *Complex Graphs and Networks*. American Mathematical Society.
21. Corrado, R. and Zollo, M. (2006). Small worlds evolving: Governance reforms, privatizations, and ownership networks in italy. *Industrial and Corporate Change*, 15(2):319–352.
22. Costa, L. F., Rodrigues, F. A., Travieso, G., and Boas, P. V. (2007). Characterization of complex networks: A survey of measurements.
23. Cowan, R. (2004). Network models of innovation and knowledge diffusion. Research Memoranda 016, Maastricht : MERIT, Maastricht Economic Research Institute on Innovation and Technology.
24. Cowan, R. and Jonard, N. (2004). Network structure and the diffusion of knowledge. *Journal of Economic Dynamics and Control*, 28:1557–1575.
25. Cvetkovic, D., Doob, M., and Sachs, H. (1995). *Spectra of Graphs: Theory and Applications*. Johann Ambrosius Barth.
26. Davis, G. and Greve, H. (1997). Corporate Elite Networks and Governance Changes in the 1980s. *American Journal of Sociology*, 103(1):1–37.
27. Davis, G., Yoo, M., and Baker, W. (2003). The Small World of the American Corporate Elite, 1982-2001. *Strategic Organization*, 1(3):301.
28. Diestel, R. (2000). *Graphentheorie*. Springer, 2nd edition.
29. Dodds, P. S., Watts, D. J., and Sabel, C. F. (2003). Information exchange and robustness of organizational networks. *Working Paper Series, Center on Organizational Innovation, Columbia Univeristy.*
30. Durrett, R. (2007). *Random Graph Dynamics*. Cambridge University Press.
31. Durrett, R. A. (2004). *Probability. Theory and Examples*. Thomson Learning.
32. Farmer, J. D., Smith, E., and Shubik, M. (2005). Economics: The next physical science?
33. Flath, D. (1992). Indirect shareholding within japan's business groups. *Economics Letters*, 38(2):223–227.
34. Freeman, L. C. (1978). Centrality in social networks conceptual clarification. *Social Networks*, 1:215–239.
35. Gabszewicz, J. (2000). *Strategic Interaction and Markets*. Oxford University Press, Center for Operations Research and Econometrics (CORE), Catholic University of Louvain.
36. Gallegati, M. and Kirman, A. P., editors (1999). *Beyond the Representative Agent*. Edward Elgar Publishing.

37. Garlaschelli, D., Battiston, S., Castri, M., Servedio, V. D. P., and Caldarelli, G. (2005). The scale-free topology of market investments. *Physica*, A(350):491–499.
38. Geipel, M. (2007). Self-organization applied to dynamic network layout. *International Journal of Modern Physics C*. forthcoming.
39. Geroski, P. A. (2000). Models of technology diffusion. *Research Policy*, 29(4-5):603–625.
40. Gigerenzer, G. and Selten, R., editors (2002). *Bounded Rationality: The Adaptive Toolbox (Dahlem Workshop Reports)*. MIT Press.
41. Godsil, C. D. and Royle, G. F. (2001). *Algebraic Graph Theory*. Springer.
42. Goyal, S. (2007). *Connections: an introduction to the economics of networks*. Princeton University Press.
43. Goyal, S. and Moraga-Gonzalez, J. L. (2001). R&d networks. *RAND Journal of Economics*, 32:686–707.
44. Granovetter, M. (1995a). *Getting a Job: A Study of Contacts and Careers*. University of Chicago Press.
45. Granovetter, M. (1995b). *Industrial and Corporate Change*, chapter Coase Revisited: Business Groups in the Modern Economy. Oxford University Press.
46. Gross, T. and Blasius, G. (2007). Adaptive coevolutionary networks - a review. eprint arXiv: 0709.1858.
47. Growiec, J., Pammolli, F., Riccaboni, M., and Stanley, H. (2008). On the size distribution of business firms. *Economics Letters*, 98(2):207–212.
48. Haken, H. (1977). *Synergetics - An Introduction; Nonequilibrium Phase Transitions and Self-Organization in Physics, Chemistry and Biology*. Springer.
49. Hausman, D. M. (2003). *Inexact and Separate Science of Economics*. Cambridge University Press.
50. Horn, R. A. and Johnson, C. R. (1990). *Matrix Analysis*. Cambridge University Press.
51. Huberman, Bernardo, A. and Hogg, T. (1995). Communities of practice: Performance and evolution. *Computational and Mathematical Organization Theory*, 1(1):73–92.
52. Jackson, M. (2008). *Social and Economic Networks*. Princeton University Press.
53. Jackson, M. and Calvo-Armengol, T. (2007). Networks in labor markets: Wage and employment dynamics and inequality. *The Journal of Economic Theory*, 132(1):27–46.
54. Jackson, M. and Rogers, B. (2005). The Economics of Small Worlds. *Journal of the European Economic Association*, 3(2–3):617–627.
55. Jackson, M. O. (2006). *Advances in Economics and Econometrics: Theory and Applications, Ninth World Congress*, volume I, chapter The Economics of Social Networks. Cambridge University Press.
56. Jackson, M. O. and Wolinsky, A. (1996). A strategic model of social and economic networks. *Journal of Economic Theory*, 71(1):44–74.
57. Jain, S. and Krishna, S. (1998). Autocatalytic sets and the growth of complexity in an evolutionary model. *Physical Review Letters*, 81(25):5684–5687.
58. Jain, S. and Krishna, S. (2001). A model for the emergence of cooperation, interdependence, and structure in evolving networks. *Proceedings of the National Academy of Sciences*, 98(2):543–547.
59. Kamien, M. I., Muller, E., and Zang, I. (1992). Research joint ventures and R&D cartels. *The American Economic Review*, 82(5):1293–1306.
60. Kirman, A. (1997). The economy as an evolving network. *Journal of Evolutionary Economics*, 7(4):339–353.
61. Kogut, B. and Walker, G. (2001). The small world of germany and the durability of national networks. *American Sociological Review*, 66(3):317–335.
62. König, M. D., Battiston, S., Napoletano, M., and Schweitzer, F. (2008). On algebraic graph theory and the dynamics of innovation networks. *Networks and Heterogeneous Media*, 3(2):201–219.
63. König, M. D., Battiston, S., Napoletano, M., and Schweitzer, F. (2009). Efficiency and stability of dynamic innovation networks. forthcoming.

64. König, M. D., Battiston, S., and Schweitzer, F. (2008). *Innovation Networks - New Approaches in Modeling and Analyzing*, chapter Modeling Evolving Innovation Networks. Springer Complexity Series.
65. Lave, J. and Wenger, E. (1991). *Situated Learning: Legitimate Peripheral Participation*. Cambridge University Press.
66. Milgram, S. (1967). The small world problem. *Psychology Today*, 2(1):60–67.
67. Newman, M. (2003). The structure and function of complex networks. *SIAM review*, 45(2):167–256.
68. Newman, M. E. J. (2007). *The New Palgrave Encyclopedia of Economics*, chapter Mathematics of networks. Palgrave Macmillan, Basingstoke.
69. Newman, M. E. J., Strogatz, S. H., and Watts, D. J. (2001). Random graphs with arbitrary degree distributions and their applications. *Physcal Review E*, 64(2):026118.
70. Newman, M. E. J., Watts, D. J., and Strogatz, S. H. (2002). Random graph models of social networks. *Proceedings of the National Academy of Sciences*, 99(90001):2566–2572.
71. Nisan, N., Roughgarden, T., Tardos, E., and Vazirani, V. V., editors (2007). *Algorithmic Game Theory*. Cambridge University Press.
72. Powell, W. W., White, D. R., Koput, K. W., and Owen-Smith, J. (2005). Network dynamics and field evolution: The growth of interorganizational collaboration in the life sciences. *American Journal of Sociology*, 110:1132–1205.
73. Reichl, L. E. (2004). *A Modern Course in Statistical Physics*. Wiley-VCH, 2 edition.
74. Riccaboni, M. and Pammolli, F. (2002). On firm growth in networks. *Research Policy*, 31(8-9):1405–1416.
75. Seneta, E. (2006). *Non-negative Matrices And Markov Chains*. Springer.
76. Slikker, M. and van den Nouweland, A. (2001). *Social and Economic Networks in Cooperative Game Theory*. Springer.
77. Stanley, R. P. (1987). A bound on the spectral radius of graphs with e edges. *Linear Algebra and its Applications*, 87:267–269.
78. Stephenson, K. and Zelen, M. (1989). Rethinking centrality: Methods and examples. *Social Networks*, 11:1–37.
79. Stoneman, P. (2002). *The Economics of Technological Diffusion*. Blackwell Publishers.
80. Van Alstyne, M. (1997). The state of network organization: A survey in three frameworks. *Journal of Organizational Computing and Electronic Commerce*, 7(2):83–151.
81. Van Zandt, T. (1997). *Contemporary Economic Development Reviewed, Volume 4: The Enterprise and its Environment*, chapter Decentralized Information Processing in the Theory of Organizations, pages 125–160. MacMillan Press Ltd.
82. Varian, H. R. (1996). *Intermediate Microeconomics: A Modern Approach*. WW Norton.
83. Vega-Redondo, F. (2006). Diffusion and growth in an evolving network. *forthcoming in International Journal of Game Theory*.
84. Vega-Redondo, F. (2007). *Complex Social Networks*. Series: Econometric Society Monographs. Cambridge University Press.
85. Venkatesh, B. and Goyal, S. (1998). Learning from neighbors.
86. Veugelers, R. (October 1998). Collaboration in R&D: An assessment of theoretical and empirical findings. *De Economist*, 146:419–443(25).
87. Wasserman, S. and Faust, K. (1994). *Social Network Analysis: Methods and Applications*. Cambridge University Press.
88. Watts, A. (2001). A dynamic model of network formation. *Games and Economic Behavior*, 34(2):331–341.
89. Watts, D. J. and Strogatz, S. H. (1998). Collective dynamics of small-world networks. *Nature*, 393:440–442.
90. Weidlich, W. (2002). *Sociodynamics - a systematic approach to mathematical modelling in the social sciences*. Taylor & Francis.
91. West, Douglas, B. (2001). *Introduction to Graph Theory*. Prentice-Hall, 2nd edition.
92. Xu, J. (2003). *Theory and Application of Graphs*. Kluwer Academic Publishers.
93. Zenou, Y. (2006). Course on networks: Theory and applications. lecture notes.

Part II
Strategic Interaction, Economic Models and Networks

Games of Coalition and Network Formation: A Survey

Marco A. Marini

Abstract This paper presents some recent developments in the theory of coalition and network formation. For this purpose, a few major equilibrium concepts recently introduced to model the formation of coalition structures and networks among players are briefly reviewed and discussed. Some economic applications are also illustrated to give a flavour of the type of predictions such models are able to provide.

1 Introduction

Very often in social life individuals take decisions within groups (households, friendships, firms, trade unions, local jurisdictions, etc.). Since von Neumann and Morgenstern's [45] seminal work on game theory, the problem of the formation of coalitions has been a highly debated topic among game theorists. However, during this seminal stage and for a long period afterward, the study of coalition formation was almost entirely conducted within the framework of games in characteristic form (cooperative games) which proved not entirely suited in games with externalities, i.e. virtually all games with genuine interaction among players. Only in recent years, a widespread literature on what is currently known as noncooperative coalition formation or endogenous coalition formation has come into the scene with the explicit purpose to represent the process of formation of coalitions of agents and hence modelling a number of relevant economic and social phenomena.[1] Moreover, following this theoretical and applied literature on coalitions, the recent paper by

M.A. Marini
Department of Economics, University of Urbino "Carlo Bo", Urbino (Italy) & CREI, Universitá Roma Tre, Rome, (Italy)
marco.marini@uniurb.it

[1] Extensive surveys of the coalition formation literature are contained in Greenberg [23], Bloch [4, 5], Yi [46, 46] and Ray and Vohra [41].

A.K. Naimzada et al. (eds.), *Networks, Topology and Dynamics*, Lecture Notes in Economics and Mathematical Systems 613,

Jackson and Wolinsky [34] opened the door to a new stream of contributions using networks (graphs) to model the formation of links among individuals.[2]

Throughout these brief notes, I survey non exhaustively some relevant contributions of this wide literature, with the main aim to provide an overview of some modelling tools for economic applications. For this purpose, some basic guidelines to the application of coalition formation in economics are presented using as primitives the games in strategic form. As far as economic applications are concerned, most of the examples presented here mainly focus, for convenience, on a restricted number of I.O. topics, as cartel formation, horizontal merger and R&D alliances.

2 Coalitions

2.1 Cooperatives Games with Externalities

Since von Neumann and Morgenstern [45], a wide number of papers have developed solution concepts specific to games with coalitions of players. This literature, known as cooperative games literature, made initially a predominant use of the characteristic function as a way to represent the worth of a coalition of players.

Definition 1. A cooperative game with transferable utility (TU cooperative game) can be defined as a pair (N,v), where $N = \{1,2,..i,..N\}$ is a finite set of players and $v : \mathbb{N} \rightarrow R_+$ is a mapping (characteristic function) assigning a value or worth to every feasible coalition, i.e. every nonempty subset of players $S \subset N$ belonging to $\mathbb{N}$, the family of nonempty coalitions $2^N \backslash \{\varnothing\}$.[3]

The value $v(S)$ can be interpreted as the maximal aggregate amount of utility members of coalition S can achieve by coordinating their strategies. In strategic environments, players' payoffs are defined on the strategies of all players and the worth of a group of players S depends on their expectations about the strategies played by the remaining players $N \backslash S$. Hence, to obtain $v(S)$ from a strategic situation, we need first to define an underlying strategic form game.

Definition 2. A strategic form game is a triple $G = \{N, (X_i; u_i)_{i \in N}\}$, in which for each $i \in N$, X_i is the set of strategies with generic element x_i, and $u_i : X_1 \times \cdots \times X_n \rightarrow R_+$ is every player's payoff function.

Moreover, henceforth we restrict the action space of each coalition $S \subset N$ to $X_S \equiv \prod_{i \in S} X_i$. Let, also, $v(S) = \sum_{i \in S} u_i(x)$, for $x \in X_N \equiv \prod_{i \in N} X_i$.[4]

[2] Myerson [36] and Aumann and Myerson [2] were among the first authors to use graphs to model cooperation between individuals. Excellent surveys of the network literature are contained in Dutta and Jackson [17] and in Jackson [28–31].

[3] Here we mainly deal with games with transferable utility. In games without transferable utility, the worth of a coalition associates with each coalition a players' utility frontier (a set of vectors of utilities).

[4] See Sect. 2.3 for an interpretation of these restrictions.

Example 1. Two-player prisoner's dilemma.

	A	B
A	*3,3*	*1,4*
B	*4,1*	2,2

Therefore, $v(N) = 6$ and $v(\{i\}) = \begin{cases} 4, & \text{if } x_j = A \\ 2, & \text{if } x_j = B \end{cases}$ for $j \neq i$.

The cooperative allocation (3; 3) can be considered stable only if every player is expected to react with strategy B to a deviation of the other player from the cooperative strategy A.

The above example shows that in order to define the worth of a coalition of players, a specific assumption on the behaviour of the remaining players is required.

2.1.1 α- and β-Characteristic Functions

The concepts of core, formally studied by Aumann [1], are based on von Neumann and Morgenstern's [45] early proposal of representing the worth of a coalition as the minmax or maxmin aggregate payoff that it can guarantee its members in the underlying strategic form game. Accordingly, the characteristic function $v(S)$ in games with externalities can be obtained assuming that outside players act to minimize the payoff of every deviating coalition $S \subset N$. In this minimax formulation, if members of S move second, the obtained characteristic function,

$$v_\beta(S) = \min_{x_{N\setminus S}} \max_{x_S} \sum\nolimits_{i \in S} u_i(x_S, x_{N\setminus S}), \tag{1}$$

denoted β-characteristic function, represents what members in S *cannot be prevented from getting*. Alternatively, if members of S move first, we have

$$v_\alpha(S) = \max_{x_S} \min_{x_{N\setminus S}} \sum\nolimits_{i \in S} u_i(x_S, x_{N\setminus S}) \tag{2}$$

denoted α-characteristic function, which represents what members in S *can guarantee themselves*, when they expect a retaliatory behaviour from the complement coalition $N\setminus S$.[5]

When the underlying strategic form game G is zero-sum, (1) and (2) coincide. In non-zero sum games they can differ and, usually, $v_\alpha(S) < v_\beta(S)$ for all $S \subset N$.

However, α- and β-characteristic functions express an irrational behaviour of coalitions of players, acting as if they expected their rivals to minimize their payoff. Although appealing because immune from any *ad hoc* assumption on the reaction of the outside players (indeed, their minimizing behavior is here not meant to represent the expectation of S but rather as a mathematical way to determine the lower

[5] Note that here players outside S are treated as one coalition, so the implicit assumption is that players in $N\setminus S$ stick together after S departure from the grand coalition N.

bound of S's aggregate payoff), still this approach has important drawbacks: deviating coalitions are too heavily penalized, while outside players often end up bearing an extremely high cost in their attempt to hurt deviators. Moreover,the little profitability of coalitional objections yield very large set of solutions (e.g. large cores).

2.1.2 Nash Behaviour Among Coalitions

Another way to define the characteristic function in games with externalities is to assume that in the event of a deviation from N, a coalition S plays *à la* Nash with remaining players.[6]

Although appealing, such a modelling strategy requires some specific assumptions on the coalition structure formed by remaining players $N\backslash S$ once a coalition S has deviated from N.

Following the Hart and Kurtz's [25] coalition formation game, two extreme predictions can be assumed on the behaviour of remaining players. Under the so called γ-assumption,[7] when a coalition deviates from N, the remaining players split up in singletons; under the δ-assumption, players in $N\backslash S$ stick together as a unique coalition.[8]

Therefore, the obtained characteristic functions can be defined as follows:

$$v_\gamma(S) = \sum_{i\in S} u_i\left(\overline{x}_S, \{\overline{x}_j\}_{j\in N\backslash S}\right), \tag{3}$$

where $\overline{x}$ is a strategy profile such that, for all $S \subset N$, $\overline{x}_S \in X_S$ and $\forall j \in N\backslash S, \overline{x}_j \in X_j$

$$\overline{x}_S = \arg\max_{x_S\in X_S} \sum_{i\in S} u_i\left(x_S, \{\overline{x}_j\}_{j\in N\backslash S}\right)$$

$$\overline{x}_j = \arg\max_{x_j\in X_j} u_j\left(\overline{x}_S, \{x_k\}_{k\in(N\backslash S)\backslash\{j\}}, x_j\right).$$

Moreover,

$$v_\delta(S) = \sum_{i\in S} u_i\left(\overline{x}_S, \overline{x}_{N\backslash S}\right),$$

where,

$$\overline{x}_S = \arg\max_{x_S\in X_S} \sum_{i\in S} u_i\left(x_S, \overline{x}_{N\backslash S}\right)$$

$$\overline{x}_j = \arg\max_{x_{N\backslash S}\in X_{N\backslash S}} \sum_{j\in N\backslash S} u_j\left(\overline{x}_S, x_{N\backslash S}\right).$$

---

[6] The idea that coalitions in a given coalition structure can play noncooperatively among them was firstly explored by Ichiishi [26].

[7] Hurt and Kurz's [25] Γ- game is indeed a strategic coalition formation game with fixed payoff division, in which the strategies consist of the choice of a coalition. Despite the different nature of the two games, there is an analogy concerning the coalition structure induced by a deviation from the grand coalition.

[8] See Chander and Tulkens [14] for applications of this approach.

In both cases, for (3) and (4) to be well defined, the Nash equilibrium of the strategic form game played among coalitions must be unique. Moreover, usually, $v_\alpha(S) < v_\beta(S) < v_\delta(S)$ for all $S \subset N$.

2.1.3 Timing and the Characteristic Function

It is also conceivable to modify the γ- or δ-assumption reintroducing the temporal structure typical of the α and β-assumptions.[9]

When a deviating coalition S moves first under the γ-assumption, the members of S choose a coordinated strategy as leaders, thus anticipating the reaction of the players in $N \backslash S$, who simultaneously choose their best response as singletons. The strategy profile associated to the deviation of a coalition S is the Stackelberg equilibrium of the game in which S is the leader and players in $N \backslash S$ are, individually, the followers. We can indicate this strategy profile as a $\widetilde{x}(S) = (\widetilde{x}_S, x_j(\widetilde{x}_S))$ such that

$$\widetilde{x}_S = \arg\max_{x_S \in X_S} \sum_{i \in S} u_i\left(x_S, \{x_j(x_S)\}_{j \in N \backslash S}\right) \tag{4}$$

and, for every $j \in N \backslash S$,

$$x_j(x_S) = \arg\max_{x_j \in X_j} u_j\left(\widetilde{x}_S, \{x_k(\widetilde{x}_S)\}_{k \in (N \backslash S) \backslash \{j\}}, x_j\right). \tag{5}$$

Sufficient condition for the existence of a profile $\widetilde{x}(S)$ can be provided. Assume that $G(N \backslash S, x_S)$, the restriction of the game G to the set of players $N \backslash S$ given the fixed profile x_S, possesses a unique Nash Equilibrium for every $S \subset N$ and $x_S \in X_S$, where X_S is assumed compact. Let also each player's payoff be continuous in each player's strategy. Thus, by the closedness of the Nash equilibrium correspondence (see, for instance, [20]), members of S maximize a continuous function over a compact set and, by Weiestrass Theorem, a maximum exists. As a consequence, for every $S \subset N$, there exists a Stackelberg equilibrium $\widetilde{x}(S)$. We can thus define the characteristic function $v_\lambda(S)$ as follows:

$$v_\lambda(S) = \sum_{i \in S} u_i\left(\widetilde{x}_S, \{\overline{x}_j(\widetilde{x}_S)\}_{j \in N \backslash S}\right).$$

Obviously, $v_\lambda(S) \geq v_\gamma(S)$. Inverting the timing of deviations and reactions, the γ-assumption can be modified by assuming that a deviating coalition S plays as follower against all remaining players in $N \backslash S$ acting as singleton leaders. Obviously, the same can be done under the δ-assumption.

2.1.4 The Core in Games with Externalities

We can test the various conversions of $v(S)$ introduced above by examining the different predictions obtained using the *core* of (N, v).

---

[9] See Currarini and Marini [15] for more details.

We first define an imputation for (N,v) as a vector $\mathbf{z} \in R^n_+$ such that $\sum_{i \in N} z_i \leq v(N)$ (feasibility) and $z_i \geq v(i)$ (individual rationality) for all $i \in N$.

Definition 3. The core of a TU cooperative game (N,v) is the set of all imputations $\mathbf{z} \in R^n_+$ such that $\sum_{i \in S} z_i \geq v(S)$ for all $S \subseteq N$.

Given that coalitional payoffs are obtained from an underlying strategic form game, the core can also be defined in terms of strategies, as follows.

Definition 4. The joint strategy $\mathbf{x} \in X_N$ is core-stable if there is no coalition $S \subset N$ such that $v(S) > \sum_{i \in S} u_i(\mathbf{x})$.

Example 2 (Merger in a linear Cournot oligopoly). Consider three firms $N = \{1,2,3\}$ with linear technology competing à la Cournot in a linear demand market. Let the demand parameters a and b and the marginal cost c, be selected in such a way that interior Nash equilibria for all coalition structures exist. The set of all feasible coalitions of the N players is

$$\mathbb{N} = (\{1,2,3\},\{1\},\{2\},\{3\},\{1,2\},\{1,3\},\{2,3\}).$$

Note that if all firms merge, they obtain the monopoly payoff $v(\{1,2,3\} = \frac{A}{4}$, where $A = (a-c)^2/b$, independently of the assumptions made on the characteristic function. These assumptions matters for the worth of intermediate coalitions. Under the α- and β-assumptions, if either one single firm or two firms leave the grand coalition N, remaining firms can play a minimizing strategy in such a way that, for every $S \subset N$, $v_\alpha(S) = v_\beta(S) = 0$. In this case, the core coincides with all individually rational Pareto efficient payoff, i.e. all points weakly included in the set of coordinates, $\mathbf{Z} = \left[\left(\frac{A}{8}, \frac{A}{16}, \frac{A}{16}\right), \left(\frac{A}{16}, \frac{A}{8}, \frac{A}{16}\right), \left(\frac{A}{16}, \frac{A}{16}, \frac{A}{8}\right)\right]$. Under the γ-assumption, we know that when, say firms 1 and 2, jointly leave the merger, a simultaneous duopoly game is played between the coalition $\{1,2\}$ and firm $\{3\}$. Hence, $v_\gamma(\{1,2\}) = \frac{A}{9}$. Similarly for all other couples of firms deviating from N. When instead a single firm i leaves the grand coalition N, a triopoly game is played, with symmetric payoffs $v_\gamma(\{i\}) = \frac{A}{16}$ (all these payoffs are obtained from the general expression $v(S) = \frac{A}{(n-s+2)^2}$ expressing firms' profits in a n-firm oligopoly). In this case, since intermediate coalitions made of two players do not give each firm more than their individually rational payoff, the core under the γ-assumption coincides with the core under the α- and β-assumptions. We know from Salant et al. [42] model of merger in oligopoly, that $v_\gamma(S) > \sum_{i \in S} v_\gamma(\{i\})$ only for $|S| > 0;8\ |N|$. This means that in the merger game the core under the γ-assumption shrinks with respect to the core under the α- and β-assumptions only for $n > 5$. Under the δ-assumption, when a single firm leaves N, a simultaneous duopoly game is played between the firm $\{i\}$ and the remaining firms $N \backslash \{i\}$ acting as a single coalition. As a result, $v(\{i\}) = \frac{A}{9}$, which is greater than $\frac{A}{12}$, the maximum payoff at least one firm will obtain in the grand coalition. Therefore, under the δ-assumption, the core is empty. Finally, note that since under the λ-assumption every single firm playing as leader obtains $v(\{i\}) = \frac{A}{12}$, in such a case the core is unique and contains only the equal split imputation $\mathbf{z} = (\frac{A}{12}, \frac{A}{12}, \frac{A}{12})$ [see Figs. 1 and 2].

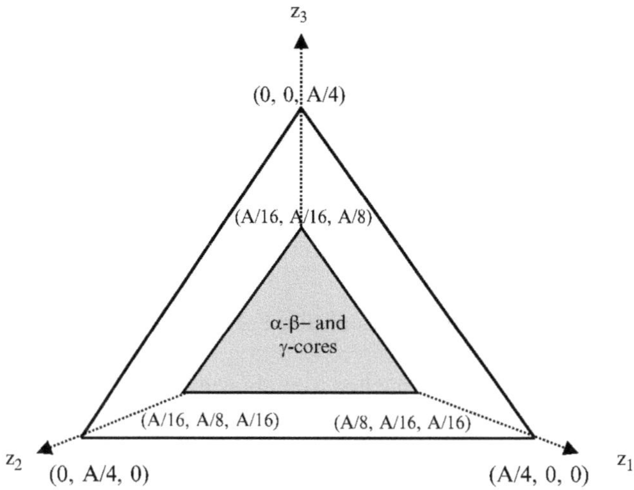

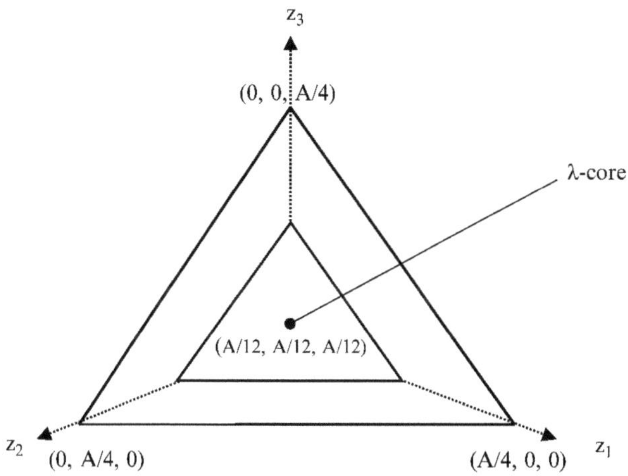

Fig. 2 Merger game: λ-core

2.2 Noncooperative Games of Coalition Formation

Most recent approaches have looked at the process of coalition formation as a strategy in a well defined game of coalition formation (see [7, 8, 47] for surveys). Within this stream of literature, usually indicated as *noncooperative theory of coalition formation* (or *endogenous coalition formation*), the work by Hurt and Kurz [25] represents the main seminal contribution. Most recent contributions along these lines include Bloch [4, 5], Ray and Vohra [40, 41] and Yi [46]. In all these works, cooperation is modelled as a two stage process: at the first stage players form coalitions, while at the second stage formed coalitions interact in a well defined strategic setting. This process is formally described by a *coalition formation game*, in which a given rule of coalition formation maps players' announcements of coalitions into a well defined coalition structure, which in turns determines the equilibrium strategies chosen by players at the second stage. A basic difference among the various models lies on the timing assumed for the coalition formation game, which can either be simultaneous (Hurt and Kurz [25], Ray and Vohra [40], Yi [46]) or sequential ([5], Ray and Vohra [41]).

2.2.1 Hurt and Kurz's Games of Coalition Formation

Hurt and Kurz [25] were among the first to study games of coalition formation with a valuation in order to identify stable coalition structures.[10] As valuation, Hurt and Kurz adopt a general version of Owen value for TU games [38], i.e. a Shapley value with prior coalition structures, that they call Coalitional Shapley value, assigning to every coalition structure a payoff vector $\varphi_i(\pi)$ in $\mathbf{R}^N$, such that (by the efficiency axiom) $\sum_{i\in N}\varphi_i(\pi)=v(N)$. Given this valuation, the game of coalition formation is modelled as a game in which each player $i\in N$ announces a coalition $S\ni i$ to which he would like to belong; for each profile $\sigma=(S_1,S_2,\dots,S_n)$ of announcements, a partition $\pi(\sigma)$ of N is assumed to be induced on the system. The rule according to which $\pi(\sigma)$ originates from σ is obviously a crucial issue for the prediction of which coalitions will emerge in equilibrium. Hurt and Kurz's game Γ predicts that a coalition emerges if and only if all its members have declared it (from which the name of "unanimity rule" also used to describe this game).

Formally:

$$\pi^{\gamma}(\sigma)=\{S_i(\sigma):i\in N\}\ ,$$

where

$$S_i(\sigma)=\begin{cases} S_i \text{ if } S_i=S_j \text{ for all } j\in S_i \\ \{i\} \text{ otherwise.}\end{cases}$$

Their game Δ predicts instead that a coalition emerges if and only if all its members have declare the same coalition S (which may, in general, differs from S). Formally:

$$\pi^{\delta}(\sigma)=\left\{S\subset N: i,j\in S \text{ if and only if } S_i=S_j\right\}.$$

It can be seen that the two rules generate different partitions after a deviation by a coalition: in the Γ-game, remaining players split up in singletons; in the Δ-game, they stick together.

Example 3. $N=\{1,2,3\}$, $\sigma_1=\{1,2,3\}$; $\sigma_2=\{1,2,3\}$; $\sigma_3=\{3\}$

$$\pi^{\gamma}(\sigma)=(\{1\},\{2\},\{3\}),$$

$$\pi^{\delta}(\sigma)=(\{1,2\},\{3\}).$$

[10] Another seminal contribution is Shenoy [43].

Note that the two rules of formation of coalitions are "exclusive" in the sense that each player of a forming coalition has announced a list of its members. Moreover, in the gamma-game this list has to be approved unanimously by all coalition members.

Once introduced these two games of coalition formation, a stable coalition structure for the game Γ (Δ) can be defined as a partition induced by a Strong Nash Equilibrium strategy profile of these games.

Definition 5. The partition π is a γ-stable (δ-stable) coalition structure if $\pi = \pi^{\gamma}(\sigma^*)$ (or $\pi = \pi^{\delta}(\sigma^*)$) for some σ^* with the following property: there exists no $S \subset N$ and $\sigma_S \in \Sigma_S$ such that

$$u_i(\sigma_S, \sigma^*_{N \setminus S}) \geq u_i(\sigma^*) \text{ for all } i \in S$$

and

$$u_h(\sigma_S, \sigma^*_{N \setminus S}) > u_h(\sigma^*) \text{ for at least one } h \in S.$$

In the recent literature on endogenous coalition formation, the coalition formation game by Hurt and Kurz is usually modelled as a first stage of a game in which, at the second stage formed coalitions interact in some underlying strategic setting. The coalition formation rules are used to derive a valuation mapping from the set of all players' announcements Σ into the set of real numbers. These payoff functions are obtained by associating with each partition $\pi = \{S_1, S_2, \ldots, S_m\}$ a game in strategic form played by coalitions

$$G(\pi) = (\{1,2,\ldots,m\}, (X_{S_1}, X_{S_2}, \ldots, X_{S_m}), (U_{S_1}, U_{S_2}, \ldots, U_{S_m})),$$

in which X_{S_k} is the strategy set of coalition S_k and $U_{S_k} : \Pi_{k=1}^{m} X_{S_k} \to R_+$ is the payoff function of coalition S_k, for all $k = 1,2,\ldots,m$. The game $G(\pi)$ describes the interaction of coalitions after π has formed as a result of players announcements in Γ.or Δ-coalition formation games.

The Nash equilibrium of the game $G(\pi)$ (assumed unique) gives the payoff of each coalition in π; within coalitions, a fix distribution rule yields the payoffs of individual members.

Following our previous assumptions (see Sect. 1.2) we can derived the game $G(\pi)$ from the the strategic form game G by assuming that $X_{S_k} = \prod_{i \in S_k} X_i$ and $U_{S_k} = \sum_{i \in S_k} u_i$, for every coalition $S_k \in \pi$. We can also assume $u_i = \frac{U_{S_k}}{|S_k|}$ as the per capita payoff function of members of S_k. Therefore, using Example 1, for the Γ-game , $u_i(x^*(\{1,2,3\}) = \frac{A}{12}$, for $i = 1,2,3$, $u_i(x^*(\{i,j\},\{k\}) = u_j(x^*(\{i,j\},\{k\}) = \frac{A}{18}$, $u_k(x^*(\{i,j\},\{k\}) = \frac{A}{9}$ and $u_i(x^*(\{i\},\{j\},\{k\}) = \frac{A}{16}$, for $i = 1,2,3$. Therefore, the grand coalition is the only stable coalition structure of the Γ-game of coalition formation. For the Δ-game, there are no stable coalition structures.

If we extend the merger game to n firms, we know that the payoff of each firm $i \in S \subset N$ when all remaining firms split up in singletons, is given by:

$$u_i\left(x^*\left(\pi^{\gamma}\left(\sigma'\right)\right)\right) = \frac{(a-c)^2}{s(n-s+2)^2},$$

where $n \equiv |N|$, $s \equiv |S|$ and $\sigma' = \left(\{S\}_{i \in S}, \{N\}_{i \in N \setminus S}\right)$. The grand coalition, induced by the profile $\sigma^* = (\{N\}_{i \in N})$, is a stable coalition structure in the Γ-game of coalition formation, if, for every $i \in N$,

$$u_i(x(\pi^\gamma(\sigma^*))) = \frac{(a-c)^2}{4n} \geq u_i\left(x\left(\pi^\gamma\left(\sigma'\right)\right)\right) = \frac{(a-c)^2}{s(n-s+2)^2}.$$

The condition above is usually verified for every $s \leq n$. Therefore, the stability of the grand coalition for the Γ-merger game holds also for a n-firm oligopoly.

2.2.2 Timing in Games of Coalition Formation

Following the literature on endogenous timing (for instance, Hamilton and Slutsky's [24]) we can add a preplay stage to the basic strategic setting (denoted *basic game*) in which players declare independently both their intention to coordinate their action with the other players as well as the timing they want to play the basic game. More specifically, every player $i \in N$ is assumed to play an extensive form game in which at stage t_0 (*coalition timing game*) announces an 2-tuple of strategies $a_i = (S, \tau) \in \mathbb{N} \times \{t_1, t_2\}$, where $\tau = \{t_1, t_2\}$ represents the time (stage 1 or 2) she intends to play the *basic game* jointly with the selected coalition $S \in \mathbb{N}$. Given the profile of announcements of the N players $a = (a_1, a_2, \ldots, a_n)$, a coalition structure $P(a) = (S_1^\tau, S_2^\tau, \ldots, S_m^\tau)$ endowed with a sequence of play of the basic game is induced, for instance, via the Hart and Kurz's unanimity rule: when a coalition of players announces both the same coalition S and the same timing, they will play the basic game of strategies simultaneously and coordinately as a coalition of players; otherwise, they will play as singletons with the timing prescribed by their own announcement. As the following example shows, the coalition formation timing rule constitutes a one-to-one mapping between the set of players' announcements and the set of feasible partitions of N.

Example 4 (Two-player). For every $i = 1, 2$ with $j \neq i$, each player's announcement set is:

$$A_i = [(\{i,j\}, t_1), (\{i,j\}, t_2), (\{i\}, t_1), (\{i\}, t_2)].$$

In this case the set of feasible partitions induced by the vector of announcement $a \in A_1 \times A_2$ includes the following six partitions:

$$(\{1,2\}^{t_1}), (\{1,2\}^{t_2}), (\{1\}^{t_1}, \{2\}^{t_1}), (\{1\}^{t_2}, \{2\}^{t_2}), (\{1\}^{t_1}, \{2\}^{t_2}), (\{1\}^{t_2}, \{2\}^{t_1}).$$

The existence of a Strong Nash equilibrium of the *coalition timing game* can be investigated. It can be shown [35] that for a symmetric strategic setting with no discount, the strategy for players of acting all together at period one constitutes an equilibrium when players'actions are strategic substitutes (in the sense of Bulow et al. [12]). Conversely, acting together at period two constitutes an equilibrium when players'actions are strategic complements.

2.3 Some Guidelines to Coalition Formation in Economic Applications

In order to compare and interpret the main predictions that endogenous coalition formation theories obtain in some classical economic problems, it can be useful to use a very simple setup in which the equal sharing rule within each coalition is not assumed but it is obtained through some symmetry assumptions imposed on the strategic form game describing the economic problem at hand. Once some basic assumptions are imposed on the strategic form games underlying the games of coalition formation, the main economic applications can be divided in a few categories: 1) games with positive or negative players-externalities; 2) games with actions that are strategic complements or substitutes; 3) games with or without coalition-synergies. According to these three features, we may have a clear picture of some of the results which can be expected from the different concepts of coalitional stability illustrated above and, in particular, of the stability of the grand coalition.[11]

We start imposing some symmetry requirements on the strategic form game G.

Assumption 1. *(Symmetric Players)*: $X_i = X \subset R$ for all $i \in N$. Moreover, for all $x \in X_N$ and all pairwise permutations $p : N \to N$:

$$u_{p(i)}\left(x_{p(1)},\ldots,x_{p(n)}\right) = u_i\left(x_1,\ldots,x_n\right).$$

Assumption 2. *(Monotone Externalities)*: One of the following two cases must hold for $u_i(x) : X_N \to R$ assumed quasiconcave:

1. Positive externalities: $u_i(x)$ strictly increasing in $x_{N\setminus i}$ for all i and all $x \in X_N$;
2. Negative externalities: $u_i(x)$ strictly decreasing in $x_{N\setminus i}$ for all i and all $x \in X_N$.

Assumption 1 requires that all players have the same strategy set, and that players payoff functions are symmetric, by this meaning that any switch of strategies between players induces a corresponding switch of payoffs. Assumption 2 requires that the cross effect on payoffs of a change of strategy have the same sign for all players and for all strategy profiles.

Lemma 1. *For all $S \subseteq N$, $\widetilde{x}_S \in \arg\max_{x_S \in X_S} \sum_{i \in S} u_i(x_S, x_{N\setminus S})$ implies $\widetilde{x}_i = \widetilde{x}_j$ for all $i,j \in S$ and for all $x_{N\setminus S} \in X_{N\setminus S}$.*

Proof. See Appendix.

An important implication of Lemma 1 is that all players belonging to a given coalition $S \subseteq N$ will play the same maximizing strategy and then will obtain the same payoff. We can thus obtain a game in valuation form from a game in partition function form without imposing a fixed allocation rule.

The next lemma expresses the fact that in every feasible coalition structure π, at the Nash equilibrium played by coalitions, when players-externalities are positive

[11] Some of the results presented here comes from Currarini and Marini [16].

(negative), being a member of bigger rather than a smaller coalition is convenient only when each member of S plays a strategy that is lower (higher) than that played by each member of a smaller coalition.

Lemma 2. *Let Assumptions 1 and 2 hold. Then for every S and $T \in \pi$, with $|T| \geq |S|$:*
(1) Under Positive Externalities, $u_s(x^*(\pi)) \geq u_t(x^*(\pi))$ if and only if $x_s \leq x_t$;
(2) Under Negative Externalities, $u_s(x^*(\pi)) \geq u_t(x^*(\pi))$ if and only if $x_s \geq x_t$.

Proof. See Appendix.

Finally, we can use a well known classification of all economic models in two classes: (1) games in which players' actions are *strategic complements*; (2) games in which players' actions are *strategic substitutes.*[12]

Definition 6. The payoff function u_i exhibits increasing differences on X_N if for all S, $x_S \in X_S$, $x'_S \in X_S$, $x_{N\backslash S} \in X_{N\backslash S}$ and $x'_{N\backslash S} \in X_{N\backslash S}$ such that $x'_S > x_S$ and $x'_{N\backslash S} > x_{N\backslash S}$ we have

$$u_i\left(x'_S, x'_{N\backslash S}\right) - u_i\left(x_S, x'_{N\backslash S}\right) \geq u_i\left(x'_S, x_{N\backslash S}\right) - u_i\left(x_S, x_{N\backslash S}\right).$$

This feature is typical of games, as price oligopoly models with differentiated goods, for which players' best-replies are upward-sloping. For these games, we can prove the following.

Lemma 3. *Let assumptions 1–2 hold, and let u_i have increasing differences on X_N, for all $i \in N$. Then for every S and $T \in \pi$, with $|T| \geq |S|$:*
(1) Positive Externalities imply $x_s \leq x_t$; (2) Negative Externalities imply $x_s \geq x_t$.

Proof. See Appendix.

Suppose now to have a game with actions that are strategic substitutes. This is the case of Cournot oligopoly and many other economic models. Suppose also that a boundary on the slope of the reaction mapping $f_S : R_{N\backslash S} \rightarrow R_S$ is imposed by the following contraction assumption.

Assumption 3. *(contraction)* Let $S \in \pi$. Then, there exists a $c < 1$ such that for all $x_{N\backslash S}$ and $x'_{N\backslash S} \in X_{N\backslash S}$

$$\left\| f_S\left(x_{N\backslash S}\right) - f_S\left(x'_{N\backslash S}\right)\right\| \leq c \left\| x_{N\backslash S} - x'_{N\backslash S}\right\|,$$

where $\|.\|$ denotes the euclidean norm defined on the space R^{n-s}.

Lemma 4. *Let assumptions 1–3 hold. Then for every S and $T \in \pi$, with $|T| \geq |S|$:*
(1) Positive Externalities imply $x_s \leq x_t$; (2) Negative Externalities imply $x_s \geq x_t$.

Proof. See Currarini and Marini [16].

[12] See, for this definition, Bulow et al. [12].

Using all lemmata presented above we are now able to compare the valuation of players belonging to different coalitions in a given coalition structure and then, to a certain extent, the profitability of deviations. However, the above analysis is limited to games in which forming a coalition does not enlarge the set of strategy available to its members and does not modify the way payoffs within a coalition originate from the strategies chosen by players in N. In fact, as assumed at the beginning of the paper, the action space of each coalition $S \subset N$ is restricted to $X_S \equiv \prod_{i \in S} X_i$. Moreover $U_S = \sum_{i \in S} u_i(x(\pi))$. The only advantage for players to form coalitions is to coordinate their strategies in order to obtain a coalitional efficient outcome. This approach encompasses many well known games *without synergies*, such as Cournot and Bertrand merger or cartel formation and public good and environmental games, but rules out an important driving force of coalition formation, i.e. the exploitation of synergies, typically arising for instance in R&D alliances or mergers among firms yielding some sort of economies of scales. Within this framework, we can present the following result.

Proposition 1. *Let assumptions 1–2 hold, and let u_i possess increasing differences on X_N, for all $i \in N$. Then the grand coalition N is a stable coalition structure in the game of coalition formation Γ derived from the game in strategic form G.*

Proof. By Lemma 3, positive externalities imply that for all π, at $x(\pi)$ larger coalitions choose larger strategies than smaller coalitions, while the opposite holds under negative externalities, and then $\frac{U_S(x^*(\pi^\gamma))}{|S|} \geq \frac{U_S(x^*(\pi^\gamma))}{|T|}$ for all $S, T \in \pi^\gamma$ with $|T| \geq |S|$. This directly implies the stability of the grand coalition in Γ. To provide a sketch of this proof, we note that any coalitional deviation from the strategy profile σ^* yielding the grand coalition induces a coalition structure in which all members outside the deviating coalitions appear as singleton. Since these players are weakly better off than any of the deviating members, and since all players were receiving the same payoff at σ^*, a strict improvement of the deviating coalition would contradict the efficiency of the outcome induced by the grand coalition. □

In games with increasing differences, players strategies are strategic complements, and best replies are therefore positively sloped. The stability of the efficient coalition structure $\pi^* = \{N\}$ in this class of games can be intuitively explained as follows. In games with positive externalities, a deviation of a coalition $S \subset N$ will typically be associated with a lower level of S's members' strategies with respect to the efficient profile $x(\pi^*)$, and with a higher level in games with negative externalities (see lemma 3 and 4 above). If strategies are the quantity of produced public good or prices (positive player-externalities), S will try to free ride on non members by reducing its production or its price; if strategies are emissions of pollutant or quantities (negative player-externalities), S will try to emit or produce more and take advantage of non members' lower emissions or quantities. The extent to which these deviations will be profitable ultimately depend on the reaction of non members. In the case of positive externalities, S will benefit from an increase of non members' production levels or prices; however, strategic complementarity implies that the decrease of S's production levels or prices will be followed by a decrease of the produced levels or prices of non members. Similarly, the increase of S's pollutant emissions or quantities will induce higher pollution or quantity levels by non members. Free riding is therefore little profitable in these games. From the above discussion, it is clear that deviations can be profitable only if best reply functions are negatively sloped, that is, strategies must be substitutes in G. However, the above discussion suggests that some "degree" of substitutability may still be compatible with stability. Indeed, if S's decrease in the production of public good is followed by a moderate increase in the produced level of non members, S may still not find it profitable to deviate from the efficient profile. Therefore, if the absolute value of the slope of the reaction maps is bounded above by 1, the stability result of proposition 1 extends to games with strategic substitutes.

Proposition 2. *Let assumptions 1–3 hold. The grand coalition N is a stable coalition structure in the game of coalition formation Γ derived from the game in strategic form G.*

Moreover, we can extend the results of proposition 1 and 2 to games with negative *coalition-externalities.*[13]

Definition 7. A game $G(\pi)$ exhibits positive (negative) coalition-externalities if, for any feasible coalition structure π and coalition $S \in \pi$, for every player $i \in S$, $u_i(x^*(\pi')) > (<)\ u_i(x^*(\pi))$ where π' is obtained from π by merging coalitions in $\pi \backslash S$.

It is clear from the above definition, that under negative coalition-externalities, $u_i(x(\pi^{\gamma}(\sigma'))) < u_i\left(x\left(\pi^{\delta}(\sigma')\right)\right)$ where $\sigma' = \left(\{S\}_{i \in S}, \{N\}_{j \in N \backslash S}\right)$ just because $\pi^{\gamma}(\sigma') = \left(\{S\}, \{j\}_{j \in N \backslash S}\right)$ and $\pi^{\delta}(\sigma') = (\{S\}, \{N \backslash S\})$. The following propositions exploits this fact.

Proposition 3. *Let assumptions 1–2 hold, and let u_i possess increasing differences on X_N, for all $i \in N$. Let also the game $G(\pi)$ exhibits negative coalition-externalities. Then the grand coalition N is a stable coalition structure in the Δ-game of coalition formation derived from the game in strategic form G.*

Proposition 4. *Let assumptions 1–3 hold. Let also the game $G(\pi)$ exhibits negative coalition-externalities. Then the grand coalition N is a stable coalition structure in the Δ-game of coalition formation derived from the game in strategic form G.*

[13] See Bloch [6] or Yi [47] for such a definition. There is not a clear relationship between games with positive (or negative) player-externalities and games with positive (or negative) coalition-externalities. However, for most well known games without synergies, both positive-player externalities (PPE) plus strategic complement actions (SC) as well as negative-player externalities (NPE) plus strategic substitute actions (SS) yield games with positive coalition-externalities. These are the cases of merger or cartel games in quantity oligopolies (NPE+SS), merger or cartel games in price oligopolies (PPE+SC) and public goods (PPE+SS) or environmental games (NPE+SS). Similarly, we can obtain Negative Coalition-Externalities in a game by associating NPE and SC as in a cartel game in which goods are complements and then the game exhibits SC.

A comparison of the above results, obtained for Hurt and Kurz's (1985) games of coalition formation, with the other solution concepts can be mentioned. It can be shown (see [46]) that for all games without synergies in which - as in the merger example - players prefer to stay as singletons to free-ride on a forming coalition – Bloch's [5] sequential game of coalition formation gives rise to equilibrium coalition structures formed by one coalition and a fringe of coalition acting as singletons. Moreover, even in a linear oligopoly merger game, Ray and Vohra's [40] *Equilibrium Binding Agreement* may or may not support the grand coalition as a stable coalition structure, depending on the number of firms in the market. When the game G is a game with synergies, a classification of the possible results. becomes even more complex. To give an illustration, we can introduce a simple form of synergy by assuming, as in Bloch's [4] and Yi's [46] R&D alliance models, that when firms coordinate their action and create a R&D alliance, they pool their research assets in such a way to reduce the cost of each firm in proportion to the number of firms cooperating in the project.[14] Let the producing cost of firms participating to a R&D alliance of s firms be $c(x_i, s_i) = (c+1-s_i)x_i$, where s_i is the cardinality of the alliance containing firm i: Let also $a > c \geq n$. As shown by Yi [46], at the unique Nash equilibrium associated with every coalition structure , the profit of each firm in a coalition of size s_i is given by

$$u_i^{\gamma}(x(\pi^{\gamma})) = \frac{\left(a-(n+1)(c+1-s_i)+\sum\limits_{j=1}^{k} s_j(c+1-s_j)\right)^2}{(n+1)^2},$$

When $\pi = \pi(\sigma')$, symmetry can be used to reduce the above expression to

$$u_i^{\gamma}(\pi^{\gamma}(\sigma')) = \frac{(a-(n-s_i+1)(c+1-s_i)+(n-s_i)c)^2}{(n+1)^2}.$$

Straightforward manipulations show that the deviation of a coalition S_i from the grand coalition in the game Γ is always profitable whenever:

$$s_i > -\frac{1}{2}n + c - \frac{1}{2}\sqrt{(n^2 - 4(nc - c^2) - 8(a-c-1)}.$$

For example, for $n = 8$, a deviation by a group of six firms ($s_i = 6$) induces a per firm payoff of $v_i^{\gamma}(\pi^{\gamma}(\sigma')) = \frac{(a-c+15)^2}{81}$ higher than the every firm's payoff in the grand coalition $v_i(\pi^{\gamma}(\sigma^*)) = \frac{(a-c+7)^2}{81}$. Therefore, it becomes more difficult to predict the stable coalition structures in Hurt and Kurz's Γ and Δ-games. In the sequential games of coalition formation [5, 41] for a linear Cournot oligopoly in which firms can form reducing-cost alliances, and each firm's $i \in S$ bears a marginal cost

$$c_i = \gamma - \theta s\,,$$

[14] This is usually classified as a game with negative coalition-externalities (see [46, 47]).

where s is the size of the alliance to which firm's i belongs, the equilibrium profit of each firm $i \in S$ is:

$$v_i(\pi) = \frac{1-\gamma}{n+1} + \theta s_i - \frac{\sum_{j \neq i} s_i^2}{n+1}.$$

Therefore, the formation of alliances induces negative externalities on outsiders, just because an alliance reduces marginal costs of participants and make them more aggressive in the market. Moreover, members of larger alliance have higher profits and then, if membership is open, all firms wants to belong to the association ([6], Bloch 2005). In the game of sequential coalition formation, anticipating that remaining players will form an association of size $(n-s)$, the first s players optimally decide to admit $s^* = (3n+1)/4$ and the unique equilibrium coalition structure results in the formation of two associations of unequal size $\pi^* = \left(\left\{\frac{3n+1}{4}\right\}, \left\{\frac{n-1}{4}\right\}\right)$.

3 Networks

3.1 Notation

We follow here the standard notation applied to networks.[15] A nondirected network (N,g) describes a system of reciprocal relationships between individuals in a set $N = \{1,2,\ldots,n\}$, as friendships, information flows and many others. Individuals are nodes in the graph g and links represent bilateral relationship between individuals.[16] It is common to refer directly to g as a network (omitting the set of players). The notation $ij \in g$ indicates that i and j are linked in network g. Therefore, a network g is just a list of which pairs of individuals are linked to each other. The set of all possible links between the players in N is denoted by $g^N = \{ij | i,j \in N, i \neq j\}$. Thus $G = \{g \subset g^N\}$ is the set of all possible networks on N, and g^N is denoted as the *complete network*. To give an example, for $N = \{1,2,3\}$, $g = \{12,13\}$ is the network with links between individuals 1 and 2 and 1 and 3, but with no link between player 2 and 3. The complete network is $g^N = \{12,23,13\}$. The network obtained by adding link ij to a network g is denoted by $g + ij$, while the network obtained by deleting a link ij from a network g is denoted $g - ij$. A *path* in g between individuals i and j is a sequence of players $i = i_1, i_2, \ldots, i_K = j$ with $K \geq 2$ such that $i_k i_{k+1} \in g$ for each $k \in \{1,2,\ldots,K-1\}$. Individuals who are not connected by a path are in different components C of g; those who are connected by a path are in the same component. Therefore, the components of a network are the distinct connected subgraphs of a network. The set of all component can be indicated as $C(g)$. Therefore, $g = \bigcup_{g' \in C(g)} g'$. Let also indicate with $N(g)$ the players who have at least one link in network g.

[15] See, for instance, Jakcson and Wolinski [34], Jackson [28] and van den Nouweland [44].

[16] Here both individuals engadged in a relationship have to give their consent for the link to form. If the relationship is unilateral (as in advertising) the appropriate model is a directed network. Also, here the intensity of a link is assumed constant.

3.2 Value Functions and Allocation Rules

It is possible to define a value function assigning to each network a worth.

Definition 8. A value function for a network is a function $v : G \to R$.

Let V be the set of all possible value functions. In some applications $v(g) = \sum_i u_i(g)$, where $u_i : G \to R$. A network $g \in G$ is defined (strongly) efficient if $v(g) \geq v(g')$ for all $g' \in G$. If the value is transferable across players, this coincides with Pareto-efficiency.[17]

Since the network is finite, it always exists an efficient network. Another relevant modelling feature is the way in which the value of a network is distributed among the individuals forming the network.

Definition 9. An allocation rule is a function $Y : G \times V \to R^N$.

Thus, $Y_i(g,v)$ is the payoff obtained by every player $i \in N(g)$ under the value function v. Some important properties of the value functions v and of the allocation rules Y can be defined.[18]

When compared to the characteristic function of cooperative games (see Sect. 1.1), here a value function v is sensitive not only to the number of players connected (in a component of g) but also to the specific architecture in which they are connected. However, v can be restricted to depend only on the number of players connected in a coalition. In a seminal contribution, Myerson [36] starts with a TU cooperative game (N,v) and overlaps a communication network g to such a framework. Myerson [36] associates a "graph-restricted value" $v^g : 2^N \to R$, assigning to each coalition S a value equal to the sum of worth generated by the connected components of players in S. Formally, players in S have links in $g(S) = \{ij \in g | i \in S,\ j \in S\}$ and this induces a partition of S into subsets of players $S(g)$ that are connected in S by g. Thus, $v^g(S) = \sum_{g' \in C^S(g)} v(g')$ for every $S \subset N$, where $C^S(g)$ indicates the set of components induced by g involving players belonging to coalition S. This value assumes that players in S can coordinate their action only within their own components.[19] Two assumptions underline this value: (1) there are no externalities between different components of a network; (2) what matters for the worth v^g is only the worth of the coalition of players which are in a component, not the type of connections existing within the coalition. Within this framework, Myerson characterizes a specific allocation rule (known as *Myerson value*) distributing the payoffs among individuals, and shows that under two axioms - fairness and component additivity - the unique allocation rule satisfying these properties is the Shapley value of the graph-restricted game (N, v^g):

[17] A network g is Pareto efficient (PE) with respect to a value v and an allocation rule Y if there not exists any $g' \in G$ such that $Y_i(g',v) \geq Y_i(g,v)$ with strict inequality for some i Note that if a network is PE with respect to v and Y for all possible allocation rules Y; it is (strong) efficient (see [28]).

[18] See Jackson and Wolinsky [34] and Jackson [29] for details.

[19] This implies a component balanced allocation rule Y.

$$Y_i(g, v^g) = \sum_{S \subset N \setminus \{i\}} \frac{|S|!(|N| - 1 - |S|)!}{|N|!} \left(v^g (S \cup \{i\}) - v^g (S)\right).$$

3.3 Networks Formation Games

3.3.1 Networks Formation in Extensive Form

Aumann and Myerson [2] propose an extensive form game to model the endogenous formation of cooperation structures. In their approach, which involves a sequential formation of links among players, bilateral negotiations take place in some predetermined order. Firstly, an exogenous rule determines the sequential order in which pairs of players negotiate to form a link. A link is formed if and only if both players agree and, once formed, cannot be broken. The game is one of perfect information and each player knows the entire history of links accepted or rejected at any time of the game. Once all links between pairs of players have formed, single players can still form links. Once all players have decided, the process stops and the network g forms and the payoff is assigned according to the Myerson value, i.e. the Shapley value of the restricted game (N, v^g). Stable cooperative structure are considered only those associated with subgame perfect equilibria of the game.

Example 5.[20] Suppose a TU majority game with $N = \{1,2,3\}$ and $v(S) = 1$ if $|S| \geq 2$ and $v(S) = 0$ otherwise. If the exogenous rule specifies the following order of pairs: $\{1,2\}$, $\{1,3\}$, $\{2,3\}$. The structure $\{1,2\}$ is the only cooperation structure supported by a subgame perfect equilibrium of the game. Neither player 1 nor player 2 have an interest to form a link with player 3, provided that the other player has not formed a link with 3. So, using backward induction, if at the final stage $\{2,3\}$ has formed, at stage 2 also $\{1,3\}$ forms and player 1 obtains a lower payoff than in a coalition with only player 2. Thus, at stage 1 player 1 forms a link with player 2 and the latter accepts. No other links are formed at the following stages.

It is possible that a subgame Nash equilibrium of the Aumann and Myerson's network formation game in extensive form does not support the formation of the complete network even for superadditive games. Moreover, no general results are known for the existence of stable complete networks even for symmetric convex games.[21]

3.3.2 Networks Formation in Strategic Form

Myerson [37] suggests a noncooperative game of network formation in strategic form.[22]

---

[20] This example is taken from Dutta et al. [19].

[21] See, for a survey of this approach, van den Nouweland [45].

[22] This game is also analyzed by Quin [39] and Dutta et al. [19].

For each player $i \in N$ a strategy $\sigma_i \in \Sigma_i$ is given by the set of players with whom she want to form a link, i.e. $\Sigma_i = (S | S \subseteq N \setminus \{i\})$. Given a n-tuple of strategies $\sigma \in \Sigma_1 \times \Sigma_2 \times \cdots \times \Sigma_n$ a link ij is formed if and only if $j \in \sigma_i$ and $i \in \sigma_j$. Denoting the formed (undirected) network $g(\sigma)$, the payoff of each player is given by $Y_i(v, g(\sigma))$ for every $\sigma \in \Sigma_N$. A strategy profile is a Nash equilibrium of the Myerson's linking game if and only if, for all player $i$and all strategies $\sigma' \in \Sigma_i$

$$Y_i(v, g(\sigma)) \geq Y_i(v, g(\sigma'_i, \sigma_{-i})).$$

We can also define a network g *Nash stable* with respect to a value function v and an allocation rule Y, if there exists a pure strategy Nash equilibrium σ such that $g = g(\sigma)$.

The concept of Nash equilibrium applied to the network formation game appears a too weak notion of equilibrium, due to the bilateral nature of links. The empty network (a g with no links) is always Nash stable for any v and Y. Moreover, all networks in which there is a gain in forming additional links but no convenience to sever existing links are also Nash stable. Refinements of the Nash equilibrium concept for the network formation process have been proposed. The *pairwise stability* introduced by Jackson and Wolinsky [34] plays a prominent role in the recent developments of the analysis of networks formation.

3.3.3 Pairwise Stability

We should expect that in a stable network players do not benefit by altering the structure of the network. Accordingly, Jackson and Wolinsky [34] defines a notion of network stability denoted *pairwise stability.*

Definition 10. A network g is pairwise stable with respect to the allocation rule Y and value function v if
(1) For all $ij \in g$, $Y_i(v, g) \geq Y_i(g - ij, v)$ and $Y_j(v, g) \geq Y_j(g - ij, v)$, and
(2) For all $ij \notin g$, if $Y_i(g + ij, v) > Y_i(g, v)$ then $Y_j(g + ij, v) < Y_j(g, v)$.

As shown by Jackson and Watts [33], a network is pairwise stable if and only if it has no *improving path* emanating from it. An improving path is a sequence of networks $\{g_1, g, \ldots, g_K\}$, where each network g_k is defeated by a subsequent (adjacent) network g_{k+1}, i.e. $Yi(g_{k+1}, v) > Yi(g_k, v)$ for $g_{k+1} = g_k - ij$ or $Y_i(g_{k+1}, v) \geq Y_i(g_k, v)$ and $Y_j(g_{k+1}, v) \geq Y_j(g_k, v)$ for $g_{k+1} = g_k + ij$, with at least one inequality holding strictly. Thus, if there not exists any pairwise stable network, then it must exists at least one cycle, i.e. an improving path $\{g_1, g, \ldots, g_K\}$ with $g_1 = g_K$. Jackson and Wolinsky [34] show that the existence of pairwise stable networks is always ensured for certain allocation rules. They prove that under the egalitarian and the component-wise egalitarian rules,[23] pairwise stable networks always exists. In particular, under

[23] The *egalitarian allocation rule* Y^e is such that $Y_i^e(g; v) = \frac{v(n)}{n}$ for all i and g. The component-wise allocation rule Y^{ce} is an egalitarian rule respecting component balance, i.e. such that $Y_i^{ce}(g; v) = \frac{v(C)}{|N(C)|}$ when $N(C)$, the set of players in component C is non empty and $Y_i^{ce}(g; v) = 0$ otherwise. See Jackson and Wolinsky [34] and Jackson [28] for details.

the egalitarian rule, any efficient network is pairwise stable. Under the component-wise allocation rule, a pairwise stable network can always be found. This can be done for component additive v by finding components C that maximize the payoffs of its players, and then continuing this process for the remaining players $N \backslash N(C)$. The network formed by all these components is pairwise stable. Another allocation rule with strong existence properties is the Myerson value. As shown by Jackson and Wolinsky [34], under Myerson's allocation rule there always exists a pairwise network for every value function $v \in V$. Moreover, all improving paths emanating from any network lead to pairwise stable networks, i.e. there are no cycles under the Myerson value allocation rule.[24]

However, as it is shown by Jackson and Wolinsky and by Jackson [28], there exists a tension between efficiency and stability whenever the allocation rule Y is component balanced and anonymous, in the sense that there does not exists an allocation rule with such properties that for all $v \in V$ yields an efficient network that is pairwise stable.

3.3.4 Further Refinements of Network Stability Concepts

As in the case of coalition formation, equilibrium concepts immune to coordinated deviations by players are also conceivable for networks (see, [18, 19, 32]). By allowing every subset of players to coordinate their strategies in arbitrary ways yields a strong Nash equilibrium for network formation games. That is, a strategy profile $\sigma \in \Sigma_N$ is a *strong Nash equilibrium* of the network formation game if there not exist a coalition $S \subseteq N$ and a strategy profile $\sigma'_S \in \Sigma_S$ such that

$$Y_i(v, g(\sigma'_S, \sigma_{N \backslash S})) \geq Y_i(v; g(\sigma)),$$

with strict inequality for at least one $i \in S$. Hence, a network g is strongly stable with respect to a value function v and an allocation rule Y, if there exists a strong Nash equilibrium σ such that $g = g(\sigma)$.

Similarly, an intermediate concept of stability, stronger than pairwise stability and weaker than strong Nash equilibrium, has been proposed [34] and denoted *pairwise Nash equilibrium*. This can be defined as a strategy profile $\sigma \in \Sigma_N$ such that, for all player i and all strategies $\sigma'_i \in \Sigma_i$,

$$Y_i(v, g(\sigma'_i, \sigma_{N \backslash \{i\}})) \geq Y_i(v, g(\sigma))$$

and there not exists a pair of agents (i, j) such that

$$Y_i(v, g(\sigma) + ij) \geq Y_i(v, g(\sigma))$$
$$Y_j(v, g(\sigma) + ij) \geq Y_j(v, g(\sigma))$$

[24] See Jackson [28] for details.

with strict inequality for at least one of the agents. Therefore, a network g is pairwise Nash stable with respect to a value function v and an allocation rule Y, if there exists a pairwise Nash equilibrium such that $g = g(\sigma)$.[25]

It can be shown that, given a value function v and an allocation rule Y, the set of strongly stable networks is weakly included in the set of pairwise Nash stable networks and that the latter set coincides with the intersection of pairwise stable networks and Nash stable networks.[26] Moreover, the set of pairwise stable networks and the set of Nash stable networks can be completely disjoint even though neither is empty.[27]

In the next section, I briefly illustrate some very simple applications of network formation games to classical I. O. models. These are taken from Bloch [9], Belleflamme and Bloch [3] as well as Goyal and Joshi [22].

3.4 Some Economic Applications

3.4.1 Collusive Networks

In Bloch [7] and in Belleflamme and Bloch [3] it is assumed that firms can sign bilateral market sharing agreements. Initially firms are present on different (geographical) markets. By signing bilateral agreement they commit not to enter each other's market.

If $ij \in g$, firm i withdraws from market j and firm j withdraws from market i. For every network g and given N firms, let $n_i(g)$ denote the number of firms in firm i' s market, with $n_i(g) = n - d_i(g)$ where $d_i(g)$ is the degree of vertex (firm) i in the network, i.e. the number of its links. If all firms are identical, firm i's total profit is

$$U_i(g) = u_i(n_i(g)) + \sum_{i,ij \notin g} u_i\left(n_j(g)\right).$$

With linear demand and zero marginal cost, under Cournot competition we obtain

$$U_i(g) = \frac{a^2}{[n_i(g)+1]^2} + \sum_{i,ij \notin g} \frac{a^2}{[n_j(g)+1]^2}.$$

If $n \geq 3$; there are exactly two pairwise stable networks, the empty network and the complete network. For $n = 2$, the complete network is the only stable network.

[25] This equilibrium concept has been adopted in applications by Goyal and Joshi [22] and Belleflamme and Bloch [3] and formally studied by Calvo-Armengol and Ilkilic [13], Ilkilic [27] and Gilles and Sarangi [21].

[26] See, for instance, Jackson and van den Nouweland [32] and Bloch and Jackson [10].

[27] See Bloch and Jackson [10, 11], for an extensions of these equilibrium concepts to the case in which transfers among players are allowed.

Note that the empty network is stable since for every symmetric firm the benefit to form a link is

$$U_i(g+ij) - U_i(g) = \frac{a^2}{n^2} - 2\frac{a^2}{(n+1)^2}$$

that, for $n \geq 3$, is negative.

For every incomplete network, $U_i(g) - U_i(g - ij) \geq 0$, requires that

$$\frac{a^2}{[n_i(g)+1]^2} - \left[\frac{a^2}{[n_i(g)+2]^2} + \frac{a^2}{[n_j(g)+1]^2}\right] \geq 0$$

and this holds only for $n_i(g) = n_j(g) = 1$, i.e. when the network is complete.

In this case,

$$U_i(g^N) - U_i(g^N - ij) = \frac{a^2}{4} - \frac{2a^2}{9} > 0.$$

Therefore, it follows that the only nonempty network which is pairwise stable is the complete network.

3.4.2 Bilateral Collaboration Among Firms

Bloch [7] and Goyal and Joshi [22] consider the formation of bilateral alliances between firms that reduce their marginal cost, as

$$c_i = \gamma - \theta d_i(g)\,,$$

where $d_i(g)$ denotes the degree of vertex i, i.e. the number of bilateral agreements signed by firm i.

Under Cournot competition with linear demand, we have each firm's profit is given by

$$U_i(g) = \left[\frac{a-\gamma}{n+1} + \theta d_i(g) - \frac{\theta \sum_j d_j(g)}{n+1}\right]^2.$$

For such a case, the only pairwise stable network turns out to be the complete network g^N (see [22]). This is because, by signing an agreement, each firm increases its quantity by $\Delta q_i = \dfrac{n\theta}{n+1}$, consequently, its profit. Moreover, when a large fixed cost to form a link is included in the model, Goyal and Joshi show that stable networks possess a specific form, with one complete component and a few singleton firms.

4 Concluding Remarks

This paper has attempted to provide a brief overview of the wide and increasing literature on games of coalition and network formation, paying a specific attention to the results which may be obtained by applying these games to some well known economic problems. It has been shown that, under reasonable assumptions mainly concerning the symmetry of players' payoffs, a number of general results can be obtained in games of coalition formation, which, in turn, can be easily applied to standard economic problems without synergies, as industry mergers and cartels, public goods games and many others. Network formation games appear as a natural extension of coalition formation games with, included, a detailed analysis of the effects of bilateral links among players. However, the issue of which network will form and which equilibrium concepts are suitable in a number of economic applications seems still largely unresolved, thus requiring further investigation. The future research agenda on the topic of network formation in social environments is certainly open to new exciting contributions.

Appendix

Lemma 1. *For all* $S \subseteq N$, $\widetilde{x}_S \in \arg\max_{x_S \in X_S} \sum_{i \in S} u_i(x_S, x_{N\setminus S})$ *implies* $\widetilde{x}_i = \widetilde{x}_j$ *for all* $i, j \in S$ *and for all* $x_{N\setminus S} \in X_{N\setminus S}$.

Proof. Suppose $\widetilde{x}_i \neq \widetilde{x}_j$ for some $i, j \in S$. By symmetry we can derive from $\widetilde{x}_S$ a new vector x'_S by permuting the strategies of players i and j such that

$$\sum_{i \in S} u_i(x'_S, x_{N\setminus S}) = \sum_{i \in S} u_i(\widetilde{x}_S, x_{N\setminus S}) \tag{6}$$

and hence, by the strict quasiconcavity of all $u_i(x)$, for all $\lambda \in (0,1)$ we have that:

$$\sum_{i \in S} u_i(\lambda x'_S + (1-\lambda)\widetilde{x}_S, x_{N\setminus S}) > \sum_{i \in S} u_i(\widetilde{x}_S, x_{N\setminus S}). \tag{7}$$

Since, by the convexity of X, the strategy vector $\left(\lambda x'_S + (1-\lambda)\widetilde{x}_S\right) \in X_S$, we obtain a contradiction. □

Lemma 2. *Let Assumptions 1 and 2 hold. Then for every* S *and* $T \in \pi$, *with* $|T| \geq |S|$*: (1) Under Positive Externalities,* $u_s(x(\pi)) \geq u_t(x(\pi))$ *if and only if* $x_s \leq x_t$*; (2) Under Negative Externalities,* $u_s(x(\pi)) \geq u_t(x(\pi))$ *if and only if* $x_s \geq x_t$.

Proof. We first prove the result for the case of positive externalities, starting with the "only if" part. By assumption 1, all members of T get the same payoff at $x(\pi)$. By definition of $x(\pi)$, the profile in which all members of T play x_t maximizes the utility of each member of T, so that

$$u_t((x_t, x_t) x_s) \geq u_t((x_s, x_s), x_s). \tag{8}$$

Suppose now that $x_s > x_t$. By assumption 1 and 2.1 we have

$$u_t((x_s, x_s), x_s) = u_{t_i}((x_s, x_s), x_s) = u_s((x_s, x_s), x_s) > u_s((x_t, x_t), x_s). \tag{9}$$

To prove the "if" part, consider coalitions T_1, T_2 and S which, as defined at the beginning of this section, are such that $|T_1| = |S|$ and such that $\{T_1, T_2\}$ forms a partition of T. By definition of $x(\pi)$, the utility of each member of S is maximized by the strategy profile x_S. Using the definition of u_s and of x_s we write:

$$u_s((x_t, x_t), x_s) \geq u_s((x_t, x_t), x_t). \tag{10}$$

By assumption 2.1, if $x_s \leq x_t$ then

$$u_s((x_t, x_t), x_t) \geq u_s((x_s, x_t), x_t). \tag{11}$$

Finally, by assumption 1 and the fact that $|T_1| = |S|$, we obtain

$$u_s((x_s, x_t), x_t) = u_{t_1}((x_t, x_t), x_s) = u_t((x_t, x_t), x_s), \tag{12}$$

implying, together with (11) and (12), that

$$u_s(x(\pi)) = u_s((x_t, x_t), x_s) \geq u_t((x_t, x_t), x_s) = u_t(x(\pi)). \tag{13}$$

Consider now the case of negative externalities (assumption 2.2). Condition (8) holds independently of the sign of the externality. Suppose therefore that $x_s < x_t$. By negative externalities and symmetry we have

$$u_t((x_s, x_s), x_s) = u_s((x_s, x_s), x_s) > u_s((x_t, x_t), x_s). \tag{14}$$

The "if" part is proved considering again coalitions T_1, T_2 and S. Again, Condition (10) holds independently of the sign of the externality. By negative externalities, if $x_s \geq x_t$ then

$$u_s((x_t, x_t), x_t) \geq u_s((x_s, x_t), x_t). \tag{15}$$

As before, we use assumption 1 and the fact that $|T_1| = |S|$ to obtain

$$u_s((x_s, x_t), x_t) = u_t((x_t, x_t), x_s), \tag{16}$$

and, therefore, that

$$u_s(x(\pi)) = u_s(x_t, x_s) \geq u_t(x_t, x_s) = u_t(x(\pi)). \tag{17}$$

□

Lemma 3. *Let assumptions 1–2 hold, and let u_i have increasing differences on X_N, for all $i \in N$. Then for every S and $T \in \pi$, with $|T| \geq |S|$: (1) Positive Externalities imply $x_s \leq x_t$; (2) Negative Externalities imply $x_s \geq x_t$.*

Proof. (1) Suppose that, contrary to our statement, positive externalities hold and $x_s > x_t$ By increasing differences of u_i *for all* $i \in N$ (and using the fact that the sum of functions with increasing difference has itself increasing differences), we obtain:

$$u_s((x_s, x_t), x_s) - u_s((x_s, x_t), x_t) \geq u_s((x_t, x_t), x_s) - u_s((x_t, x_t), x_t). \tag{18}$$

By definition of x_s we also have:

$$u_s((x_t,x_t),x_s) - u_s((x_t,x_t),x_t) \geq 0. \tag{19}$$

Conditions (18) and (19) directly imply:

$$u_s((x_s,x_t),x_s) - u_s((x_s,x_t),x_t) \geq 0. \tag{20}$$

Referring again to the partition of T into the disjoint coalitions T_1 and T_2, an application of the symmetry assumption 1 yields:

$$u_s((x_s,x_t),x_s) = u_{t_1}((x_s,x_t),x_s); \tag{21}$$

$$u_s((x_s,x_t),x_t) = u_{t_1}((x_t,x_t),x_s).$$

Conditions (20) and (21) imply

$$u_{t_1}((x_s,x_t),x_s) \geq u_{t_1}((x_t,x_t),x_s). \tag{22}$$

Positive externalities and the assumption that $x_s > x_t$ imply

$$u_{t_2}((x_s,x_t),x_s) > u_{t_2}((x_t,x_t),x_s). \tag{23}$$

Summing up conditions (22) and (23), and using the definition of T_1 and T_2, we obtain:

$$u_t((x_s,x_t),x_s) > u_t((x_t,x_t),x_s), \tag{24}$$

which contradicts the assumption that x_t maximizes the utility of T given x_s.

The case (2) of negative externalities is proved along similar lines. Suppose that $x_s < x_t$. Conditions (20) and (21), which are independent of the sign of the externalities, hold, so that (22) follows. Negative externalities also imply that if $x_s < x_t$ then (23) follows. We therefore again obtain condition (24). □

References

1. Aumann R (1967) A survey of games without side payments. In: Shubik M (ed) Essays in mathematical economics. Princeton University Press, Princeton, pp 3–27
2. Aumann R, Myerson R (1988) Endogenous formation of links between players and coalitions: an application of the shapley value. In: Roth A (ed) The shapley value. Cambridge University Press, Cambridge, pp 175–191.
3. Belleflamme P, Bloch F (2004) Market sharing agreements and stable collusive networks. Int Econ Rev 45:387–411.
4. Bloch F (1995) Endogenous structures of associations in oligopolies. Rand J Econ 26: 537–556.
5. Bloch F (1996) Sequential formation of coalitions with fixed payoff division. Games Econ Behav 14:90–123.

6. Bloch F (1997) Non cooperative models of coalition formation in games with spillovers. In: Carraro C, Siniscalco D (eds) New directions in the economic theory of the environment. Cambridge University Press, Cambridge
7. Bloch F (2002) Coalition and networks in industrial organization. The Manchester School 70:36–55
8. Bloch F (2003) Coalition formation in games with spillovers. In: Carraro C (ed) The endogenous formation of economic coalitions. Fondazione Eni Enrico Mattei Series on Economics and the Environment, Cheltenham, UK. Elgar, Northampton, MA
9. Bloch F (2004) Group and network formation in industrial organization. In: Demange G, Wooders M (eds) Group formation in economics: networks, clubs and coalitions. Cambridge University Press, Cambridge
10. Bloch F, Jackson MO (2006) Definitions of equilibrium in network formation games. Int J Game Theory 34:305–318
11. Bloch F, Jackson MO (2007) The formation of networks with transfers among players. J Econ Theory 133:83–110
12. Bulow J, Geanokoplos J, Klemperer P (1985) Multimarket oligopoly: strategic substitutes and complements. J Polit Econ 93:488–511
13. Calvó-Armengol A, Ilkilic R (2004) Pairwise stability and Nash equilibria in network formation. Universitat Autonoma de Barcelona (unpublished)
14. Chander P, Tulkens H (1997) The core of an economy with multilateral externalities. Int J Game Theory 26:379–401
15. Currarini S, Marini MA (2003) A sequential approach to the characteristic function and the core in games with externalities. In: Sertel M, Kara A (eds) Advances in economic design. Springer, Berlin
16. Currarini S, Marini MA (2006) Coalition formation in games without synergies. Int Game Theory Rev 8(1):111–126
17. Dutta B, Jackson MO (2003) On the formation of networks and groups. In: Dutta B, Jackson MO (eds) Networks and groups: models of strategic formation. Springer, Heidelberg
18. Dutta B, Mutuswami S (1997) Stable networks. J Econ Theory 76:322–344
19. Dutta B, van den Nouweland, Tijs AS (1998) Link formation in cooperative situations. Int J Game Theory 27:245–256
20. Fudenberg D, Tirole J (1991) Game theory. MIT, Cambridge, MA
21. Gilles RP, Sarangi S (2004) The role of trust in costly network formation. Virginia Tech (unpublished)
22. Goyal S, Joshi S (2003) Networks of collaboration in oligopoly. Games Econ Behav 43:57–85
23. Greenberg J (1994) Coalition structures. In: Aumann RJ, Hart S (eds) Handbook of game theory, vol 2. Elsevier, Amsterdam
24. Hamilton J, Slutsky S (1990) Endogenous timing in duopoly games: stackelberg or cournot equilibria. Games Econ Behav 2:29–46
25. Hart S, Kurz M (1983) Endogenous formation of coalitions. Econometrica 52:1047–1064
26. Ichiishi T (1983) A social coalitional equilibrium existence lemma. Econometrica 49:369–377
27. Ilkilic R (2004) Pairwise stability: externalities and existence. Universitat Autonoma de Barcelona (unpublished)
28. Jackson MO (2003) The stability and efficiency of economic and social networks. In: Koray S, Sertel M (eds) Advances in economic design. Springer, Heidelberg. Reprinted in Dutta B, Jackson MO (eds) Networks and groups: models of strategic formation. Springer, Heidelberg
29. Jackson MO (2005a) A survey of models of network formation: stability and efficiency. In: Demange G, Wooders M (eds) Group formation in economics: networks, clubs, and coalitions. Cambridge University Press, Cambridge
30. Jackson MO (2005b) The economics of social networks. In: Blundell R, Newey W, Persson T (eds) Proceedings of the 9th World Congress of the Econometric Society, Cambridge University Press, Cambridge, August 2005
31. Jackson MO (2007) Social and economic networks. Princeton University Press, Princeton

32. Jackson MO, van den Nouweland A (2005) Strongly stable networks. Games Econ Behav 51:420–444
33. Jackson MO, Watts A (2002) The evolution of social and economic networks. J Econ Theory 196(2):265–295
34. Jackson MO, Wolinsky A (1996) A strategic model of social and economic networks. J Econ Theory 71:44–74
35. Marini M (2007) Endogenous timing with coalitions of agents. CREI, Roma Tre (unpublished)
36. Myerson R (1977) Graphs and cooperation in games. Math Oper Res 2:225–229
37. Myerson R (1991) Game theory: analysis of conflict. Harvard University Press, Cambridge, MA
38. Owen R (1977) Value of games with a priori unions. In Hein R, Moeschlin O (eds) Essays in mathematical economics and game theory. Springer, New York, pp 76–88
39. Quin CZ (1996) Endogenous formation of cooperative structures. J Econ Theory 69:218–226
40. Ray D, Vohra R (1997) Equilibrium binding agreements. J Econ Theory 73:30–78
41. Ray D, Vohra R (1999) A theory of endogenous coalition structures. Games Econ Behav 26:286–336
42. Salant SW, Switzer R, Reynolds J (1983) Losses from horizontal merger: the effects of an exogenous change in industry structure on Cournot–Nash equilibrium. Q J Econ 98:185–99
43. Shenoy (1979) On coalition formation: a game theoretical approach. Int J Game Theory 8:133–164
44. van den Nouweland A (2005) Models of network formation in cooperative games. In: Demange G, Wooders M (eds) Group formation in economics: networks, clubs, and coalitions. Cambridge University Press, Cambridge
45. von Neumann J, Morgenstern O (1944) Theory of games and economic behaviour. Princeton University Press, Princeton
46. Yi S-S (1997) Stable coalition structure with externalities. Games Econ Behav 20:201–237
47. Yi S-S (2003) The endogenous formation of economic coalitions: the partition function approach. In: Carraro C (ed) The endogenous formation of economic coalitions. Fondazione Eni Enrico Mattei Series on Economics and the Environment, Cheltenham, UK. Elgar, Northampton, MA

Network Formation with Closeness Incentives

Berno Buechel

Abstract Closeness centrality is an index that has been widely used to assess the strength of an agent's position in a network of relationships. We study the formation of networks in a strategic setting, where every agent tries to optimize his closeness centrality. We investigate how the curvature of the benefit function (decreasing vs. increasing marginal returns) affects the set of stable networks and compare the results to the well-known connections model of Jackson and Wolinsky (JET 71, 1996). It turns out that our model can "replicate" the connections model in the sense that each result is translatable from one model into the other and the sets of stable networks coincide for certain specifications. We conclude that the two models incorporate the same kind of linking behavior and that grouping these "closeness-type" models means a first step in organizing network formation models by the type of incentives.

1 Introduction

Positions in social networks play a predominant role for economic outcomes. For example, consider a network of R&D collaborations in a technology-based industry. Companies which occupy a very "central" position are considered to better acquire and exploit knowledge that finally promotes their performance (e.g. [15]). In the field of social network analysis there is a long and rich history of studying benefits of network structures in various contexts. Beyond describing case studies, measures were developed that quantitatively assess the "merit" of certain network positions.[1]

B. Buechel
Institute of Mathematical Economics, Economic Behavior and Interaction Models
33501 Bielefeld, Germany
berno@wiwi.uni-bielefeld.de

[1] The underlying assumption of these approaches is that there are some structural features of networks (also called network statistics) that always have an impact on the opportunities or well-being of the agents, be it firms, people, or other units of decision.

A.K. Naimzada et al. (eds.), *Networks, Topology and Dynamics*,
Lecture Notes in Economics and Mathematical Systems 613,

This paper considers the problem from a different perspective by asking how network structures can be affected by agents that strive for beneficial positions.[2] As the impetus of each individual's linking behavior we use one of the most customary centrality indices: closeness centrality [8]. Closeness captures the idea that it is beneficial for an agent to have short paths to many agents in the network. Applications range from performance of organizational units [17], over web-based communities, to status in school classes.[3]

We model a situation of two-sided link formation, based on the framework introduced by Jackson and Wolinsky [12]. An important example therein is the connections model they presented, which was intensively studied thereafter (see, e.g. [1]). The benefits of the connections model represent information transmission with some decay. As the decay is based on the length of communication paths, the benefits of the connections model are also considered as a "closeness-like" centrality index [3]. Since closeness and connections are clearly similar in spirit, the question arises, to what extent the results for the connections model persist when using closeness centrality instead.

Experience shows that in network formation models minor changes may have major effects. E.g. [6] and [11] both study network formation based on the same concept ("structural holes"), but use a different specification, which leads to quite different results. Based on these considerations this paper investigates robustness of the connections model: *How do the stable networks in the closeness model differ from those in the connections model and what is in general the significance of the curvature of the benefit function?* It turns out that the closeness model in its linear version "replicates" the results of the (symmetric) connections model, meaning that both payoff functions incorporate the same behavior.

In the next section we will introduce the model. Section 3 provides basic results. Section 4 compares the linear closeness model to the connections model.

2 Model

Before motivating and defining the closeness model, we introduce the necessary definitions.

2.1 Framework

Let $N = \{1,\dots,n\}$ be a (finite, fixed) set of *agents/players*, with $n \geq 3$. A *network/graph* g is a set of unordered pairs, $\{i,j\}$ with $i,j \in N$. This set represents

[2] This idea can also be found in Rogers [16], who models the formation of weighted graphs using an index of social influence.

[3] Freeman [8] clarifies that closeness measures one dimension of centrality, while there are other dimensions, i.e. closeness does not sufficiently capture the intermediary role of some network positions.

who is linked to whom in a non-directed graph, i.e. $\{i, j\} = ij \in g$ means that player i and player j are linked under g. Let g^N be the set of all subsets of N of size two and G be the set of all possible graphs, $G = \{g : g \subseteq g^N\}$. A network can be seen as a (irreflexive, symmetric) binary relation on the player set. It can be represented by a matrix of zeros and ones called adjacency matrix.

Let $L_i(g)$ be the set of links that player i is involved in, that is $L_i(g) = \{ij \in g : j \in N\}$, and $l_i(g)$ its cardinality, called *degree*. An isolate is a player with degree zero and a pendant is a player with degree one (this structure is called a *loose end*).

A *circle* of size K is a sequence of K distinct players $(i_1, \dots, i_K)$ such that $\{i_k, i_{k+1}\} \in g \; \forall k \in \{1, \dots, K\}$, where $i_{K+1} := i_1$.

A *path* between two players i and j is a sequence of distinct players $(i_1, \dots, i_K)$ such that $i_1 = i$, $i_K = j$, and $\{i_k, i_{k+1}\} \in g \; \forall k \in \{1, \dots, K-1\}$. The (geodesic) *distance* between two players is the length of their shortest path(s), where the length is the number of links in the sequence. Formally, we can define the distance between two players ($d_{ij}(g)$) in a graph g by the corresponding adjacency matrix $A(g)$: $d_{ij}(g) := min\{k \in \mathbb{N} : A^k(g)_{ij} \geq 1; M\}$. If two players cannot reach one another (there does not exist a path connecting them), we define their distance as M, a number that is bigger than the feasible distances (see Sect. 2.2).

A graph is called *connected*, if there exists a path between any two players in the graph. A set of connected players is called *component*, if they cannot reach agents outside this set. A link is called *critical*, if its deletion increases the number of components in a graph. A graph is called *minimal*, if all links are critical. A *tree* is a connected network that is minimal.

Game Theoretic Framework

We base our model on a game-theoretic framework introduced by [14], [12] and [1]. Without defining the game explicitly, we take the "shortcut" of working with preferences and directly applying a stability concept.

For each player $i \in N$ a utility function represents his preferences over the set of possible graphs, $u_i : G \to \mathbb{R}$.

We work with the most basic equilibrium concept due to [12]: a network is considered as "stable" if no link will be added or cut (by two, respectively one player). Formally, a network g is pairwise stable (PS) or just *stable* if:

(1) $\forall ij \in g, \quad u_i(g) \geq u_i(g \backslash ij)$ and $u_j(g) \geq u_j(g \backslash ij)$
(2) $\forall ij \notin g, \quad u_i(g \cup ij) > u_i(g) \Rightarrow u_j(g \cup ij) < u_j(g)$

2.2 Closeness

The unique feature of the model presented here is the benefit function, which will be based on closeness centrality. Closeness incorporates the idea that an agent prefers networks, in which his average distance is short. Closeness is used in a wide variety of applications. We can identify the following basic arguments why closeness can be beneficial:

- *The higher your closeness, the smaller the distance to an arbitrary node in the network.* E.g. think of some researcher having a revolutionary idea. In the (co-author) network of this field of research, people with high closeness are likely to find out about the result much earlier.
- *The higher your closeness, the higher the expected spillovers from other agents.* As short geographic "distances" lead to external economies of scale, e.g. cars production in Detroit or Silicon Valley (electronics), it is plausible that short (network) distances have a similar effect, especially in times of well-developed information and communication technologies.
- *The higher your closeness, the higher your status.* In the friendship network of a school class one can assess the popularity of pupils by their closeness.
- *The higher your closeness, the better you can shape the community.* Networks with shorter paths facilitate quick diffusion of information/innovation. Agents with high closeness, therefore, can better spread their ideas.
- *The higher your closeness, the better you are informed.* The idea here is that accuracy of knowledge decreases with distances.

The listed arguments are not necessarily valid throughout all applications; the list only reflects how the importance of closeness can be justified. Our model is not dependent on the empirical validity of these motivations. The fact that some researchers or businessmen claim that closeness is desirable, provides enough justification to study network formation based on closeness incentives.

Definition of Closeness

We can generally define closeness such that benefits of an agent i gained by network g are decreasing with the (geodesic) distances of i to all other agents. To measure closeness there are some more details to look at.

To handle pairs that cannot reach one another, one can either restrict attention to the set of connected graphs, which would be a harsh assumption in a network formation game, or it must be defined what the distance of unconnected agents is. Here we define it as M. When i and j are connected, their distance is in $[1, n-1]$, hence let $M > n-1$.[4]

In the literature on centrality it is standard to normalize an index between 0 and 1. We follow this convention by defining *closeness* of an agent i as the following affine transformation of his average distance $\frac{\sum_{j \in N} d_{ij}(g)}{n-1}$:

$$Close_i(g) = \frac{M}{M-1} - \frac{\sum_{j \in N} d_{ij}(g)}{(M-1)(n-1)}.$$

[4] It is often convenient to define $M = n$. In this paper, however, we will keep it as a parameter.

There is another operationalization which is more prominent in the literature: the closeness definition according to Freeman [8]: $FrClose_i(g) := \frac{n-1}{\sum_{j \in N} d_{ij}(g)}$.[5] The author's trade-off here was: while Freeman's version (inverse distances) is much more customary, our closeness definition (reverse distances) more naturally separates the measurement of a structural feature of a network (network statistic) from its evaluation (by keeping the units, as argued in [18]). In the next subsection we will see that this choice does not restrict generality, e.g. if people strive just for Freeman-closeness, this is a special case of our model.

2.3 Model Specification

Our model is based on three major assumptions on individual behavior:

1. *The agents take linking decisions in respect to their degree and their closeness, where closeness is beneficial and links are costly.* To get a pure model we exclude all other aspects. One can think of any decision about adding or cutting links as a proposed exchange of average distance vs. degree: You can buy closeness by adding links; you can save costs by passing on closeness.
2. *The utility of a player is composed in an additive way by costs and benefits.* This assumption is not very restrictive as utility functions that are not additive separable may be transformed into this form. But it is a very convenient assumption: As the cross-derivatives are zero, the assumption uncouples the effects on utility coming from a change in closeness and a change in degree.
3. *The players are homogeneous in respect to preferences.* It is an interesting question to ask how networks evolve when players differ in their preferences (see, e.g. [9]). As applications of our model are very different in nature, however, we put emphasis on the different contexts that influence everybody's choice, not on the difference between agents (as also argued in [5]).

By introducing a (non-decreasing, twice differentiable) benefit function $b : [0,1] \to \mathbb{R}$ and a (non-decreasing, twice differentiable) cost function $c : [0, n-1] \to \mathbb{R}$, we can put all assumptions together to what we call the *closeness model*:

All agents $i \in N$ decide about links according to preferences that can be represented by $u_i(g) = b(Close_i(g)) - c(l_i(g))$.

Although concave and convex cost functions are reasonable – concave costs represent the combination of fix costs and variable costs; convex costs represent the scarcity of resources (e.g. time) – we will restrict attention to linear cost functions

[5] In the original version Freeman closeness is only defined for connected graphs. The extension to all networks works with the definition of the distance of unconnected players (as M).

$c^{linear}(g) = \bar{c}l_i(g), \;\; where \;\; \bar{c} \in (0,\infty)$.[6] The justification is that a concave or convex cost function would induce similar behavior as the benefit function does when transformed with the inverse function. So these aspects are assumed to be absorbed by the benefit function.

For the benefit function we will distinguish three cases: concave shape, convex shape and linear shape.[7] The first one represents decreasing marginal returns. Formally, $\forall x, x', \Delta > 0$ a concave benefit function implies $b(x+\Delta) - b(x) \geq b(x'+\Delta) - b(x')$ whenever $x \leq x'$ (by the mean value theorem). Convexity implies increasing marginal returns: just let $x' \leq x$.

Remark 1. By taking the following convex benefit function $f(x) = [M - x(M-1)]^{-1}$, the benefits are equivalent to Freeman-closeness (with linear evaluation), because $f(Close_i(g)) = FrClose_i(g)$.

The *marginal costs* $\bar{c}$ are constant and serve as the parameter for our model. The marginal benefits depend on the network g and on the shape of the benefit function. Let $\beta_i^{ij}(g)$ denote the *marginal benefit* that link ij (either added or cut) means to player i in graph g. That is, $\beta_i^{ij}(g) := b(Close_i(g \cup ij)) - b(Close_i(g \setminus ij))$.

When players take linking decisions, they compare marginal costs and marginal benefits: in graph g player i is eager to form a link to j ($ij \notin g$) iff $\beta_i^{ij}(g) > \bar{c}$ and i wants to cut a link with k ($ik \in g$) iff $\beta_i^{ik}(g) < \bar{c}$.[8]

3 General Results

This section provides boundaries (thresholds of the parameter) for stable networks in the closeness model and addresses how they can be affected by the curvature of the benefit function.

3.1 Connectedness and Loose Ends

To have a shorter notation, we substitute two often needed units of closeness:

1. $T1 := \frac{1}{(n-1)(M-1)}$. This is the smallest possible change in closeness, as it corresponds to a shift in distance of 1. It occurs when two players, who were at distance two, form a link and only the distance between these two changes, e.g. because they are already directly linked to everybody else.

[6] In fact, this assumption restricts preferences to be quasi-linear in degree.

[7] In [10] the role of concave/convex benefits is nicely elaborated. [13] analyzes decreasing marginal returns in a similar model, but with one-sided link formation.

[8] When marginal benefits are equal to marginal costs, the player is indifferent. In this case he does not cut the link, respectively does not initiate the new link (but agrees when asked).

2. $T2 := \frac{1}{(n-1)}$. This is the change in closeness of a player that links with an isolate. As his distance shifts from M to 1, his closeness increases by $\frac{M-1}{(n-1)(M-1)} = T2$.

The following results provide two characteristics of all stable networks.

Proposition 1. *In a closeness model with linear costs and concave benefits the following holds:*

(1) If $\bar{c} < b(1) - b(1 - T2)$, all stable graphs are connected.
(2) If $\bar{c} > b(T2) - b(0)$, no stable graph exhibits loose ends.

Proof. (1) Take any unconnected graph g. Take any player i and let $Close_i(g) =: x$. Linking with somebody of another component leads to a shift in closeness of at least $T2$. Because $x + T2 \leq 1$ and $b(\cdot)$ concave, it holds that $b(x+T2) - b(x) \geq b(1) - b(1-T2)$. By assumption the marginal costs are lower, such that i wants to form this link. As in any unconnected graph there exist two players who are not connected, they will alter the network structure, which makes g unstable.

(2) Take any network g with at least one pendant and let i be his (only) neighbour. Denote $Close_i(g) =: x$. Cutting the link to the pendant means a shift in closeness of $T2$. Because $x \geq T2$ and $b(\cdot)$ concave, it holds that $b(x) - b(x-T2) \geq b(T2) - b(0)$. By assumption the marginal costs are higher. Therefore, i will cut the link, which makes g unstable. □

The intuition behind the result is that the thresholds of (1) and (2) are just the minimal and the maximal marginal benefit that a link to an isolated node can mean.[9]

If the benefit function is not concave but convex, these two thresholds just switch roles, as stated by the following proposition.

Proposition 2. *In a closeness model with linear costs and convex benefits the following holds:*

(1) If $\bar{c} < b(T2) - b(0)$, all stable graphs are connected.
(2) If $\bar{c} > b(1) - b(1 - T2)$, no stable graph exhibits loose ends.

The proof is analogue to the proof of Proposition 1. Connectedness and non-existence of pendants heavily restrict the candidates for stable networks.

3.2 Existence

With the assumption of a convex benefit function, there is a very simple – admittedly not a very elegant – way of proving existence of stable graphs.

[9] I.e. the threshold in (2) is the marginal benefit of a new link in the empty graph $\beta_i^{ij}(g^{empty})$; and the threshold in (1) is the marginal benefit that cutting a link means to the center of a star $\beta_c^{ci}(g^{\star})$.

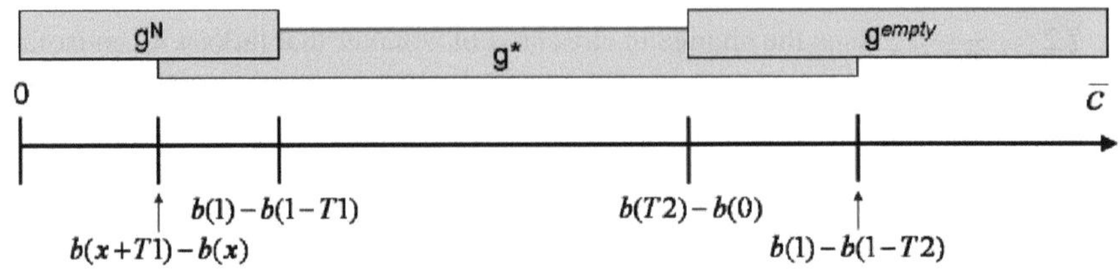

Fig. 1 Existence of stable networks for convex benefits

Proposition 3. *In a closeness model with linear costs and convex benefits the following holds: for any parameter value there exists at least one stable network.*

Proof. To show that for any marginal costs $\bar{c} \in (0,\infty)$ there exists a stable network, we take for low costs the complete graph, for high costs the empty network, and in the medium range the star. It is easy to verify that:

- The complete graph is stable if $\bar{c} \leq \beta_i^{ij}(g^N) = b(1) - b(1-T1)$. Remember that *T1* is the shift in closeness when distance increases by 1.
- The empty network is stable if $\bar{c} \geq \beta_i^{ij}(g^{empty}) = b(T2) - b(0)$.
- A star is stable if $b(x+T1) - b(x) \leq \bar{c} \leq \min\{b(1) - b(1-T2), b(x) - b(0)\}$, where $x := \frac{M}{M-1} - \frac{2n-3}{(M-1)(n-1)}$ is the closeness of a peripheral player (pendant). To verify the result, note that this condition precludes all possible deviations: (a) no peripheral players add a link $\bar{c} \geq b(x+T1) - b(x)$; and (b) the center does not cut a link $\bar{c} \leq b(1) - b(1-T2)$; and (c) no peripheral player cuts a link $\bar{c} \leq b(x) - b(0)$.

To prove existence for any marginal cost $\bar{c}$, it remains to show that (1) the lower bound of the star is below the upper bound of the complete network and (2) the upper bound of the star is above the lower bound of the empty network (see Fig. 1). 1. $b(x+T1) - b(x) \leq b(1) - b(1-T1)$ follows from $x + T1 \leq 1$ and convexity of $b(\cdot)$. And 2. $b(1) - b(1-T2) \geq b(T2) - b(0)$ follows from convexity of $b(\cdot)$; and $b(x) - b(0) \geq b(T2) - b(0)$ follows from $b(\cdot)$ increasing and $x \geq T2$. □

Figure 1 shows the idea of the proof: For any marginal cost, we can give a trivial example for a pairwise stable network.[10]

Remark 2. Figure 1 also contains the thresholds for Proposition 2 (on the right). In the case of concave benefits these two thresholds not only switch positions, but also switch their roles as stated in Proposition 1.

Besides these trivial examples (empty, complete, star) there are many more stable networks (which will be addressed in Sect. 4).

---

[10] For concave benefits the thresholds shift such that these trivial graphs do not span the whole parameter space. So in the case of concavity there are two "gaps" for which we could neither prove existence nor non-existence; for all other parameter values, existence is assured.

3.3 Pairwise Nash Stability

Besides pairwise stability there are other equilibrium concepts for network formation models, most of which are refinements of (PS). One of the most used stems from a non-cooperative framework and is called pairwise nash stability (PNS) (see, e.g. [2]). We can directly define it by just strengthening condition (1) of (PS): A network g is *pairwise nash stable* (PNS) if:

(1) $\forall i \in N,\ \forall l \subseteq L_i(g) \quad u_i(g) \geq u_i(g \backslash l)$.
(2) $\forall ij \notin g \quad u_i(g \cup ij) > u_i(g) \Rightarrow u_j(g \cup ij) < u_i(g)$.

In the closeness model, (PNS) is not always a proper refinement of (PS):

Proposition 4. *In a closeness model with linear costs and concave benefits the set of pairwise stable networks* $[PS]$ *and the set of pairwise nash stable networks* $[PNS]$ *coincide.*

One direction of the result follows directly from the definitions: $[PNS] \subseteq [PS]$. The other direction is more intriguing. Because of its length we omit the proof here and just present the main ideas[11]:

Calvó-Armengol and Ilkiliç [7] show that $[PNS]$ and $[PS]$ coincide, if the utility function $u(\cdot)$ satisfies a property called $\alpha - convexity$ in current links. Moreover, if costs and benefits are additively separable and marginal costs are constant, it is enough to show that the benefit function satisfies $\forall i \in N, \forall g \in G, \forall l \subseteq L_i(g)$,

$$\beta_i^l(g) \geq \sum_{ij \in l} \beta_i^{ij}(g), \tag{1}$$

where $\beta_i^l(g) := b_i(Close_i(g)) - b_i(Close_i(g \backslash l))$ denotes the marginal benefit that the deletion of the links (in l) means to some player i.

In essence, the condition says that the deletion of some of player i's links hurts him weakly more than the sequential deletion of these links, one at the time. For constant marginal costs it is intuitive that this is the condition requiring that deviations of cutting more than one link are only utility improving, if deviations of cutting just one link are, which is sufficient for $[PS] = [PNS]$.

To show that condition (1) holds in a closeness model with concave benefits, we need two steps: one step shows that the shift in closeness on the left-hand side of (1) cannot be smaller than the shift in closeness on the right-hand side. The other step exploits decreasing marginal returns (which guarantee, roughly, that multiple small reductions of closeness are not evaluated as severely as one big reduction).

The proof of Proposition 4 clarifies the role of the benefit function for the stability of networks: it is a genuine feature of the model that cutting one link at a time shifts closeness (weakly) less than cutting them at once. The concavity of the benefit function is just used to preserve this feature.

---

[11] The complete proof can be requested by the author.

This section showed how the curvature of the benefit function shifts thresholds for stable graphs. In the next section we study a special case, in which multiple thresholds coincide (to what we call "transition points").

4 The Linear Closeness Model

In the *linear closeness model*, we assume all players to have a linear cost function and a linear benefit function.[12] Without restriction of generality, we represent any player's preferences by $u_i^{linear}(g) = Close_i(g) - \bar{c}l_i(g)$. Note that by taking the id-function as benefit function, we mingle in this section what we distinguished before: the closeness of an agent and his benefit derived from closeness.

4.1 Transition Points

The first proposition is a corollary of Propositions 1 and 2, as the linear benefit function is a special case of both, concave and convex benefits functions.

Proposition 5. *Let again* $T2 := \frac{1}{(n-1)}$. *In the linear closeness model the following holds:*

(1) For $\bar{c} < T2$, *all stable graphs are connected.*
(2) For $\bar{c} > T2$, *no stable graph exhibits loose ends.*

Excluding pendants implies for the stable networks: (a) they cannot be minimal (i.e. a tree); (b) there exists at least one circle if the graph is non-empty; and (c) if the graph is connected, then it must contain at least n links. Observe that in this result two thresholds coincide: $b(1) - b(1-T2) = b(T2) - b(0) = T2$. This is also true for the next transition point.

Proposition 6. *Let again* $T1 := \frac{1}{(n-1)(M-1)}$. *In the linear closeness model the following holds:*

(1) For $\bar{c} < T1$, *the unique stable network is the complete network.*
(2) $T1 \leq \bar{c} \leq T2$, *a star shaped graph is stable, but not necessarily unique.*

Proof. Remember that *T1* is the shift in closeness when distances shift by 1.

(1) The minimal increase in benefit that a new link can lead to for both its owners is $T1$; because a new link reduces at least the distance to the other player from 2 to 1. So, if costs are strictly lower than this, it follows immediately that nobody wants to cut a link in any graph (stability of complete graph) and any two players, who are not directly linked, will add a link (uniqueness).

(2) Shown in proof of Proposition 3 □

---

[12] As a consequence, the linear closeness model differs from Freeman-closeness (with linear evaluation), but it is equivalent to Freeman-Closeness with a certain concave benefit function.

Costs below *T1* are considered as very small; costs above *T2* are considered as very high. However, *T2* is not necessarily a threshold for uniqueness (of the empty graph): There is a third transition point (which can be bigger than *T2*).

Let *T3* be the maximal marginal benefit that a non-critical link can mean to both its owners. We claim that $T3 = \frac{n-1}{4(M-1)}$.[13] $T3$ occurs in the line graph, where the pendants form a link, respectively in a circle graph as the marginal benefit of cutting a link.

Proposition 7. *In the linear closeness model the following holds: (1) For* $\bar{c} > T3$ *every stable graph is minimal or empty. (2) If* $T3 \geq T2$,[14] *then for* $\bar{c} > T3$ *the unique stable graph is the empty network.*

Proof. (1) If a non-empty network is not minimal, then there must be at least one non-critical link. By the definition of *T3*, networks with such links cannot be stable in this cost range. (2) The empty graph is stable because $\bar{c} \geq T2$. For uniqueness note that any non-empty graph must contain either loose ends or circles. By Proposition 5 we can exclude all graphs with loose ends for $\bar{c} > T2$. By (1) we can exclude all graphs with circles for $\bar{c} > T3$. □

The transition points organize the equilibria in the parameter space. For very small costs and for very high costs, there are only trivial stable networks. In the medium cost range we can find a multitude of stable networks. Figure 2 shows one example of a stable network in the linear closeness model for *n=14*, *M=n* and medium costs.

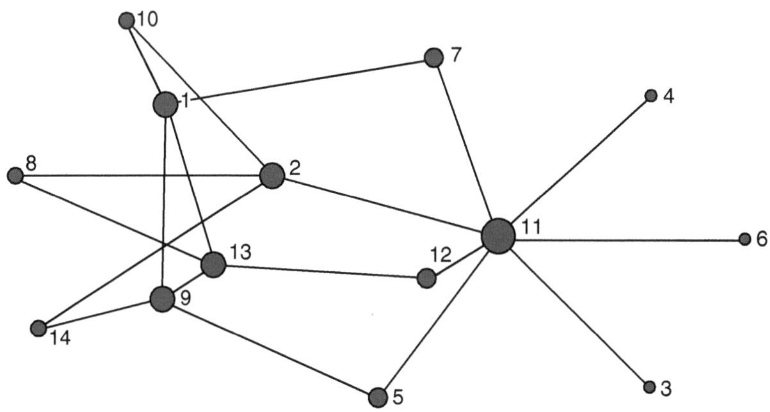

Fig. 2 Example of a stable network (the size and the position of a node indicates its closeness)

[13] The derivation of the value for $T3$ can be requested by the author.

[14] Mostly we will assume that M is such that $T3 \geq T2$ holds. We treat the exception of $T3 \leq T2$ in the next subsection as Proposition 8.

4.2 Comparison to Connections Model

In the famous example of the (symmetric) connections model, basically the following benefit is used: $Connections_i(g) = \sum_{j \in N \setminus i} \delta^{d_{ij}(g)}$, where $\delta \in (0,1)$.[15] So every reachable agent is of value, but this diminishes with distance. Like in the closeness model agents gain from short paths to other nodes. But there is also a difference: In the connections model agents benefit from having many nodes close to them; while in the closeness model agents benefit from having a small average distance.

While the motivation of the two models is similar, the results turn out to be almost identical.

Observe first that Propositions 5 and 6 correspond directly to the results of the connections model, where $T1 \hat{=} \delta - \delta^2$ and $T2 \hat{=} \delta$.

For n not too big, a computer can enumerate all networks and check for stability.[16] We did this for $n = 8$ with the connections model (taking $\delta = 0.5$ and $\delta = 0.8$), and for the closeness model once with the convex benefit function according to Freeman and once taking the linear closeness model (with $M = n$).

For $n = 8$ there are 12,346 different isomorphic graphs. In the linear closeness model only 45 of them are stable for some parameter range (greater than 0).[17] As depicted in Table 1, those 45 networks are not identical to the 63 stable networks with convex benefit function (Freeman), but overlap to some extent. The stable networks of the linear closeness model and the connections model overlap more heavily.

All of the above models are driven by similar linking behavior, which we call "closeness-type" incentives: there is high incentive to link to agents who are at high distance (or in another component) and there is low incentive to keep links that do not shorten some paths significantly. Interestingly, the differences within these models stem from specification details – be it increasing (instead of constant) marginal returns or level of decay – rather than from the choice of the model (connections

Table 1 Stable networks in the linear closeness model and related models for n=8

Number of stable networks (for some cost range)	Total	Also stable in linear closeness model
Freeman closeness	63	29
Connections $\delta = 0.5$	29	26
Connections $\delta = 0.8$	45	45

[15] By convention, here $M = \infty$ (see [12]).

[16] I thank Vincent Buskens for programming the routines to find all the stable networks for the various centrality measures.

[17] That is: we did not count the networks which are "stable" for only one point in the parameter space, e.g. the networks which are only stable if $\bar{c} = T1$.

vs. closeness). In this sense the connections model turns out to be "robust" and we conclude that the linear closeness model incorporates the same behavior as the connections model with some decay.

4.3 Trees

A very special case of the connections model occurs when the decay is very small or zero. Then distances become irrelevant and benefits only depend on the size of an agent's component. In this context the stable networks are trees, as non-critical links are worthless.

It turns out that the closeness model can also replicate this feature of the connections model by setting M sufficiently large[18]:

Proposition 8. *In the linear closeness model for marginal costs in the range $T3 < \bar{c} < T2$ the following holds:*

(1) All stable networks are trees.
(2) All trees are stable.

Proof. (1) Trees are characterized as minimal graphs that are connected. For $\bar{c} < T2$ all stable graphs are connected, as shown in Proposition 5(1). For $\bar{c} > T3$ all stable graphs are minimal as shown in Proposition 7(1).

(2) A graph is stable, if (a) nobody cuts a link and (b) no two players add a link. (a) As a tree is minimal, cutting a link leaves two components (unconnected groups of players). The more agents there are in the other component, the higher the loss of benefits. The highest incentive to cut is always given by the neighbor of a pendant; he loses closeness of $\frac{1}{n-1} = T2$. By assumption, marginal costs are lower than this (minimal marginal benefit), therefore no agent in a tree will cut a link. (b) Adding a link to a tree is an addition of a non-critical link (it is a property of trees to be maximally acyclic graphs). For $\bar{c} > T3$ this cannot be favorable for both (by the definition of *T3*). □

Remark 3. Note that many trees are also stable, when costs are below *T3* (see [4]).

This section showed that in the linear closeness model multiple thresholds (for stable networks) coincide. Given this feature, we can replicate the connections model with or without decay.

5 Concluding Remarks

We introduced a network formation model based on incentives to optimize closeness centrality. We analyzed how the set of stable networks depends on the curvature of the benefit function and compared the results to the connections model. It turns out

---

[18] Letting $M > \frac{1}{4}(n-1)^2 + 1$ assures that $T2 > T3$.

that the linear closeness model represents the same kind of behavior as the connections model, because each result is translatable from one model into the other and the sets of stable networks coincide for some parameter settings.

By grouping these "closeness-type" models, we have made a first step in organizing network formation models by the type of incentives. Accordingly, Buechel and Buskens [4] investigate the emerging networks in this setting and compare the results to incentives of a different type. Further research should clarify which structural patterns of networks emerge in which context and discuss the welfare implications.

The main limitations of our model – as in most models of strategic network formation – are the strong assumptions on behavior: Our agents are endowed with complete information and high rationality.

Acknowledgments I would like to thank Bastian Westbrock, Dinko Dimitrov, Andreas Brunhart, Tim Hellmann, Evan Shellshear and an anonymous referee for insightful comments. I gratefully acknowledge financial support from the international research training group "Economic Behavior and Interaction Models" funded by the German Science Foundation.

References

1. Bala V, Goyal S (2000) A noncooperative model of network formation. Econometrica 68:1181–1229
2. Bloch F, Jackson M (2006) Equilibrium definitions in network formation games. Int J Game Theory 34:305–318
3. Borgatti S, Everett M (2006) A graph-theoretic perspective on centrality. Soc Networks 28:466–484
4. Buechel B, Buskens V (2008) The dynamics of closeness and betweenness. Bielefeld University, Mimeo
5. Burger M, Buskens V (2006) Social context and network formation: experimental studies. ISCORE 234, Utrecht University
6. Buskens V, Van de Rijt A (2008) Dynamics of networks if everyone strives for structural holes. Am J Sociol (forthcoming)
7. Calvó-Armengol A, Ilkiliç R (2007) Pairwise-stability and nash equilibria in network formation games. Working paper (2007-01), Univ. Autnoma de Barcelona
8. Freeman L (1979) Centrality in social networks: conceptual clarification. Soc Networks 1(3):215–239
9. Galeotti A, Goyal S, Kamphorst J (2006) Network formation with heteregeneous players. Games Econ Behav 54:353–372
10. Goyal S, Joshi S (2006) Unequal connections. Int J Game Theory 34:319–349
11. Goyal S, Vega-Redondo F (2008) Structural holes in social networks. J Econ Theory (forthcoming)
12. Jackson M, Wolinsky A (1996) A strategic model of social and economic networks. J Econ Theory 71:44–74
13. Kamphorst J (2007) Network formation with decreasing marginal benefits of information. Utrecht School of Economics, Mimeo
14. Myerson RB (1991) Game theory: analysis of conflict. Cambridge University Press, Cambridge

15. Powell WW, Koput KW, Smith-Doerr L (1996) Interorganizational collaboration and the locus of innovation: networks of learning in the biotechnology. Adm Sci Q 41:116–145
16. Rogers B (2006) A strategic model of network status. Working paper (2006-01), Caltech, Pasadena
17. Song S, Nerur S, Teng J (2007) An exploratory study on the roles of network structure and knowledge processing orientation in work unit knowledge management. ACM SIGMIS Database 38:8–26
18. Valente TW, Foreman RK (1998) Integration and radiality: measuring the extent of an individual's connectedness and reachability in a network. Soc Networks 20:89–105

A Dynamic Model of Segregation in Small-World Networks

Giorgio Fagiolo, Marco Valente, and Nicolaas J. Vriend

Abstract Schelling [19–22] considered a simple model with individual agents who only care about the types of people living in their own local neighborhood. The spatial structure was represented by a one- or two-dimensional lattice. Schelling showed that an integrated society will generally unravel into a rather segregated one even though no individual agent strictly prefers this. We make some steps to generalize the spatial proximity model to a proximity model of segregation. That is, we examine models with individual agents who interact "locally" in a range of network structures with topological properties that are different from those of regular lattices. Assuming mild preferences about with whom they interact, we study best-response dynamics in random and regular non-directed graphs as well as in small-world and scale-free networks. Our main result is that the system attains levels of segregation that are in line with those reached in the lattice-based spatial proximity model. That is, Schelling's original results seem to be robust to the structural properties of the network. In other words, mild proximity preferences coupled with adjustment dynamics can explain segregation not just in regular spatial networks but also in more general social networks.

G. Fagiolo
Laboratory of Economics and Management, Sant'Anna School of Advanced Studies, Pisa, Italy
giorgio.fagiolo@sssup.it

M. Valente
Faculty of Economics, University of L'Aquila, L'Aquila, Italy
mv@business.auc.dk

N.J. Vriend
Department of Economics, Queen Mary, University of London, London, UK
n.vriend@qmul.ac.uk

A.K. Naimzada et al. (eds.), *Networks, Topology and Dynamics*,
Lecture Notes in Economics and Mathematical Systems 613,

1 Introduction

Segregation has been widely recognized as a critical issue, from both a socio-political and a public-economic perspective, in many Western countries. In the USA, segregation has dominated the political debate for some time in the second half of the last century. More recently, segregation issues are increasingly becoming one of the main points in the political agenda of the majority of European countries, and this trend is likely to be reinforced by the geo-political turmoils due to the events following 9/11 and the ongoing enlargement process of the EU.

The main problem faced by countries trying to reduce segregation is that we still do not know how to attain this goal. Indeed, the plethora of integration policies that have been implemented in the last decades turned out to be almost completely ineffective. In particular, all policies aimed at changing individual preferences towards multiculturalism (e.g., by promoting people openness and tolerance with respect to diversity) did not substantially improve integration [3]. Therefore, gaining a better knowledge of the forces underlying the dynamics leading to segregated societies seems crucial today as it was in the second half of the 20th century.

Exactly in those years, Schelling [19–22] studied a simple model of segregation with individual agents who only care about the types of people living in their own local neighborhood. The spatial structure was represented by a one- or two-dimensional lattice. Schelling showed that an integrated society will generally unravel into a rather segregated one even though no individual agent strictly prefers this. Rather, segregation seemed due to the spontaneous dynamics of the economic forces, with all individuals following their incentives to move in the most attractive locations. The preferences considered in the spatial proximity model are said to be mild, as everybody would be happy in a perfectly integrated society.

More recently, Pancs and Vriend [17] examined the robustness of Schelling's spatial proximity model. They showed that the model can be further simplified (rendering the individual preferences even more salient as an explanatory variable of segregation), and that these proximity preferences may be even more extreme in favor of integration. This focus on mild individual preferences or preferences that even favor integration is not to say that institutional constraints or racism may not hinder integration. But what the model shows is that even without such obstacles one should perhaps expect segregation. It seems that any integration policy must be based on a good understanding of these spontaneous dynamics.

Both original Schelling's model and Pancs and Vriend's robustness analyses explore segregation dynamics on regular (one- or two-dimensional) lattices. In other words, they both study the emergence of segregation in a geographical space. Indeed, lattices are widely employed in local-interaction models because they can be considered as a first approximation of geographical space [8]. The idea that people care about their spatial proximity can be justified by the fact that this is where people mow their lawn, where their children play outside, where they do their shopping, and where they park their car. The social environment is, however, not limited to this spatial proximity. People also interact through networks of friends, relatives, and colleagues, and through virtual communities on the internet. And they are likely to have preferences with whom they do this, just as they have preferences about their spatial proximity.

This suggests that segregation need not necessarily occur at the spatial (neighborhood) level: one might conceive people who are socially segregated despite being spatially integrated. This appeared to be the case with some of the recent terror suspects in the Netherlands and the UK. Therefore, a better understanding of the phenomenon of segregation in more general network structures seems desirable.

In this paper, we generalize Schelling's *spatial* proximity model to a proximity model of segregation where individual agents interact "locally" in a range of social network structures with topological properties that are different from those of regular lattices. Among all network structures alternative to regular lattices, we explore in particular small-world networks, which have been found to be a good proxy of real-world social interaction structures [2]. We stick to standard assumptions as far as types and preferences are concerned, and we study the ensuing best-response dynamics in two setups. In the first one (global-move setup) agents that are not satisfied with their current state can choose uniformly at random any empty location in the whole network (i.e., move arbitrarily far away in the social space). In the second setup (local-move), they are bound to choose one of the available locations in their social neighborhood (if any).

The rest of the paper is organized as follows. Section 2 discusses in more detail the classes of networks that we consider in our analysis. In Sects. 3 and 4 we present the model in its global- and local-move variants, and we discuss its implementation. Section 5 introduces the index that we employ to measure segregation in social networks. Simulation results are in Sect. 6, which also contains a sensitivity analysis of the parameter space. Finally, Sect. 7 concludes.

2 Social Networks and Small Worlds

The last fifteen years have witnessed an incredible outburst of empirical studies on natural, social and economic networks [1, 16, 23, 24]. More specifically, the bulk of contributions has focused on the structural and topological properties of empirically-observable networks such as the Internet and the WWW, airline connections, scientific collaborations and citations, trade and labor-market contacts, friendship and other social relationships, business relations and R&D partnerships, cellular, ecological and neural networks.

The main message of this vast literature is that most real-world networks belong to a particular class of structures that display neither the intrinsic spatial regularity of lattices, nor the disorder of random graphs [i.e., networks where any two agents are neighbors, independently of all the others, with some given probability, see 5]. To see why, let us begin with some basic definitions.

It is well-known that the simplest mathematical description of a network can be given in terms of an undirected graph $G = (n, A)$, where n is the number of nodes (individuals) and A is a $n \times n$ symmetric matrix whose generic element a_{ij} is equal to 1 if nodes i and j are linked by an edge (i.e., they are neighbors, either in a geographical or a social space), and 0 otherwise. Different networks can be taxonomized according to their structural and topological properties [18, 23, 24]. The most salient characteristics of a network can be summarized by the distributions of three statistics: (1) degree of a node; (2) clustering of a node; (3) shortest-path length between any two nodes. The degree of a node is simply the number of neighbors it has. Lattices are regular graphs because all nodes have the same degree. In random graphs node degrees are heterogeneous and symmetrically distributed around the average degree, which is proportional to the probability that any two nodes are neighbors. The clustering of a node is instead the likelihood that any two neighbors of that node are themselves neighbors. For each node i, this can be easily computed by counting the number of triangles with i as one vertex (and dividing this number by the total number of triangles that i could have formed given its degree). Obviously, lattices are in general much more clustered than random graphs, as their nodes are typically distributed in tightly connected clusters where any two neighbors are also neighbors by construction.[1] Finally, the shortest path length between any two nodes (i, j) is defined as the minimum number of links that one has to traverse to get from i to j. This measure has been popularized as the "degrees of separation," see Watts [26]. Again, lattices are extreme cases where this measure is generally high, as any two nodes far away in the lattice can reach each other only by travelling through all nodes that are in between. More precisely, the average distance between any two edges increases as $\sqrt{N}$, much faster than in random graphs, where it only increases as lnN.

Recent empirical studies [see 1, 16] have shown that real-world social networks are neither regular lattices nor random graphs, but lie in between. Indeed, they belong to the class of "small worlds" [13, 15, 25]. This type of networks preserves the high clustering level displayed by lattices, while exhibiting a smaller average shortest-path length, which only increases as lnN as in random graphs. This means that individuals embedded in real-world social networks tend to form tightly connected local clusters (of friends, relatives, business partners, etc.) as happens in geographical space. However, these local clusters are also frequently connected among them by shortcuts that allow any two agents who are arbitrarily far away in the social space to reach each other in a few steps [actually only six, on average, in many cases; see 15, for a review].

Networks belonging to the small-world class strongly differ, however, as to the shape of their degree distributions [2]. A first sub-class, which we will label as "Watts-Strogatz" (WS) in what follows, exhibits a quasi-symmetric degree distribution, centered around the average $d > 0$ (and tails possibly decaying exponentially fast as in the Gaussian distribution). To the second type of small-world networks belong the so-called "scale-free" networks (henceforth SF), i.e., networks whose degree distribution is right-skewed and decays with a power-law tail. Therefore, in WS small-world networks most of the nodes have the same degree. On the contrary,

[1] This may not be the case, however, for some particular choices of the metrics (e.g., the Von-Neumann one) and a relatively small interaction radius (e.g., equal to one). More on that below.

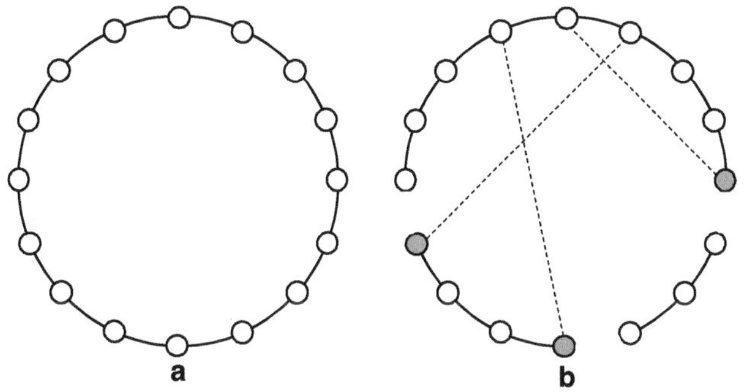

Fig. 1 Rewiring procedure for WS small-world graphs. Panel (**a**): we start from a circle where each node is connected with two neighbors. Panel (**b**): the graph after three nodes (*in grey*) have successfully rewired one of their links. *Dashed lines* depict rewired links

Fig. 2 An example of a WS graph obtained from $M = 100$ nodes originally lying on a circle and interacting with two neighbors only ($r = 1$). Rewiring probability $\beta = 0.2$. Node colors depend on degree after rewiring. *Black nodes*: $d \geq 3$. *Grey nodes*: $d = 2$. *White nodes*: $d = 1$

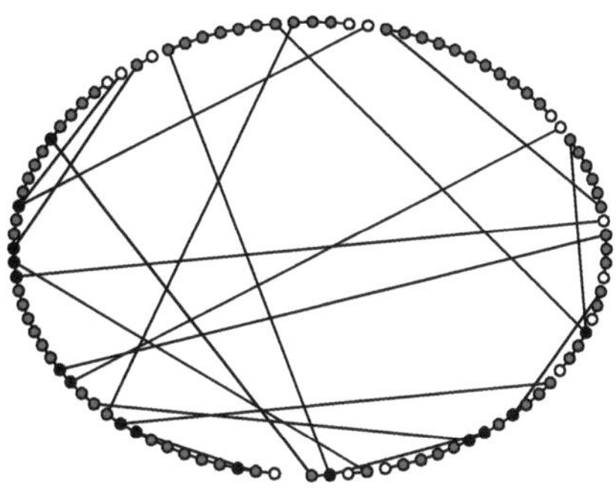

SF networks are characterized by a few nodes holding many partners (i.e., the hubs) and many nodes holding a few partners.

WS and SF small-world networks also differ in their generating mechanism [6]. Think of a given graph G as the (long-run) equilibrium of some stochastic dynamic model of node and edge dynamics. The most simple way to generate WS graphs runs as follows [27]. Start with a regular lattice (e.g., a circle), where each node has two neighbors – one on the left, one on the right, as in Fig. 1, panel (a). At each $t = 0$, pick a node (say i) at random and rewire with probability $\beta \in [0,1]$ one of its two links (say the one with j) to another node in the graph (say $h \neq i$) chosen at random. Figure 1, panel (b) shows the graph after three successful rewirings (nodes that have successfully rewired are shown in grey). If one repeats this procedure for a large number of times (avoiding the case that more than one link connects the same couple of nodes), the resulting graph is a small world, provided that β is sufficiently small (typically between 0.01 and 0.3), see Fig. 2 for an example. Notice how a small fraction of nodes (in black) hold more than two links, with some of their links being to nodes located arbitrarily far from them on the circle, while the other nodes (in grey or white) who kept only one or two links are usually linked to their direct neighbors only. This mechanism allows one to span the space of a wide variety of networks, from lattices ($\beta = 0$) to random graphs ($\beta = 1$).

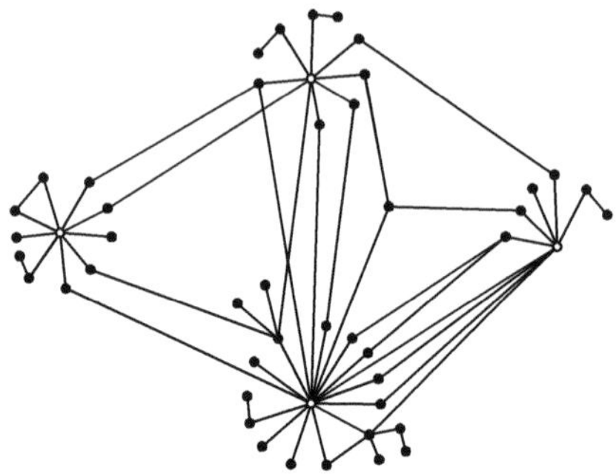

Fig. 3 An example of a SF graph obtained applying a preferential attachment algorithm. The initial population size is $M_0 = 16$. All initial nodes have degree $d = 4$. Final population size is $M = 100$. We only plot a subset of all nodes for simplicity. *White nodes* are the "hubs"

SF networks can instead be generated as the limit of a growth process known as "preferential attachment" [4]. According to this algorithm, one starts from an initial graph G_0 (e.g., a small lattice) and adds a new node to the graph in each step. The newly added node makes new connections with existing nodes, where the probability of connecting with any existing node is proportional to the current degree of the latter. As this process goes on, better-connected nodes attract more and more entrants (i.e., the rich get richer). The resulting (limit) graph can be shown to be a small world with a power-law degree distribution. The underlying assumption of this setup is that any node can hold at no cost any arbitrarily large number of nodes (as network size increases). As Fig. 3 shows for an instance of a SF graph obtained by applying the above procedure for $n = 100$, a few "hubs" (in white) holding a large number of links coexist with many nodes (in black) connected with a small number of other nodes, and possibly with the hubs.

To sum up: Recent empirical works have robustly highlighted that small-world (WS and SF) networks ubiquitously emerge in many social contexts. Therefore, small-world networks seem to be the most natural candidate to test the robustness of Schelling's spatial proximity model when agents are placed in more general types of networks. In the following section we shall present an extension of the basic Schelling's model that explores this direction.

3 The Model

Consider a society composed of N agents who can locate themselves in one of the $M \geq N \geq 3$ available locations. Each location can contain at most one agent. Locations can be connected or not. We model locations and connections through a graph G composed of M nodes and a collection of non-directed edges linking any pair of nodes. Edges are described by the (symmetric) $M \times M$ matrix $W = \{w_{kh}\}$, where $w_{kk} = 0 \; \forall k = 1, \ldots, M$ and $w_{kh} = w_{hk} = 1$ if and only if there is an edge connecting nodes k and h, and zero otherwise. We define the "neighborhood" V_k (or the "interaction group") of a node k as the set of nodes that node k is linked to

$$V_k = \{h \in I_M : w_{kh} = w_{hk} = 1\}, \tag{1}$$

where $I_M = \{1, \ldots, M\}$.

We suppose that each node is empty (i.e., it does not contain an agent) with probability $\theta \in (0,1)$, while it is occupied with probability $1-\theta$. Therefore, on average, there are $N=(1-\theta)M$ agents in the society. Each agent can be one of two types, say -1 and $+1$. Time is discrete, and time ticks are labeled by $t=0,1,2,\ldots$.

Agents have standard, binary, Schelling-type preferences: they are happy if and only if the relative frequency of agents of their own type is greater or equal than 0.50 in their neighborhood. More formally, if node i is occupied by an agent of type $s \in \{-1,+1\}$ at time t

$$u_{it} = u_{it}(s) = \begin{cases} 1, & \text{if } x_{it}(s) \geq 0.5, \\ 0, & \text{otherwise,} \end{cases} \tag{2}$$

where $u_{it} = u_{it}(s)$ is the utility of agent i (of type s) at time t and $x_{it}(s)$ is the current relative frequency of agents (i.e., filled nodes) of type s in V_i.[2]

The initial state of the system is characterized by: (1) an instance of the network structure, i.e., a graph $G_0 = \{I_M, W_0\}$ (more on that below); (2) an allocation of agents and types across the M available nodes. The initial allocation of agents and types across the M nodes is drawn uniform randomly. Thus, at $t=0$, each node $i \in I_M$ will be either empty or occupied. If it is occupied, this will be either a -1 or a $+1$ agent, each with probability 0.5. Thus, in the society there will be, on average, $N/2$ agents of type -1 and $N/2$ agents of type $+1$.

The dynamics runs as follows. At each $t>0$, an agent, say k, is drawn at random (and independently) from $I_N = \{1,\ldots,N\}$. As far as the behavior of the chosen agent is concerned, we shall explore two models:

- *Global-Move (GM) Model*: Agent k computes the utility that he could earn at each available (i.e., empty) node in the *whole network* G_0 (including in the list his current node).
- *Local-Move (LM) Model*: Agent k computes the utility that he could earn at each available (i.e., empty) node in his *neighborhood* V_k *only* (including in the list his current node).

Then, in both LM and GM models, agent k chooses the node that provides the highest achievable utility level (i.e., either one of the empty nodes or his current location). Ties are resolved by randomizing among all nodes providing the same maximal utility level. Notice that we assume no inertia in the agents' choices. That is, agents' current locations do not bias their choices (e.g., because of moving costs). The GM model also assumes that agents can move to any empty node in the network, i.e., there are no information or moving constraints or costs (see Sect. 7 for a discussion). Hence, the LM model can be justified by the presence of either a moving cost or some information costs preventing agents to observe anything that is outside their current neighborhood.

[2] In line with Pancs and Vriend [17], we assume that the utility associated to an empty neighborhood is zero.

4 Implementation

The initial network G_0 is chosen at random to belong to the small-world class. However, to benchmark our analysis against Schelling's one and Pancs and Vriend's results, we also study the behavior of the model in the case where initial graphs are two-dimensional lattices. More precisely, we experiment with Von-Neumann (VN) and Moore (M) two-dimensional, boundary-less lattices (i.e., torii). It is well-known that neighborhoods V_k in two-dimensional lattices are completely defined up to the choice of a metric (specifying how to compute the distance between any two nodes) and an interaction radius r. Let (x_h, y_h) the coordinates of node h in the lattice. In the VN case the distance between nodes (k, k') is given by

$$\delta_{VN}(k,k') = |x_k - x_{k'}| + |y_k - y_{k'}|, \tag{3}$$

while in the Moore case it is equal to

$$\delta_M(k,k') = \max\{|x_k - x_{k'}|, |y_k - y_{k'}|\}. \tag{4}$$

Therefore, if one defines

$$V_k(r) = \{h = 1, \ldots, M : \delta_\bullet(k,h) \leq r\}, \tag{5}$$

it is easy to see that in VN lattices all nodes have a degree $d_{VN} = 2r(r+1)$, while in Moore lattices one has $d_M = 4r(r+1)$, see Fig. 4 for an illustration.

Initial small-world networks are instead generated using the following procedures:

- *WS Graphs*: We start from a two-dimensional boundary-less lattice with VN neighborhoods for a certain $r \geq 1$. Then, each edge (h,k) is independently rewired to a randomly chosen node, say k', outside $V_h(r)$ with some probability $\beta \in (0,1)$. In case of rewiring, the edge (h,k) is deleted and replaced by the

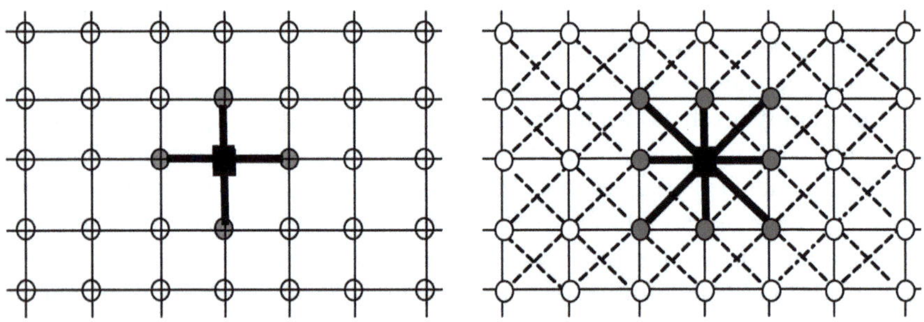

a 2D-VN: r=1 **b** 2D-M: r=1

Fig. 4 An example of neighborhood shapes with two-dimensional Von-Neumann (2D-VN) and Moore (2D-M) lattices for $r = 1$

new edge (h,k'). This yields a symmetric degree distribution, centered around $2r(r+1)$. In the benchmark results presented below, we employ $\beta = 0.2$ and then we study what happens when β is tuned in the unit interval.

- *SF Graphs*: We employ a standard "preferential attachment" procedure, starting with M_0 nodes linked through a 2D-VN lattice with $r = 1$ (and thus an initial degree $d = 4$). One node at a time is added until a size M is reached. In any step, the additional node is allowed to form 4 links. Each new link is formed by choosing one of the existing nodes with a probability proportional to its current degree.

The model contains a small number of free system- and network-specific parameters. System parameters are M (number of nodes) and θ (average percentage of empty nodes). Network specific parameters characterize – given the class of networks to be implemented – the set of possible networks from which the one actually in place will be drawn. VN and Moore lattices are characterized by their degree. WS graphs are parameterized by β and the degree d of the underlying lattice (before rewiring). Finally, SF graphs depend on the initial population size M_0. Simulations reveal that average degree d and M_0 are linked by the following (approximate) relation:

$$d \simeq 0.00003 \cdot M_0^3 - 0.0062 \cdot M_0^2 + 0.3485 \cdot M_0 + 3.1916. \tag{6}$$

Notice that d grows for $M_0 \leq 39$ and decreases for $M_0 \geq 40$. Hence, in both lattices and small-world graphs, the only common network-specific parameter to be considered is the (average) node degree d. WS graphs can also be studied for different β levels.

In the lattice-case, the initial graph is automatically defined once one specifies the degree d. In small worlds, given a choice for the network class and for the network-specific parameters of that class (e.g., d and β for a WS graph), each time we draw G_0 uniformly at random from the set of all possible graphs belonging to that class and with the given network-specific parameters.

5 Measuring Segregation in Networks

A number of indices have been suggested in the literature to measure segregation when the agents are located on generic networks [see, e.g., 7, 10–12, 14, and references therein]. Here, we will employ *Freeman's segregation index* (FSI) [11, 12]. The rationale underlying the computation of the FSI is that if a given agent-attribute (in our case the type $+1$ or -1) does not matter for social relationships (i.e., for the link structure as described by G_0), then the links among the agents should be distributed randomly with respect to that attribute. Therefore, suppose we observe a given allocation of agent types across the M nodes, connected through the network G_0. Let us, then, split the agents in two groups according to their type and, *for each type*, let us count the number of cross-group links (i.e., the number of links connecting any pair of agents of different types), as well the number of within-group links (i.e., the number of links connecting any pair of agents of the same type). This gives us a 2×2 contingency table whose generic entry l_{xy} gives us the number of links between type-x and type-y agents in G_0. Similarly, one can compute the expected contingency table for a random allocation of agent types on G_0. The difference between the number of cross-group ties expected by chance and the number of observed ties (divided by expected ones) gives us the FSI. The index ranges between -1 and 1, with the highest segregation level obtained when there are no cross-group links in place.

We also check our results against a number of alternative segregation indices, such as the "spectral segregation index" [7], those proposed in Fershtman [10] and Freeman [12], and some of the indices originally developed in the lattice-case [see 17], e.g., the average *mix deviation* index. As we discuss in Sect. 7, our main results are not qualitatively altered if one considers these alternative segregation measures. Therefore, in what follows we will mainly focus on FSI as our measure of segregation in networks.

6 Results

In this section, we explore the behavior of our model for a society of $M = 100$ nodes. Our study will take the form of a Monte Carlo (MC) analysis. The procedure is as follows. For each choice of network class and network-specific parameters we generate a number of independent runs. For each run, where necessary, we randomly select a specific instance of the network class, and we generate an initial allocation of agents and types across the network uniformly at random. We, then, let the best-response dynamics run, and collect system statistics when either FSI or the configuration of types across the M nodes have reached a steady-state. This typically happens well before $T = 50{,}000$ time-steps with probability one. We independently repeat this exercise 1,000 times, computing the Monte Carlo (MC) average and standard deviation of FSI. Since across-run variability turns out to be very small (across-run standard deviations are of an order of magnitude of 10^{-5}) and MC distributions appear to be symmetric, we report below MC averages of FSI only.

The main questions we initially address are:

Q1 Are segregation levels in WS and SF small-world networks (as measured by FSI) different from those attained in a society where individuals live in lattices?
Q2 Does the answer to Q1 differ when individuals behave according to a GM or a LM model?

Let us begin with a rough comparison of segregation levels. Figures 5 and 6 show average FSI levels for lattices and small-world graphs in both the GM and LM models. Although segregation seems to be slightly larger in lattices than in WS and SF graphs, overall levels in both lattices and small worlds remain quite large according to FSI (which, we recall, records values close to one only in extreme cases).

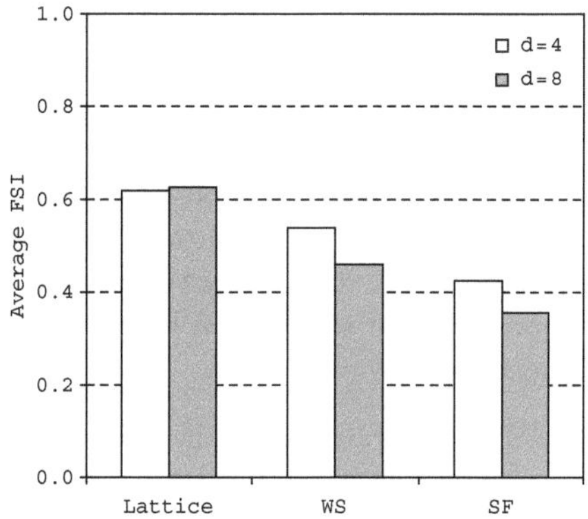

Fig. 5 GM Model. Average FSI levels in lattices, WS and SF graphs for average degrees $d = 4, 8$. Parameters: $M = 100$, $\theta = 0.3$. Note: $d = 4$: VN-lattice; $d = 8$: M-lattice

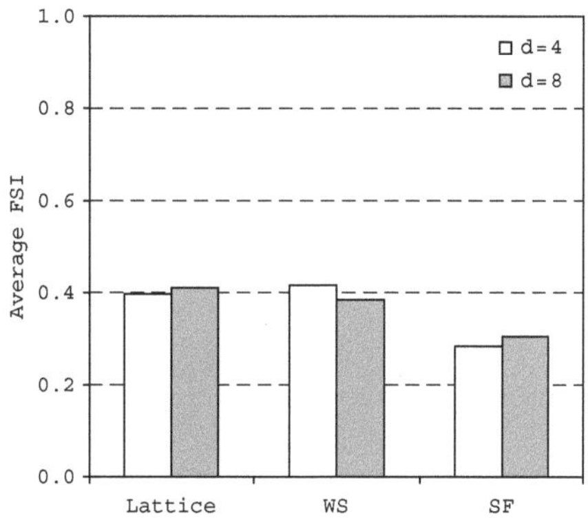

Fig. 6 LM Model. Average FSI levels in lattices, WS and SF graphs for average degrees $d = 4, 8$. Parameters: $M = 100$, $\theta = 0.3$. Note: $d = 4$: VN-lattice; $d = 8$: M-lattice

Schelling's results seem to be confirmed when one moves from spatial to social segregation in the GM model.

This result was somewhat expected. When one leaves a lattice world to move in the small-world realm, two important features change. First, the average path length tends to decrease; second, and most important here, neighborhood sizes become heterogeneous (less in WS graphs, more in SF ones). In the GM model agents can freely move in any empty location of the network, no matter their neighborhood size. Hence, it would have been surprising if segregation levels would have substantially changed.

When one comes to the LM model, however, heterogeneity in neighborhood sizes might have some impact on segregation levels. Our simulations instead show that this is not the case, see Fig. 6. Segregation levels in WS and SF networks remain comparable to those in lattices: Schelling's results seem even more robust. Notice also that segregation levels decrease when one moves from a GM to a LM model. In fact, agents in a LM world tend to explore a smaller number of options and the ensuing dynamics turn out to be more "sticky." As a result, high levels of segregation can be attained less easily by the system.

The above results indicate that both Q1 and Q2 have a simple, common answer: "Not very much." If any, some overall decrease in segregation levels is observed in SF networks.[3] However, SFI differentials are not so large to draw statistically-significant implications (more on that in Sect. 7).[4]

These findings are robust to a sensitivity analysis across system and network-specific parameters [9]. However, inspection of Figs. 5 and 6 suggests that segregation levels do exhibit some variation with average degree, hinting to some parameter dependence of segregation levels. We can then formulate the following additional questions:

Q3 Do segregation levels in WS and SF small-world networks (as measured by FSI) change with average degree and percentage of empty nodes (θ)?
Q4 Do segregation levels in WS networks change with the rewiring parameter (β)?

Let us begin with Q3. Figures 7 and 8 show how segregation levels change with average degree and the percentage of empty nodes in WS and SF graphs for the

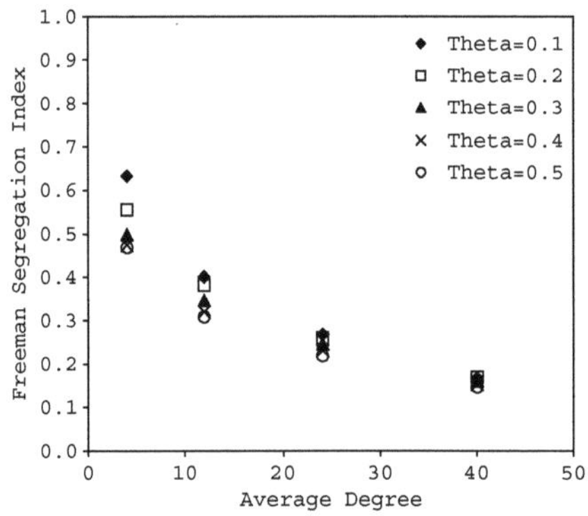

Fig. 7 GM Model. Average FSI levels in WS graphs vs. average degree for different levels of θ and $M = 100$

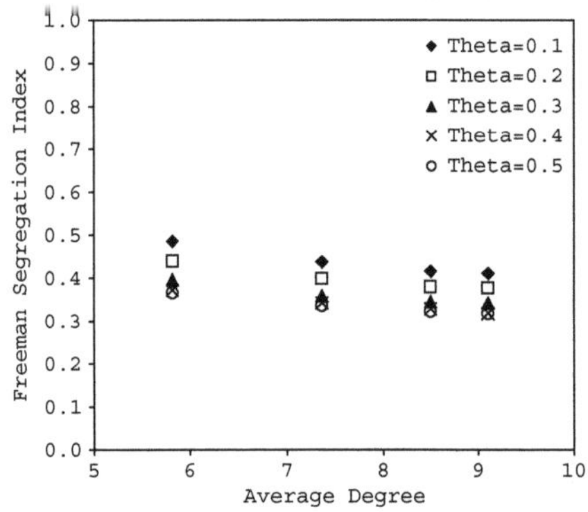

Fig. 8 GM Model. Average FSI levels in SF graphs vs. average degree for different levels of θ and $M = 100$

---

[3] Segregation levels in WS graphs are smaller than in lattices in the GM model only.

[4] Simulations also show that the average values of FSI found throughout our analysis are significantly larger than average FSI values obtained in purely-random allocations of networks and types (when best-response dynamics is not made at work), see Fagiolo et al. [9] for details.

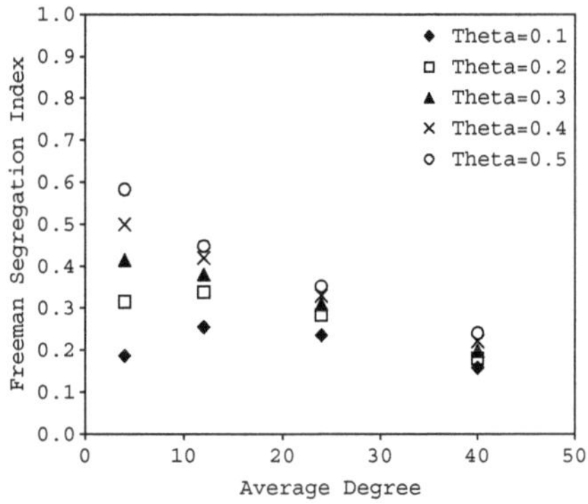

Fig. 9 LM Model. Average FSI levels in WS graphs vs. average degree for different levels of θ and $M = 100$

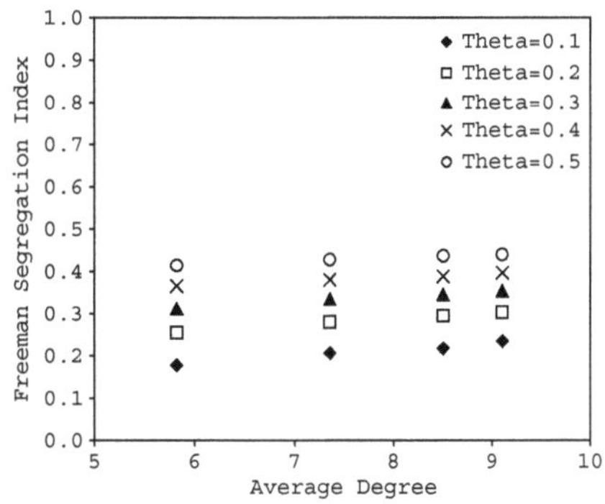

Fig. 10 LM Model. Average FSI levels in SF graphs vs. average degree for different levels of θ and $M = 100$

GM model. We see that FSI levels are decreasing with the average degree for any value of θ. Very high segregation levels are attained by the system when the society is poorly connected and there is a small percentage of empty nodes. As the connectivity increases, segregation becomes somewhat less pronounced, but even in very connected societies, segregation levels remain significantly high. Furthermore, segregation tends to decrease in the GM model as the percentage of empty nodes increases, as agents have more degrees of freedom to move around.

The above results substantially change in the LM model, see Figs. 9 and 10. Recall that, in the LM model, degree heterogeneity (which increases as one goes from WS to SF graphs) does now play some role. While more connected WS societies are less segregated, this is not so in SF networks, where segregation mildly increases with average degree. As a result, topological properties of SF networks seem to have an impact on segregating dynamics in our model. Despite more heterogeneity seems to slightly decrease segregation levels given the same connectivity, switching from a less degree-heterogeneous to a more degree-heterogenous society changes the way in which connectivity affects segregation levels.

Finally, Figs. 11 and 12 show segregation levels attained in a WS network for different values of the rewiring probability in both the GM and the LM model.

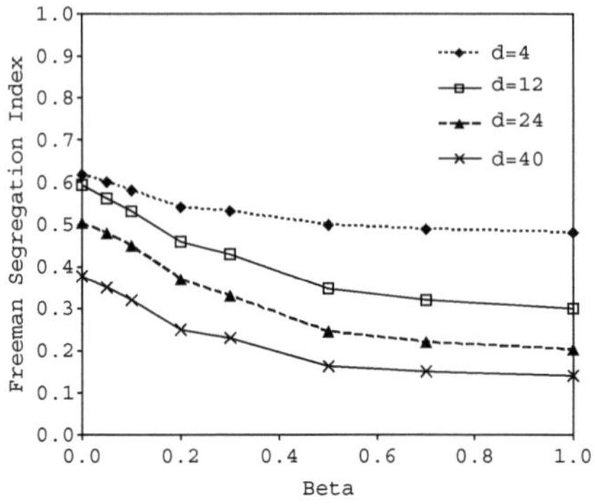

Fig. 11 GM Model. Average FSI levels in WS graphs vs. rewiring probability for different average degrees. Parameters: $\theta = 0.3$ and $M = 100$

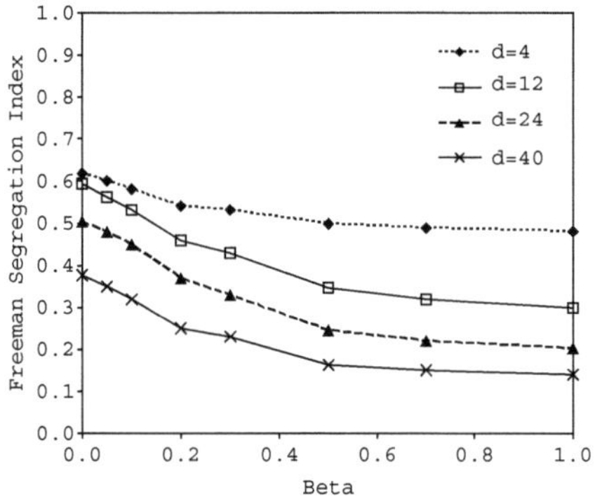

Fig. 12 LM Model. Average FSI levels in WS graphs vs. rewiring probability for different average degrees. Parameters: $\theta = 0.3$ and $M = 100$

It is easy to see that FSI levels are only mildly decreasing with β. In the SF case, segregation levels are substantially stable. This implies that, as one interpolates between lattices ($\beta = 0$) and purely random graphs ($\beta = 1$), segregation levels remain quite stable.

7 Concluding Remarks

In his seminal contributions, Schelling [19–22] studied a proximity dynamic model of spatial segregation where individuals lived on one- or two-dimensional lattices. He showed that a perfectly integrated society would evolve into a segregated one even though no individual agent would have strictly preferred that outcome in his local neighborhood.

In this paper we have argued that segregation might occur not only in the geographical space, but also in more general social networks. Empirical evidence indicates that in the real-world such networks are neither lattices nor random graphs, but rather belong to the class of small worlds. Building upon this evidence, we have presented a dynamic model of segregation where individuals interact in small-world social networks. The model sticks to Schelling's original formulation as far as individual preferences are concerned and studies best-response dynamics as in Pancs and Vriend [17]. We consider two specifications of the general model, one in which agents can move arbitrarily far away in the social space from their current location (global move model), the other wherein agents can only move in their current social interaction group (local move model).

When one replaces lattices with small worlds, the degree distribution becomes heterogeneous. This heterogeneity is relatively low in Watts-Strogatz (WS) small worlds, while becomes more relevant in scale-free (SF) small worlds. Our main result is that this increase in degree heterogeneity does not dramatically affect Schelling's findings: segregation levels remain comparable to those attained in a lattice world in both a local- or global-move model. We also perform a sensitivity analysis on the parameter space. Simulations show that in the global-move model more connected societies are less segregated. However, when a local-move model is considered, this result does not hold anymore: higher connectivity can imply either smaller or larger segregation levels, depending on the heterogeneity of degrees. The higher heterogeneity, the more likely is that strongly connected societies are also more segregated.

The above results are robust to a series of extensions and modifications [see 9, for details]. These include: (1) segregation measures alternative to the FSI [e.g., the spectral segregation index proposed in 7]; (2) additional network structures such as regular or random graphs; (3) average percentages of empty nodes (θ) larger than 50%; (4) network size (M).

Furthermore, similar results are obtained if one introduces some "inertia" in the picture. Suppose that an agent located in node i is drawn at random from $I_N = \{1,\ldots,N\}$. With inertia, this agent stays put if there is no vacant location that he would strictly prefer to his current location. The idea of inertia is based on the implicit modelling assumption of some small costs of moving (smaller than the smallest possible difference in satisfaction between any two locations, but otherwise arbitrarily small). Notice that under the *inertia* rule, satisfied agents will never move.

Many interesting issues remain to be explored. First, agents in our model jump from its current location to an available one without being affected by the topology of the network. This implies that average path length has no effect whatsoever in the dynamic process leading to segregation. Therefore, the current formulation of our model does not fully exploit a fundamental difference existing between lattices and small worlds, i.e., the fact that in small worlds average path length tends to decrease. By incorporating into agents' behavioral rules an appropriate algorithm governing the path they follow to travel from their current node to the newly selected one, one might attempt to explore the role played by average path length in the picture.

Finally, our results indicate that degree heterogeneity does affect segregation levels. In particular, when one switches from WS to SF networks, segregation seems to generally decrease and the way in which connectivity affects segregation substantially changes. Therefore, a deeper understanding of the behavior of the model in SF graphs seems desirable.

Acknowledgements We thank Yannis Ioannides, Giulia Iori, Alan Kirman, Akira Namatame, workshop participants of iNeck (Strasbourg), WEHIA 2005 (Colchester), SCE 2005 (Washington DC) and Complexity 2006 (Aix en Provence) as well as the theory reading group at QM (London) for their comments. Part of this work was done while G.F. was visiting the Bureau d' Economie Théorique et Appliquée (BETA), Université Louis Pasteur, Strasbourg, France. G.F. gratefully acknowledges financial support by the CNRS (Centre National de la Recherche Scientifique), France. All usual caveats apply.

References

1. Albert R, Barabási A (2002) Statistical mechanics of complex networks. Rev Mod Phys 74:47–97
2. Amaral L, Scala A, Barthélemy M, Stanley H (2000) Classes of small-world networks. Proc Natl Acad Sci 97:11149–11152
3. Baldwin T, Rozenberg G (2004) Britain 'must scrap multiculturalism'. The Times, April 2003
4. Barabási A, Albert R (1999) Emergence of scaling in random networks. Science 286:509–512
5. Bollobás B (1985) Random graphs. Academic, New York
6. Dorogovtsev S, Mendes J (2003) Evolution of networks: fom biological nets to the Internet and WWW. Oxford University Press, Oxford
7. Echenique F, Fryer R (2005) On the measurement of segregation. Working Paper No. 11258, NBER
8. Fagiolo G (1998) Spatial interactions in dynamic decentralized economies: a review. In: Cohendet P, Llerena P, Stahn H, Umbhauer G (eds) The Economics of networks. Interaction and behaviours. Springer, Berlin
9. Fagiolo G, Valente M, Vriend N (2007) Segregation in networks. J Econ Behav Organ 64: 316–336
10. Fershtman M (1997) Cohesive group detection in a social network by the segregation matrix index. Soc Networks 19:193–207
11. Freeman L (1972) Segregation in social networks. Sociol Methods Res 6:411–427
12. Freeman L (1978) On measuring systematic integration. Connections 2:9–12
13. Milgram S (1967) The small world problem. Psychol Today 2:60–67
14. Mitchell J (1978) On freeman's segregation index: an alternative. Connections 2:9–12
15. Newman M (2000) Models of the small world: a review. arXiv:cond-mat/ 0001118v2
16. Newman M (2003) The structure and function of complex networks. SIAM Rev 45:167–256
17. Pancs R, Vriend N (2007) Schelling's spatial proximity model of segregation revisited. J Public Econ 91:1–24
18. Pastos-Satorras R, Vespignani A (2004) Evolution and structure of the Internet. Cambridge University Press, Cambridge
19. Schelling T (1969) Models of segregation. Am Econ Rev 59:488–493
20. Schelling T (1971) Dynamic models of segregation. J Math Sociol 1:143–186
21. Schelling T (1971) On the ecology of micromotives. The Public Interest 25:61–98
22. Schelling T (1978) Micromotives and macrobehavior. W.W. Norton, New York
23. Scott J (2000) Social network analysis: a handbook. Sage, London
24. Wasserman S, Faust K (1994) Social network analysis. methods and applications. Cambridge University Press, Cambridge
25. Watts D (1999) Small worlds. Princeton University Press, Princeton
26. Watts D (2003) Six degrees: the science of a connected age. W.W. Norton, New York
27. Watts D, Strogatz S (1998) Collective dynamics of 'small-world' networks. Nature 393:440–442

Interdependent Preferences

Ahmad K. Naimzada and Fabio Tramontana

Abstract Consumer preferences can be influenced by the choices of other consumers in many ways. For example, through imitative (or *bandwagon*) and escaping (or *snob*) behaviors. Consumption choices are adopted with regard to a reference group. In this paper we propose a model in which agents can be not only locally connected to their reference group but, in some degree, also globally connected to all agents. The result is a network structure in which consumers are the nodes, and links are given by local and global interactions. In particular, links between agents belonging to different social groups can only be given by the presence of global interaction whereas local interaction creates a link between agents of the same class. The first aim of the paper is the building of a general model. The second one is the study, in a particular simplified case, of the dynamic behaviours of the model arising with different weights given to the local or to the global interactions. The main result of the paper is given by the determination of the conditions for the coexistence of different types of attractors (fixed points, periodic cycles and chaotic attractors) and by the numerical study of the global bifurcations that change the qualitative nature of the attractors and the structure of the basins of attraction.

1 Introduction

There is a classical body of literature in the field of the effects of experience on choices: see, e.g. [1–10]. In a well known paper Benhabib and Day [11] proposed a rational choice model in a stationary environment (without price dynamics) with

A.K. Naimzada
Università degli Studi di Milano-Bicocca, Piazza dell'Ateneo Nuovo 1, 20126 Milan, Italy
ahmad.naimzada@unimib.it

F. Tramontana
Università Politecnica delle Marche, Piazzale Martelli 8, 60121 Ancona, Italy
f.tramontana@univpm.it

A.K. Naimzada et al. (eds.), *Networks, Topology and Dynamics*, Lecture Notes in Economics and Mathematical Systems 613,

preferences depending endogenously on past experience that can lead to chaotic behaviour. Instead, in [12] is developed a model of consumer choice with interdependent preferences characterized by imitative behavior and sluggish transmission of changes in the consumption of a group of individuals onto the preferences of another individual. In the present article, in contrast with most of this literature, we look at the consequences of the endogenous change of preferences by past consumption activity also of other consumers. We extend the investigation of Benhabib and Day in two main directions: (1) we build a model in a discrete-time framework with individual preferences depending on past consumption behaviour of other individuals to whom the agent is related in one way or another, constructing demand schedules with the presence of imitative (or bandwagon) and escaping (or snob) effects; (2) we permit the local and global interaction of heterogeneous agents from the point of view of preferences. In our analysis we look at role played by the presence of heterogeneous agents and by the nature of interaction among them. The main result is the rich variety of dynamic behaviours (in/stability, bifurcations, complex dynamics) offered by the model when the degree of heterogeneity is relatively high and the degree of local interaction is not too high with respect to the global one. Even when all agents are of the bandwagon type we find multiple equilibria with low and high levels of consumption. The paper is organized as follows. In the second section we develop the model. In the third section we show and clarify the structure of interaction. In the fourth section we give some results, through numerical simulations, about the existence of steady states and the main bifurcations of the dynamic system. In the fifth section we complete the numerical analysis of the model investigating its global properties. Finally, the main results are summarized in the concluding section.

2 The General Model

Let us consider a market of two goods, named x and y. The preferences of the consumers are represented by a Cobb–Douglas utility function:

$$U(x,y) = x^{\alpha} y^{1-\alpha}, \tag{1}$$

where $\alpha \in [0,1]$ is the elasticity of the good x and measures the consumer's preference for it. In each period consumers are endowed with an income m, so the budget constraint takes this form:

$$m = P_x x + P_y y, \tag{2}$$

where P_x and P_y are the prices of the goods x and y respectively. The rational choice (obtaining by solving the optimization problem $x = \arg\max U\left(x, \frac{m-P_x x}{P_y}\right)$) is given by the following quantities of goods bought by the representative consumer:

$$x^* = \frac{\alpha m}{P_x} \qquad \text{and} \qquad y^* = \frac{(1-\alpha)m}{P_y}. \tag{3}$$

In the classical model, the parameter α is fixed, reflecting the absence of social interaction. Instead, we assume that the preference of the consumer varies over time, reflecting the social consumption experience. This means that each individual is embedded in some type of neighbourhood (in terms of location, income, education, etc.) and the consumption of the neighbours influences his own consumption. In particular, we consider a single period lagged interdependence, i.e. the acquisition of information about the behaviour of other agents requires time. We propose the following general specification:

$$\alpha_{t+1}^{i} = \sum_{j=1}^{N} w^{i,j} f^{i,j}(x_t^j), \tag{4}$$

where α_{t+1}^{i} is the parameter weight in the utility function of the agent i and $f^{i,j}(x_t^j)$ is the way the preference of the agent i at the time $t+1$ is influenced by the consumption of the agent j in t. In general, each consumer i, can be influenced in different ways by different neighbours, so $f^{i,j}(x_t^j)$ needs not be equal to $f^{i,k}(x_t^k)$ for $j \neq k$. Finally, $w^{i,j}$ represents the weight of the influence of the consumption of the agent j on the choices of i. The neighbourhood of i is formed by the $K \leq N$ agents whose influence on i is strictly positive. If $w^{i,j} = 0$ then the agent j is not a neighbour of i, so his/her consumption has no (direct) influence on i (maybe j could influence another agent in the neighbourhood of i, so j can influence i in an indirect way).

Using (4) we can see the network structure formed by the consumers. We call two consumers i and j *connected*, if one of them influences the other. The connection could be only a *one way connection* ($w^{i,j} = 0$ but $w^{j,i} \neq 0$ or the opposite) or a *two way connection* ($w^{i,j} \neq 0$ and $w^{j,i} \neq 0$). When two consumers don't influence each other there are no links between them (this happens when $w^{i,j} = 0$ and $w^{j,i} = 0$). Links do not only have a direction (who influences who) but also a thickness, given by the weight of the influence when it is positive. The function (4) has these global properties:

$$0 \leq \alpha^{min} \leq f^{i,j}(x_t^j) \leq \alpha^{max} \leq 1 \qquad \forall i,j\,, \tag{5}$$

which means that the preference parameter α has values inside the closed interval $[\alpha^{min}, \alpha^{max}]$ and, by definition, can not be higher then 1 or negative. If α^{max} is strictly lower then 1, then the good y has to be bought in a minimum quantity (y is *indispensable*); whereas if α^{max} is strictly positive then is the good x that has to be bought in a minimum quantity (x is *indispensable*). According to the shape of $f^{i,j}$, is possible to characterize the behaviour of different kinds of consumers depending on the ways in which the consumption of an agent can influence the behaviour of another agent. We consider two different consumers.

3 The Model with Bandwagon and Snob Consumers

We assume that, at each time step, economic agents consider, in taking their present decisions, the past social consumption experience in one of the following two ways: by *imitating* or by *escaping*. The first is the case in which people desire to adopt and imitate the behaviour of others, in other words to join the crowd. This is the so-called *bandwagon effect* and means that the more an agent sees of a good, the more he prefers it. In this case f must be increasing in the consumption of the neighbours and tending asymptotically to α^{max}, which is the maximum value for α.

When people have the desire for distinction or to disassociate from common masses, we speak of the *snob effect*, which means that the more an agent sees of a good, the less he prefers it. In this case f must be decreasing in the consumption of the neighbours and tending asymptotically to α^{min}. According to the definition, the change of preferences for a bandwagon agent is such that $\frac{\partial \alpha_{t+1}}{\partial x_t^j} > 0$ for each neighbour j of the bandwagon agent, whereas $\frac{\partial \alpha_{t+1}}{\partial x_t^j} < 0$ for each neighbour j of the snob agent.

Until now we have characterized the nodes of our network by saying that they could be of two different types: bandwagons and snobs. The next step consists in putting links into the network structure – that is, we have to identify who does influence the bandwagons' choices in the way we have showed in this section – the same for the snob group. We have also to give an intensity to each connection that represents the thickness of the links.

4 Local and Global Interactions

We introduce an (exogenously given) parameter Ω that represents the share of snob agents in the population (normalized to 1). If $0 < \Omega < 1$ snobs and bandwagons coexist, and variations of Ω modulate the share of snob consumers within the population.

We assume that there are two spaces for social interaction. First of all, we have a *strict local neighbourhood* (in our case, the same class) and then we have a *global social space*. If we consider only two kinds of consumers, we can analyze the evolution of the consumption of the representative bandwagon agent and of the representative snob agent. In this way, the local interaction for a bandwagon agent in period $t+1$ is given by the influence of the consumption of the representative bandwagon agent in period t (x_t^b), which is also his own consumption in the previous period because. In our simple model, all the agents of the same kind always consume the same quantities of goods as they have the same preferences. The same happens for the representative snob agent. In order to characterize the global interaction we can specify, in each period, the consumption of the representative agent of the whole population, given by the average consumption. Considering both these kinds of interactions that influence the preferences, we have the following dynamic system in the consumption of the representative snob agent x^s and the representative bandwagon agent x^b:

$$T(x^b, x^s): \begin{cases} x_{t+1}^b = [\gamma f^b(x_t^b) + (1-\gamma) f^b(\Omega x_t^s + (1-\Omega) x_t^b)] \frac{m^b}{P} \\ x_{t+1}^s = [\gamma f^s(x_t^s) + (1-\gamma) f^s(\Omega x_t^s + (1-\Omega) x_t^b)] \frac{m^s}{P} \end{cases}, \qquad (6)$$

where $\gamma \in [0,1]$ is the weight of local interaction, $\Omega x_t^s + (1-\Omega)x_t^b$ is the average consumption in period t and x^b (x^s) and m^b (m^s) are consumption and income of the representative bandwagon (snob) agent respectively. The price of y is normalized to 1, so we have $P_x = P$ and $P_y = 1$. At each time period the consumption of y is univocally given by the relation $y_t = m - Px_t$.

What does the dynamical system (6) imply about the connections among consumers? We can summarize the main properties of the network structure implicit in (6) in the following way:

1. There exist only three kinds of links in terms of thickness. The first kind of link connects bandwagon agents with the other bandwagons, the second kind connects snobs consumers with the other snobs and the last kind connects agents of different types.
2. Each agents is more influenced by consumers belonging to the same group.

To understand better points 1 and 2, we have to characterize the weight of the connection between pairs of agents. Let us consider the bandwagon representative agent. The local interaction has a weight equal to γ and it involves the $1-\Omega$ bandwagon agents, so the weight is $\gamma/(1-\Omega)$ for each one. The weight of the global interaction is $(1-\gamma)$ and is equally distributed among all the agents. To sum up, the consumption of each bandwagon agent influences the preference of the representative bandwagon agent with a weight given by $\gamma/(1-\Omega)+(1-\gamma)$ because its influence is both local and global.

The weight of the influence of each snob agent, instead, is simply $(1-\gamma)$, so the link between one bandwagon and another bandwagon is thicker then the one connecting the bandwagon with the snob agent. In the same way, the link between a snob agent and another snob agent has got a total weight equal to $\gamma/\Omega + (1-\gamma)$.

In Fig. 1 we have the case with two snobs and two bandwagons. We can see the thicker link between agents belonging to the same group. When there is the same number of snob and bandwagon agents in the population, we have only two kind of links, but if the share differs, the link between two snobs will also differ from the link between two bandwagons. In order to obtain some explicit results we have to introduce a particular form of the interaction functions f.

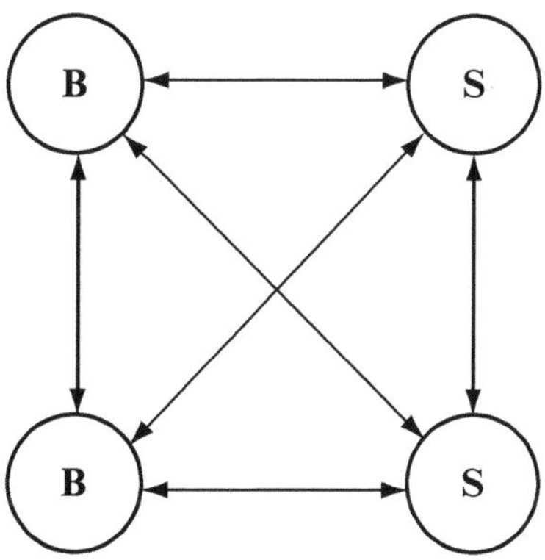

Fig. 1 The network structure connecting bandwagons and snobs

5 Results

The dynamical system (6) is formed by a pair of very complicated nonlinear equations and we are not able to characterize analytically the steady states of the system. We can only observe the results given by a high number of simulations for some particular specifications of the functions $f^b(x)$ and $f^s(x)$ that characterize the way in which the preferences of the two social groups are influenced by the consumption of the neighbours.

A fairly simple choice for the function that represents the preferences of the bandwagons is given by

$$f^b(x) = 1 - \frac{b}{b + x^{\rho_b}}, \tag{7}$$

where $b > 0$ and $\rho_b > 0$ are parameters reflecting the intensity of the influence of previous period behaviours on the preferences, i.e. "to what degree the agent is a bandwagon consumer". For the snobs we can use

$$f^s(x) = \frac{a}{a + x^{\rho_s}}, \tag{8}$$

where $a > 0$ and $\rho_s > 0$ reflect the intensity of the influence of previous period behaviours on the preferences of a snob agent, i.e. "to what degree the agent is a snob consumer". If ρ_b and ρ_s are both high valued, then the degree of heterogeneity is high too. The graphs of these functions are represented in Fig. 2.

We can see that the function associated with the bandwagon increases in neighbours' consumptions, whereas for the snobs the relation is the opposite. This confirm that the two functions suitably characterize the evolution of the preferences of the two groups. In particular, as $f^b \in [0,1[$ and $f^s \in]0,1]$, we consider the case in which no good is indispensable.

We are interested in the effects of variations of the parameters that express the share of different types of agents in the population (Ω) and the weight of local interaction (γ). Let us start with the role of the share of snobs in the population.

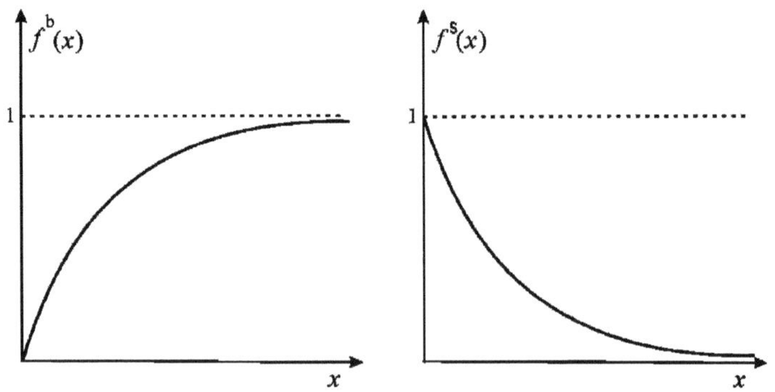

Fig. 2 Bandwagon effect and snob effect

5.1 The Role of the Share Parameter Ω

The parameter Ω regulates the share of snobs in the population. It varies between 0 and 1. If $\Omega = 1$ then the whole population is made up by snob agents. In this case we have only one dynamic equation – the one characterizing the evolution of the consumption of the representative snob agent:

$$x_{t+1} = s(x_t) = \frac{a}{a + x_t\rho}\frac{m}{P}, \tag{9}$$

where the apex "*s*" has been omitted because there remain only one kind of agents in the population.

The one-dimensional map (9) is decreasing, hence there exists only one fixed point x^*, whose value is increasing in m, a and ρ and decreasing in P. The fixed point can lose stability via flip bifurcation if the value of ρ is sufficiently high – that is, if the agent is *snob enough*. In that case we obtain a 2-period cycle whose points are characterized by extreme values of the consumption. In one period the agent consumes a quantity close to 0 of the good whereas in the next period he/she uses almost all income for that good (see Fig. 3). It is well known that this kind of map can not have more than one fixed point and period attractors with periodicity higher than two.

When there are only bandwagons in the population ($\Omega = 0$), the model reduces to this one-dimensional map:

$$x_{t+1} = b(x_t) = \left(1 - \frac{b}{b + x_t\rho}\right)\frac{m}{P}. \tag{10}$$

In this case, if the agent is *bandwagon enough* – that is, when the parameter ρ is sufficiently high – the map (10) has three fixed points. In fact, an increase in the value of ρ gives an S-shape to $b(x_t)$, like in Fig. 4. Usually the lower fixed point (the one corresponding to consumption close to 0) and the higher (the one with consumption close to the maximum budget available) are locally stable, whereas the intermediate fixed point is unstable. This means that when the model passes from a situation

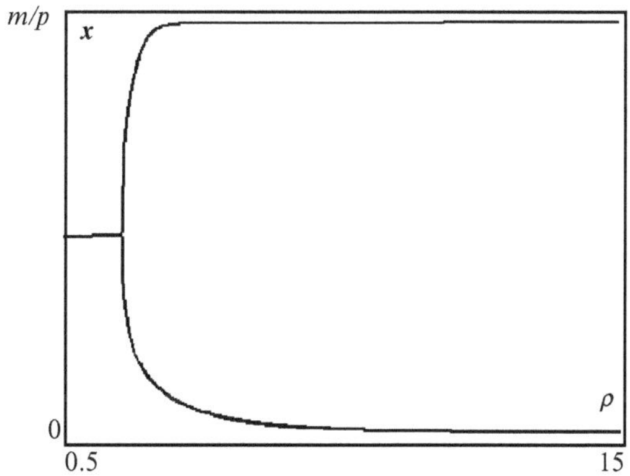

Fig. 3 Only snob people. The bifurcation diagram is obtained using the parameters $a = 0.9$, $m = 2.87$ and $P = 1.4$

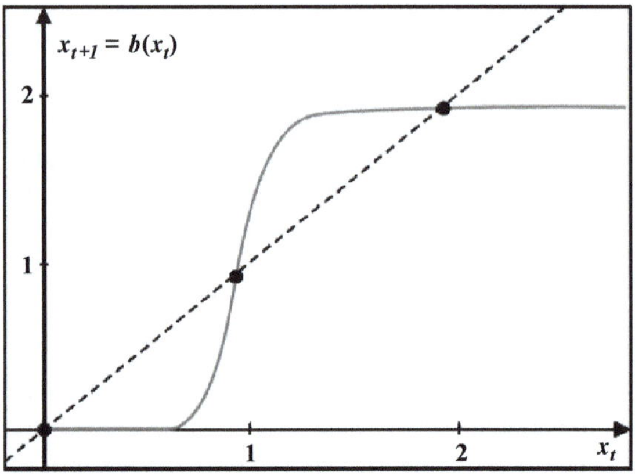

Fig. 4 Only bandwagon people. The picture is obtained using the parameters $b = 0.47$, $\rho = 11.24$, $m = 2.7$ and $P = 1.4$

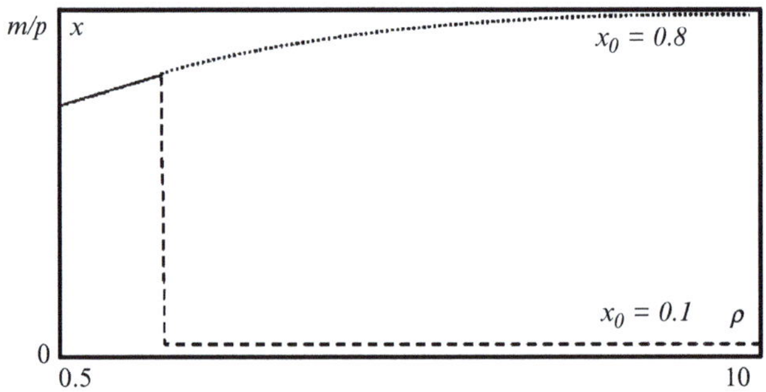

Fig. 5 Only bandwagon people. The bifurcation diagram is obtained using the parameters $b = 0.47$, $m = 2.7$ and $P = 1.4$. Where the curve is smooth the consumption converges to the unique fixed points. For the values of the parameter in which there are two curves (*red and blue*) there are two locally stable fixed points: starting from an initial condition close to 0 the bifurcation diagram is the higher (*dotted*) one whereas the lower (*dashed*) one corresponds to an initial condition close to m/P

with only one stable fixed point to another with two stable fixed points, the initial condition becomes crucial in order to know which kind of long run consumption habit will characterize the behaviour of a population entirely formed by bandwagon agents. In Fig. 5 we have used two different initial conditions. When the parameter ρ is sufficiently high we obtain convergence to different fixed points. From these extreme cases we can conclude that when the representative snob consumer is *snob enough* is possible that the dynamic process does not converge to a fixed point but to a periodic cycle (a 2-cycle). Instead, when the representative consumer is a sufficiently emphasized bandwagon consumer, we can have situations of multistability. In these cases we need to know the initial condition in order to characterize the final outcome of the dynamic process. We can suppose that combining these two results (that is, the degree of heterogeneity is high) makes it possible to obtain the coexistence of more than one attractor, which are more complicated than fixed points. We will come back to this point in the Sect. 5.3.

5.2 The Role of Local and Global Interaction

For local interaction we intend the interaction among agents belonging to the same group (snobs with snobs and bandwagons with bandwagons). Instead, the global interaction expresses interaction with all the other agents, without considering to which group they belong. When both the interactions are present, then agents are more influenced by agents of the same social class. This could be the case in which the two groups live in different places and agents have social interactions especially with people like them. In the extreme case $\gamma = 1$, the groups have no interactions at all with agents belonging the other group. In taking their consumption decisions they only consider the past choices of the members of the same groups. The system (6) is reduced in a couple of independent difference equations that express the consumption of the representative snob agent and the representative bandwagon agent:

$$x_{s,t+1} = s(x_{s,t}) \tag{11}$$

$$x_{b,t+1} = b(x_{b,t}) \tag{12}$$

that are exactly the equations already analyzed in the previous subsection, with the difference that in this case the two groups coexist even if they don't influence each other.

The extreme case $\gamma = 0$ represents the situation in which bandwagon and snob are influenced in the same way by all the other consumers, because the interaction is only global. The map (6) can be redesigned in this way:

$$T(x^b, x^s) : \begin{cases} x^b_{t+1} = b(\Omega x^s_t + (1-\Omega)x^b_t \\ x^s_{t+1} = s(\Omega x^s_t + (1-\Omega)x^b_t \end{cases} . \tag{13}$$

Increasing the value of γ, agents give an increasing importance to the consumers belonging to the same group, until the extreme case in which they are only influenced by themselves and do not consider, in taking their decisions, the choice of the other class of consumers (when $\gamma = 1$). It is hard to analyze bidimensional maps like (6) and (13) because the equations have strong nonlinearities. In the next section we use both analytical and numerical tools.

5.3 The General Case

When the parameters Ω and γ don't assume extreme values we have the general model in which snobs and bandwagons coexist and the interaction is both local and global. In this case, the fixed points of the map (6) are given by the intersections of these functions:

$$T^*: \begin{cases} x^b = F(x^s) = \left\{ \left[\frac{(1-\gamma)a}{\frac{P}{m^s}x^s - \gamma\left(\frac{a}{a+(x^s)^{\rho_s}}\right)} - a \right]^{\frac{1}{\rho_s}} - \Omega x^s \right\}^{\frac{1}{1-\Omega}} \\ x^s = G(x^b) = \left\{ \left[\frac{(1-\gamma)b}{(1-\gamma) - \frac{P}{m^b}x^b + \gamma\left(1 - \frac{b}{b+(x^b)^{\rho_b}}\right)} - b \right]^{\frac{1}{\rho_b}} - (1-\Omega)x^b \right\}^{\frac{1}{\Omega}} \end{cases} . \tag{14}$$

Before proceeding with the analysis of $F(x^s)$ and $G(x^b)$, we note that in the equations are present exponential expressions with real exponent. When the base is negative, $F(x^s)$ and $G(x^b)$ are not defined, there are no fixed points at all and the system always diverges.

The system of equations (14) is too complicated to be completely analyzed analytically, but something can be said about some properties of the solutions. Considering $F(x^s)$, we can note that it is defined for $x^s > \hat{x}^s$, such that:

$$0 < \hat{x}^s < \frac{m^s}{P} \tag{15}$$

and it has got these properties:

$$\lim_{x^s \to \hat{x}^s} F(x^s) = +\infty \qquad \text{and} \qquad \lim_{x^s \to +\infty} F(x^s) = -\infty \,. \tag{16}$$

To understand better the shape of $F(x^s)$, we need now to analyze the sign of the derivative $F'(x^s)$[1] (which is always negative), so we can draw approximately $F(x^s)$ as you can see in Fig. 6.

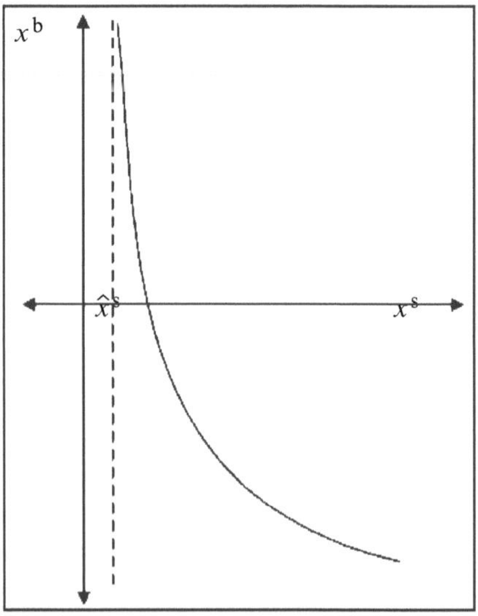

Fig. 6 The shape of $F(x^s)$

[1] The expression of the derivative $F'(x^s)$ is reported in the appendix.

Now we need to analyze $G(x^b)$, which is defined in the interval $[0, \hat{x}^b[$ where:

$$(1-\gamma)\frac{m^b}{P} < \hat{x}^b < \frac{m^b}{P}. \tag{17}$$

Moreover, we have that $G(0) = 0$ and the function owns a vertical asymptote:

$$\lim_{x^b \to \hat{x}^b} G(x^s b) = +\infty. \tag{18}$$

The derivative $G'(x^b)^2$ is not easy to analyze because it may change sign more than one time, depending on the values of the parameters, with the only rule that $G'(0) < 0$. Numerically we found that if the bandwagon consumers are such that ρ_b is not too low (we have seen that this condition implies that agents have an accentuated imitative behaviour) and the share of snobs in the population Ω is not too high (we know that for $\Omega = 0$ three fixed points coexist), $G(x^b)$ assumes the shape represented in Fig. 7a.

The shapes of $F(x^s)$ and $G(x^b)$ are such that they may intersect three times (Fig. 7b). More complicated dynamics occur when bandwagon and snobs are almost balanced (Ω close to 0.5), as we can see in the bifurcation diagram shown in Fig. 8. Looking at the bifurcation diagram in Fig. 8, obtained with initial conditions close to the fixed point with low consumption for bandwagon people, we can see that increasing the value of Ω produces a cascade of flip bifurcations leading to chaos. When the value of Ω further increases, the process changes direction and a backward flip (or *period halving*) bifurcation occurs. In fact we have seen that if there are too many snobs in the population there is no more multistability. The most complicated situation we can find is a cycle of period two. This proves that in the general case (i.e. when both classes coexist and there are both global and local interactions) is possible to obtain convergence to a cycle with periodicity higher than two (Fig. 9a) – even to chaos (Fig. 9b).

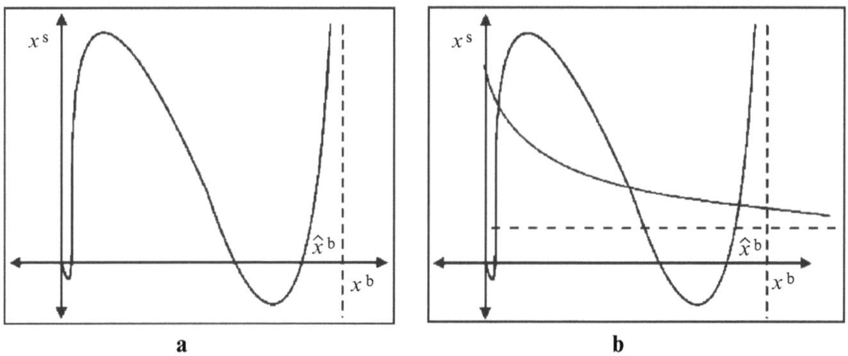

Fig. 7 $F(x^s)$ and $G(x^b)$

[2] The expression of the derivative $G'(x^b)$ is reported in the appendix.

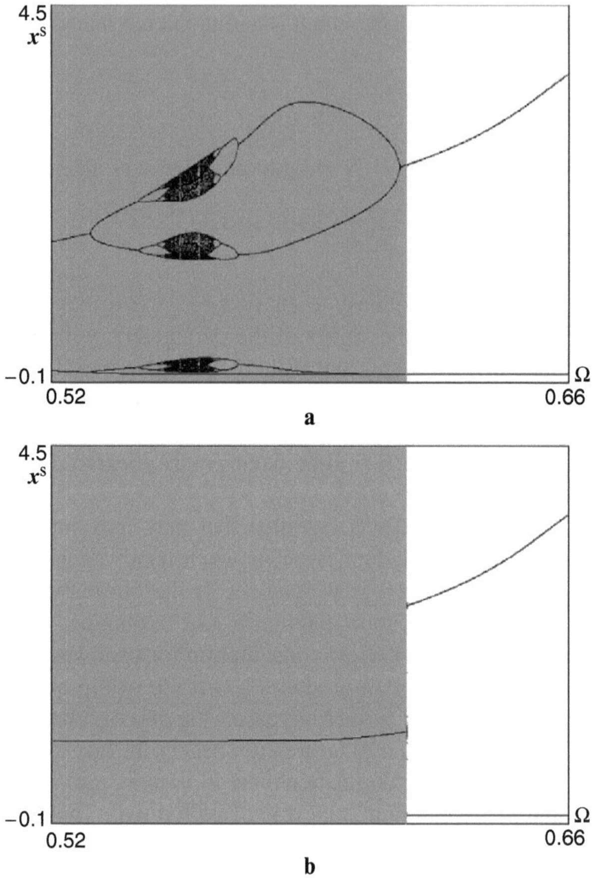

Fig. 8 The bifurcation diagram (**a**) is obtained starting from the point $x^b = 0.2$; $x^s = 0.8$, varying the value of Ω on the range 0.52, 0.66 and keeping fixed this set of parameters: $\gamma = 0.212$, $m^s = 28,32$, $m^b = 26.32$, $a = 2.628$, $b = 4.631$, $\rho^b = 5,604$, $\rho^s = 14,824$ and $P = 7,656$. For the diagram (**b**) we used the initial condition $x^b = 3$; $x^s = 0.8$. On the *white side* there is no difference between the two initial conditions

Figure 9a,b confirm that the fixed points are particularly different with respect to the consumption of the bandwagon agent, whereas in the periodic points (or in the points of a chaotic attractor) we have an higher variance in the consumption of the snob agent.

The role played by the parameter that regulates the weight of the local interaction (γ) is similar to the role played by Ω. In fact, if it assumes a high value, we only observe convergence to a fixed point or to a 2-cycle, whereas for lower values of the parameter we see high periodicity cycles and chaos (Fig. 10).

If we repeat such numeric analysis, we observe other combinations of exogenous parameters, and if ρ^b and ρ^s are high enough we obtain similar results. With low

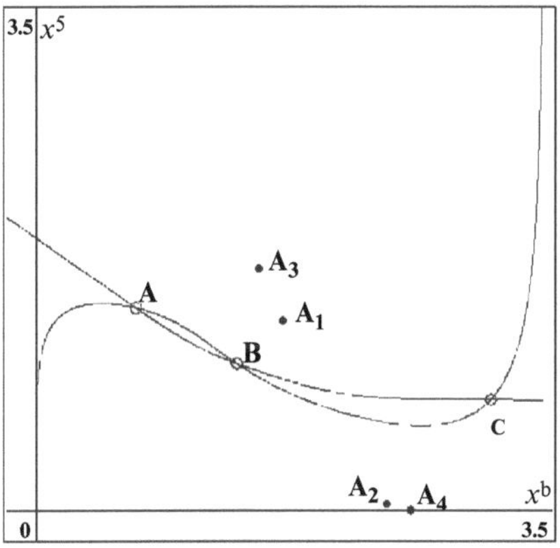

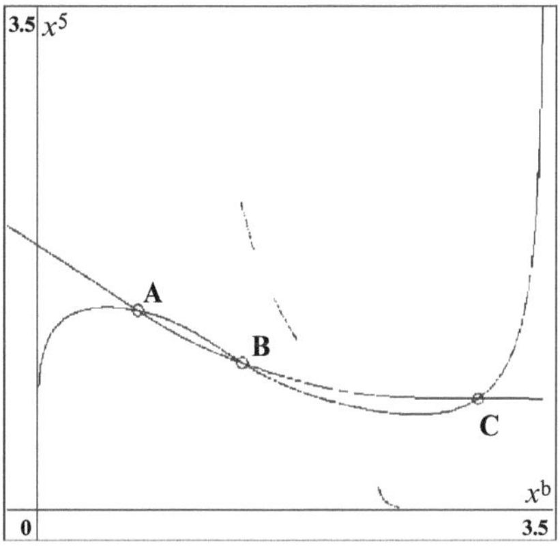

Fig. 9 The initial condition used is $x^b = 0.2$; $x^s = 0.8$ and we keep fixed this set of parameters: $\gamma = 0.212$, $m^s = 28,32$, $m^b = 26.32$, $a = 2.628$, $b = 4.631$, $\rho^b = 5,604$, $\rho^s = 14,824$ and $P = 7,656$. In (**a**) $\Omega = 0.537$ and in (**b**) $\Omega = 0.5565$

values of ρ^b and ρ^s the situation becomes quite simple with only one globally stable fixed point. Indeed, in this case the distinction between snobs and bandwagons is not very marked.

6 Global Analysis

In situations of multistability, is important to characterize the set of initial conditions which leads to every coexistent attractor. These sets are called basins of attraction of the attractors. Let us consider a situation in which a stable fixed point coexists with a chaotic (or periodic with a high period) attractor (Fig. 11).

According to what we said in the previous subsection, this is one of the most interesting situations because the attractors are different in many ways. In the stable fixed point C, the bandwagons' consumption is higher than the snob's one. The points of the chaotic attractor (indicated by arrows), instead, are such that in some periods the snobs will consume more than the bandwagons. The chaotic attractor has also the property that the variance of the snobs' consumption is larger than the variance of the consumption of the other group. These differences among the attractors emphasize the importance of a global analysis of the dynamical system. In the situation in Fig. 11a the grey area represents the basin of attraction of the fixed point whereas the white area is the basin of the chaotic attractor (whose points are in the zone indicated by the arrows). Roughly speaking, we can see that starting from a point close to the fixed point A the system converges to the chaotic attractor and if the initial condition is located close to the fixed point C then the system converges to C. The situation is not always so simple. In Fig. 11b we can see that the structure

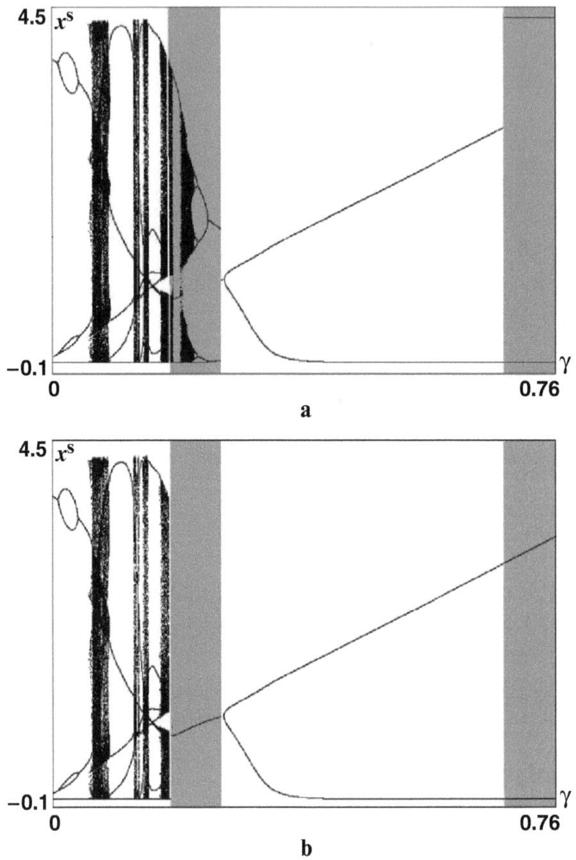

Fig. 10 In (**a**) the initial condition is $x^b = 0.2$; $x^s = 0.8$, whereas in (**b**) we used $x^b = 3$; $x^s = 0.8$. The bifurcation diagram is obtained varying the value of γ on the range 0–0.76 and keeping fixed this set of parameters: $\Omega = 0.554$, $m^s = 28.32$, $m^b = 26.32$, $a = 2.628$, $b = 4.631$, $\rho^b = 5.604$, $\rho^s = 14.824$ and $P = 7{,}656$. In the *grey-shaded* portion the diagrams are different, so there is coexistence

of the basins of attraction is more complicated, and even starting close to A there exists the possibility of convergence to the fixed point C.

It is not easy to characterize the parameters that easily permit a situation like the one represented in Fig. 11b. We can only conclude that in this kind of nonlinear model and in presence of multistability the situation could be quite complicated and a global analysis becomes necessary for completely understanding it. We intend to go into the question of the emergence of the disconnected basins of attraction from a mathematical point of view in a separated paper.

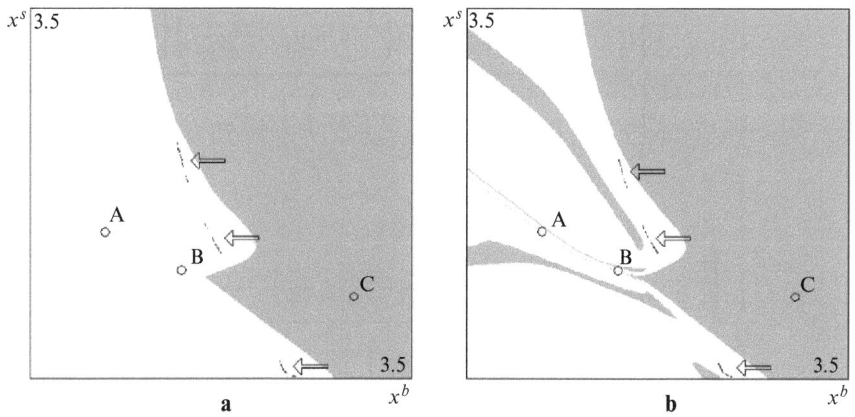

Fig. 11 Basins of attraction. The figures are obtained using this set of parameters: $\gamma = 0.212$, $m^s = 28.32$, $m^b = 26.32$, $a = 2.628$, $b = 4.631$, $\rho^b = 5.604$, $\rho^s = 14.824$, $P = 7{,}656$, $\Omega = 0.56$ in (**a**) and $\Omega = 0.554$ in (**b**)

7 Conclusions

In this work, starting from the simple model of consumer choice and introducing local and global interaction and heterogeneous agents, we have shown that endogenous preferences determine the emergence of different phenomena. The first major achievement is the coexistence of three steady states characterized by varying degrees of stability. In particular, there is an always unstable intermediate steady state that separates the basins of attraction of different attractors. There are intermediate values for the two parameters that govern the degree of local/global interaction and the snob/bandwagon share of agents that give rise to the occurrence of more complicated dynamics, cycles and chaos (in presence of relatively high degree of heterogeneity). The switch to these dynamics happens around the two extreme fixed points through period-doubling bifurcations. Moreover, we have shown a further complication involving the emergence of disconnected basins of attraction. In our future research we are interested in introducing a third kind of consumers who are not influenced by the others (*Independent Consumers*). We think that the presence of this class of consumers could stabilize the consumption of bandwagons and snobs.

Appendix

Derivatives of $F(x^s)$ and $G(x^b)$:

Using software for symbolic calculus (Mathematica), we know that the derivatives $F'(x^s)$ and $G'(x^b)$ may be written as follows:

$$F'(x^s) = -\frac{1}{1-\Omega}\left\{\frac{\left[\frac{P}{m^s}+\frac{a\rho_s\gamma(x^s)^{-1+\rho_s}}{(a+(x^s)^{\rho_s})^2}\right]\left[a\left(-1+\frac{1-\gamma}{\frac{Px^s}{m^s}-\frac{a\gamma}{a+(x^s)^{\rho_s}}}\right)\right]^{-1+\frac{1}{\rho_s}}}{\left[P\left(\frac{Px^s}{m^s}-\frac{a\gamma}{a+(x^s)^{\rho_s}}\right)^2\right]\frac{1}{a(1-\gamma)}}+\Omega\right\} \tag{19}$$

$$G'(x^b) = \frac{1}{\Omega}\left\{-1-\frac{b(1-\gamma)\left[b\left(-1+\frac{m^b(b+(x^b)^{\rho_s})(-1+\gamma)}{(x^b)^{\rho_s}(-m^b+Px^b)+b(Px^b+m^b(-1+\gamma))}\right)\right]^{\frac{1-\rho_b}{\rho_b}}}{\left[P\left(-1+\frac{Px^b}{m^b}+\frac{b\gamma}{b+(x^b)^{\rho_b}}\right)^2\right]\left[-\frac{P}{m^b}+\frac{b\gamma\rho_b(x^b)^{-1+\rho_b}}{(b+(x^b)^{\rho_b})^2}\right]^{-1}}+\Omega\right\} \tag{20}$$

from which we obtained the properties given in Sect. 5.3.

References

1. Day RH (1971) Rational choice and economic behavior. Theory Decis 1:229–251
2. Day RH, Kennedy PE (1970) Recursive decision systems: an existence analysis. Econometrica 38:666–681
3. Duesenberry JS (1949) Income, saving, and the theory of consumer behavior. Harvard University Press, Cambridge
4. Frank RH (1985) Choosing the right pond: human behavior and the quest for status. Oxford University Press, Oxford
5. Georgescu-Roegen N (1950) The theory of choice and the constancy of economic laws. Q J Econ 64:125–138
6. Georgescu-Roegen N (1971) The entropy law and the economic process. Harvard University Press, Cambridge
7. Leibenstein H (1950) Bandwagon, snob, and veblen effects in the theory of consumers' demand. Q J Econ 64:183–207
8. Modigliani F (1948) Fluctuations in the income-savings ratio: a problem in income forecasting. Stud Income Wealth 11:371–443
9. Pareto V (1964) Cours D'Economie Politique. Librairie Droz, Geneva
10. Veblen T (1899) The theory of the leisure class: an economic study of institutions. Allen & Unwin, London
11. Benhabib J, Day RH (1981) Rational choice and erratic behavior. Rev Econ Stud 48:459–471
12. Iannaccone LR (1989) Bandwagons and the threat of chaos: interpersonal effects revisited. J Econ Behav Organ 11:431–442

Co-Evolutive Models for Firms Dynamics

Giulia Rotundo and Andrea Scozzari

Abstract This paper considers the Bak–Sneppen (B&S) Self-Organized Criticality model originally developed for species co-evolution. We focus both on the original application of the model on a lattice, and on scale-free networks. Stylized facts on firms size distribution are also considered for the application of the model to the analysis of firms size dynamics. Thus, the B&S dynamics under the uniform, Normal, lognormal, Pareto, and Weibull distributions is studied. The original model is also extended by introducing weights on links connecting species, and examining the topology of the resulting Minimum Spanning Tree (MST) of the underlying network. In a system of firms a MST may evidence the template of the strongest interactions among firms. Conditions that lead to particular configurations of a MST are investigated.

1 Introduction

In the framework of Econometrics, the availability of large electronic databases has led to an increasing number of statistical analysis of raw data. Numerical studies on the size of firms often are concerned with the detection of the probability distribution that best fits the data sets. It is remarkable that same probability distributions for the size of firms are validated on large data sets encompassing several different industry sectors and long time extension. There is now wide empirical evidence indicating that the distribution of the degrees of the nodes in many networks

G. Rotundo
Faculty of Economics, University of Tuscia, Viterbo, Italy
giulia.rotundo@uniroma1.it

A. Scozzari
Faculty of Economics, University of Rome "La Sapienza", Rome, Italy
andrea.scozzari@uniroma1.it

A.K. Naimzada et al. (eds.), *Networks, Topology and Dynamics*, Lecture Notes in Economics and Mathematical Systems 613,

representing system of firms follows a power law. Other probability distributions are validated only on constrained subsets of data, but they result invariant on quite long time intervals [2, 3, 11, 17]. Although each single firm can experience fluctuations in its size, at a first analysis the collective evolution of firms, traced through their sizes, can be represented by a stationary process. Hence, the contribution of many single units to observables at the macrolevel has opened the way to complex systems approaches mainly based on stochastic models in discrete time and space state. The emergence of such common behavior from several data sets was explained by the fact that most firms share the same kind of hierarchical managing structure and internal organization [2, 3]. Furthermore, the collective behavior of groups of agents has been often interpreted as a mechanism for social contacts, as the basis for opinion switching relevant for financial applications [20, 29, 45], as well as in many other fields [6, 31–33]. These studies point out the relevance of closeness of agents in term of their similarity in the preferential direction for information spreading. Moreover, spatial closeness plays the major role in neighbors definitions in crowd behavior [43], thus validating models that mainly consider local dynamics driven by short range interactions. In particular, percolation models are well suitable for the modelling of such phenomena.

Bak and Sneppen (B&S) have introduced a simple model in the class of percolation models, addressing to each unit (agent) the role of species in an evolutionary context [10]. The B&S model is of intrinsic interest, since it is one of the simplest models giving rise to Self-Organizing Criticality (SOC) behavior. Species co-evolve to a stationary state and exhibit "intermittent dynamics", that is, species undergo long periods of little changes, called stasis, which are punctuated by sudden bursts of activities called avalanches (which are correlated with extinction events). The original formulation of the model on d-dimensional lattices with the usual boundary conditions has been already extensively studied [10, 39]. However, as argued by Watts and Strogatz [49], most real-life networks are neither perfectly ordered, nor completely disordered, but fall under the category of "small-world" networks, which interpolate smoothly between the two extremes [1, 49]. Such networks are characterized by a high degree of local order, yet appear disordered on a large-scale because of the presence of shortcuts in the networks. Because of their wide applicability, there have recently been numerous papers characterizing the properties of such networks. Scale-free networks provide a good example of small-world networks. Extension of the B&S co-evolutive dynamics on scale-free networks has evidenced different stasis dynamics of the involved species depending on their number of first neighbors [34, 35]. In this paper, we consider the application of the B&S model to firms size co-evolution by referring to probability distributions of the firms size drawn from empirical literature. We study this dynamics on lattice and scale-free networks.

The dynamics of the B&S model, at its stationary phase, seems to be well suitable for firms size modelling, since it accomplishes the permanence of the same firm's size probability distribution on long time intervals. We study the B&S dynamics by considering the same probability distributions that were already validated in the literature for the description of the firms size. In particular, the B&S dynamics seems to be suitable for modelling the extinction process of the firms, whose sizes are well fitted by a Weibull distribution.

In the original B&S model applied to lattice networks, links only serve to drive the dynamics and to define the neighborhood of each species. A link between two species represents a dependence between them, whose straightforward interpretation in natural evolution models represents a prey-predator relationship.

In this paper we introduce weights on links connecting the nodes, representing species (e.g., firms), of the network. Weights are intended to give a measure of the closeness among nodes, and their natural meaning range, for instance, from the amount of commercial interchange among firms, to the closeness of management teams, to the intensity of technological innovation, and so forth [3, 4, 42].

Furthermore, weights on links allow to devise subnetworks (or spanning subnetworks) pointing out the strongest relationships between species (e.g., firms). Among different classes of spanning subnetworks, we consider here the Spanning Trees, and, in particular, we are interested in finding a Minimum weight Spanning Tree (MST) of the underlying network. This is due to by both the existence of efficient algorithms for finding a MST even on huge networks, and the ease of interpreting its *geometry* from an economic viewpoint. The geometry of an MST is important in the *strong disordered limit* [19], and it remains unaltered on random graphs even if the distribution of disorder is made very broad [19]. The main reason can be addressed to the ordering of links weights, and any probability distribution that does not alter the order of the weights gives rise to the same MST configuration. Numerical studies on the geometry of Minimum Spanning Trees, when random uncorrelated weights are assigned to the links or edges of a network, have been provided for square and cubic lattices [19] and for scale-free networks [28, 36, 47]. The latter ones show universality of the spanning trees geometry highlighted in the scale-free structure of the MST itself. In fact, on non-sparse scale-free networks, the MST nodes degree distribution follows a power law with a degree exponent close to the one of the original network, and independent from the weight distribution. The B&S nodes co-evolution can be extended to links weights co-evolution. When applying the B&S dynamics on networks representing system of firms, we obtain not only a co-evolution of the firms size but also a evolution of the MSTs describing the evolution of the strongest relationships between firms. Furthermore, we are also able to investigate the conditions that give rise to different MST structures, with the aim to provide support for policies decisions.

We are aware of the fact that the B&S model is a toy model, but it has the advantage to hold some mathematical tractability, it can serve as a first approximation of collective behavior of market agents, and of the comprehension of properties and limits of the simple interactions between agents. Therefore, it can constitute a starting point for the development of more complex models.

The paper is organized as follows. The next section reviews the econometric literature about the distributions of the size of the firms. Section 3 shows the B&S model and its extensions, and finally Sect. 4 reports some results on our MST approach.

2 Empirical Studies on the Size of Firms

Several studies have been made on the detection of skew distributions of firms sizes and on the validity of Gibrat's law of proportionate effect for the growth rate in order to explain the empirically observed distribution of the firms size. This law states that the expected increment of a firm size in each period is proportional to the current size of the firm, and that the growth rate of each firm is independent from its size (Gibrat's law in weak form). Therefore, under the hypothesis that the growth rates are identically independent distributed, the distribution of the log of a firm's size tends to the lognormal distribution for $T \to \infty$, i.e. on sufficiently large time interval [16].

Mainstream econometric literature on firms size is aimed at showing the limits of the Gibrat model, and new growth rates and firms size distributions are proposed for fitting data. The studies about size and growth rate of firms differ for the hypotheses tested and for the data sets that were used. In the literature, data from Census and COMPUSTAT data bases are mainly analyzed. Census data give information about small firms, that are crucial for understanding the impact of social dynamics at the individual level. The volume of sales is used as a proxy for a firm size, and in some studies other fundamental variables like the total assets, sales and the number of employees are used as a complementary variables in order to check the validity and robustness of the results. Literature focuses mostly on the Pareto distribution as well as on the lognormal distribution as an alternative for the size of firms. The discussion is not purely an academic exercise. Right skewness implies that most firms have a size just below the average, and that there are few huge firms and some others very small. The detection of the proper distribution allows to explain differences in the reaction of the market to external shocks, like natural catastrophes, or the impact on some economy of exogenous economic factors. Computer aided simulations of economic systems show that, in the case of lognormally distributed data, shocks are absorbed, while in the case of Pareto distribution, correlations internal to the system can amplify the external shocks leading to strong oscillations of the entire system and to run the risk of a collapse [14]. These studies can help both in driving the best policies for economy development and in detecting the maximum charge of bad events (taxes, wars, natural catastrophes etc.) that can be supported without a complete crash. The next three subsections review the literature presenting some other studies on the distribution of the size of the firms supporting different probability distributions [11].

2.1 *Econometric Analysis Supporting the Gibrat's Law*

The distributions of the firms sizes in industrial countries are highly skewed, that is, a small number of large firms coexists along with a large number of small firms. The presence of right skewness supports both the Gibrat's law and the Pareto distribution.

Some studies [24, 25] inquiry the independence between growth rate and size. In [24] it is shown that the lognormal distribution hypothesis holds for UK firms larger than eight employers. Later, the same authors report that the size of the distribution of the UK companies is close to the lognormal, although the hypothesis of lognormality can be statistically rejected [25]. The test of the suitability of other distributions shows that the Pareto distribution performs the fit worse than the lognormal one at the upper tail. Other studies report that the fit of the lognormal distribution to size data is quite close to the mean, but it performs less on the tails. Families of functions, that include the lognormal one as a particular function, and that take into account a power law decay of tails, have also been developed. The goodness of statistical fit allows for some compromise. The weak form of Gibrat's law has been shown to be compatible with power law under further hypotheses. As an example, the first model is the Simon's model [16] where the Gibrat's law is combined with an entry process to obtain a Levy distribution for the firms size. Particular assumptions like the validity of the detailed balance, that states the time-reversal symmetry for the growth rate, show that Gibrat's law and Pareto–Zipf's law hold for firms larger than a fixed threshold [21]. This property is not valid in general [27], but the behavior of the largest companies is important because it influences the entire economy. Therefore, such analysis is useful for driving economic policies at the Country level. On the other hand, districts constitute small worlds with a prevalence of small sized industries, so that policies for district developments will be different from those based on the common behavior of big firms, and need a finer analysis.

2.2 Econometric Analysis Supporting the Pareto Distribution

Histograms of companies sizes exhibit skewness. In some data set, skewness has been shown to be robust over time [8]. It even lived through large-scale demographic transitions in the work forces and widespread technological changes. Finer analysis have shown that skewness grows during growing phases of the economy and decreases during recessions [22], thus being an indicator of such economic cycles. A characteristic that emerges is that, although the position of individual firms in a size distribution does depend on the definition of size, the shape of the distribution does not. The main concern is to select the best fit to data histograms. Although in older studies [24, 25] the lognormal hypothesis received great attention, in recent papers the main results indicate power law for firm size and Laplace law for firms growth rates [16]. The two results are strictly connected. In fact, it can be shown, under proper hypotheses, that the logarithm of a Pareto random variable follows an exponential distribution, and that the difference of two exponential random variables results in a Laplace distribution [40].

The power law behavior seems to be valid also for parameters that are common in most developed Countries. The results reported in [15] can be interpreted as the existence of a significant range of the world GDP distribution where countries share a common size-independent average growth rate. Further particular hypotheses like entry and exit of companies from the market, provide results that contradict the Gibrat's law. As an example in [2, 3, 44] the exponential distribution for the growth rate of firms has been found to hold for the 20 years 1974–1993 of COMPUSTAT publicly-traded United States manufacturing firms, whilst the variance of the growth rate should grow with the size of the firm. A model is also proposed which offers a possible explanation for the power law relationship between firms size and the variance of growth rate [46], showing a power law dependence of the variance of the growth rate conditioned to the size of the firms. It has been shown that such kind of dependence may rise from a hierarchical management organization provided with a disobedience probability.

2.3 Econometric Analysis Supporting the Weibull Distribution

A comparison with the distribution of the extinction rate of species introduces immediately the comparison between the B&S model for species co-evolution and the studies presented in Di Guilmi et al. [17]. In the former case the species, corresponding to nodes of a lattice in the B&S model, are the extinguished ones. The changing in the nodes' value in the B&S dynamics, indicates the extinction of a specie (firm) that will be replaced by a new specie represented by that node. The classification of firms by size is best fitted by a Weibull function. A best fit Weibull parameters table, reported in [17], considers data about the extinction rate of firms in eight OECD countries, and divides data into six classes by number of employees. In our application, we perform simulations with the B&S model referring to each of the six classes.

3 The Dynamic Model

In its original formulation, at each time $t = 0,\ldots,T$, the d-dimensional B&S dynamic model considers L^d species organized in a simple regular lattice with the usual boundary conditions in dimension d. Each species, represented by each node of the lattice, is fully described at time t by its fitness, $f_i^d(t)$, $i = 1,\ldots,L^d$, drawn at time 0 from a uniform distribution in $[0,1]$. Therefore, we are considering a network where to each node is assigned a value (fitness) while each link or edge of the network simply represents the connection between two nodes. In financial applications the values $f_i^d(t) \in [0,1)$ can be chosen for representing firm fitness [5, 7], as well as prices or opinions [12, 41, 50]. At each time step, the B&S model selects the node with the minimum fitness and changes its fitness and those of its $2d$ adjacent nodes by randomly generating new values from a uniform distribution in $[0,1)$.

One of the key problems related to the B&S evolution model is to compute the limit distribution for the values of the nodes at a stationary *regime*, as the time of the system grows to infinity. Computer simulations show that for a time long enough, under the B&S dynamics the maximum of minima of fitnesses are above a critical threshold f_c, apart from some periods, called avalanches, where they fall below f_c [9, 10, 23, 26]. In the one-dimensional case (i.e., $d = 1$) the limit (marginal) distribution is uniform on $(f_c, 1)$, with $f_c \sim 0.667$. These results were confirmed theoretically through the mean field approximation [18, 39, 48]. Many other properties of the B&S model were obtained after a change of distributions: from uniform to exponential.

3.1 The B&S-Exponential Model

The B&S-exponential model is defined to be the model obtained from the B&S-uniform one by substituting the hypothesis $f_i^d(t) \sim U[0,1), i = 1,\dots,L^d$, by $f_i^d(t) \sim D[0,\infty), i = 1,\dots,L^d$, where $D[0,\infty)$ is the exponential distribution in $[0,\infty)$.

Intermediate results in the proof assessing the existence of the critical threshold f_c and the behavior of the joint distribution of $f_i^d(t)$, $i = 1,\dots,L^d$, are provided in [37, 38] using the exponential setup. The results were reported to the original B&S model through the following lemma based on some remarks in [37, 38].

Lemma 1. *The B&S-exponential model has the same dynamics properties of the B&S-uniform one.*

Proof. We follow the rules for random numbers generation. Let x be a random number sampled by a random variable uniformly distributed in $[0,1)$. The function $q : [0,1) \to [0,\infty)$, such that $q(x) := -ln(1-x)$ transforms x into a random number $y = q(x)$ sampled by a random variable exponentially distributed in $[0,\infty)$. If at times $t = 0,\dots,T$, $q(\cdot)$ is applied to the $f_i^d(t)$, $i = 1,\dots,L^d$, then it provides a one-to-one mapping between the fitnesses of the B&S-uniform and the B&S-exponential model. Moreover, the dynamics of the evolution of the fitnesses is still based on the minimum value of the fitnesses. Actually, $q(\cdot)$ is a monotone function, therefore at time t, the transformation applied on the $f_i^d(t)$, $i = 1,\dots,L^d$, maintains the ordering of the values of the nodes, and the evolution rule selects the same node and its neighbors both in the uniform and in the exponential setup. □

After Lemma 1, the function $q(\cdot)$ changes the values of the fitness of each node, but it does not change the dynamics of the B&S-uniform model. Furthermore, the results on the threshold in the B&S-exponential model can be easily suited to the original B&S model by the following remark.

Remark 1. If the B&S-exponential model has threshold f_c, then the B&S-uniform model has threshold $q(f_c) = 1 - e^{-f_c}$.

3.2 Extension of B&S Model to Other Probability Distributions

Let us consider the B&S model where the uniform distribution has been substituted by using a random variable X, defined on an arbitrary interval I, with cumulative probability distribution $F_X(x)$.

Theorem 1. *Let f_c be the threshold for the B&S-uniform model. Consider now the B&S model where the uniform distribution has been replaced by a random variable X with cumulative probability distribution $F_X(x)$. Then, the resulting B&S model has threshold $F_X(f_c)$.*

Proof. Given a random variable U uniformly distributed over the interval $(0,1)$, $X = F_X^{-1}(U)$ (provided F is invertible). Hence, $F_X(X)$ is uniformly distributed in $(0,1)$. Since $F_X(X)$ is continuous, monotone and non decreasing it is order preserving. Therefore, $F_X(X)$ maps the B&S model with any probability distribution to the B&S model with uniform distribution and $F_X(f_c)$ holds. □

Theorem 2. *If the B&S model has limit distribution given by the product of uniform distributions above f_c, then the B&S model where the uniform distribution is replaced by a random variable X exponentially distributed has limit distribution given by the product of exponential distributions above $F_X(f_c)$.*

Proof. The proof follows from Theorem 1 by considering the transformation $F_X(\cdot)$ on the fitnesses, where $F_X(\cdot)$ is the cumulative distribution function of a random variable X exponentially distributed. □

Firstly, in this paper we are interested in studying the B&S evolution model by referring to the distributions mostly used for describing the size of the firms. Therefore, here we consider the uniform, Normal, lognormal, Pareto, log-Pareto, Exponential, and Weibull distributions. Let us consider the B&S model in which the uniform distribution is substituted by one of the above distributions. The following remarks hold.

Remark 2. If the B&S-uniform model has threshold f_c, then the B&S model using Normal, lognormal, Pareto, log-Pareto, Exponential, Weibull, distributions, has thresholds $F_X(f_c)$, where $F_X(x)$ is the cumulative distribution function of a random variable X following the Normal, lognormal, Pareto, log-Pareto, Exponential, Weibull distribution, respectively.

Remark 3. If the B&S-uniform model has limit distribution of the fitnesses given by the product of uniform distributions, then the B&S model using the lognormal, Pareto, log-Pareto, Weibull distributions has limit distribution given by the product of the lognormal, Pareto, log-Pareto, Weibull distribution, respectively.

3.3 Extension of the B&S Model: The Case of Scale-Free Networks

The B&S model on lattices can be generalized by referring to arbitrary finite connected networks, like, for instance, small world networks [30] and scale-free (SF) networks. SF networks have been recognized to describe several real growing networks, and, at the same time, have proved to show very peculiar properties for diffusion properties. In a system of firms this serves for modelling the impact of external factors, as the spreading of innovation, the external modification of demand and supply and so forth. In particular, the diffusion properties corresponding to the fault tolerance property can provide the maximal amount of changes that a system can bear before having a deep drastic change in the firms organization. It is then natural to ask whether and to what extent the topology of these complex networks would affect the results obtained in classical evolution models like the B&S one. It results that the critical thresholds continue to exist only on a subset of SF networks [34].

4 Minimum Spanning Tree

We extend the B&S model by assigning weights to each link of the underlying network (lattice or scale-free networks). In our application, each node-firm value (fitness) represents the size of a firm, while a link provides the connection between two firms. Depending on the application, a weight associated to a link or edge may represent the quote of participation of a firm into another [42], the intensity of technological innovation [4], or the tightness of management structure [3]. In networks or graphs applications, given a graph, a customary problem is to find out the relevant relationships between nodes. This is often accomplished by finding certain spanning subgraphs of the given graph. Different classes of subgraphs can be considered, each providing different properties about the closeness of the nodes. In this paper we refer to trees. A spanning subtree of a network gives the minimum way of connecting all the nodes of the graph. Among all the possible spanning subtrees we will look for the one (or the ones since there may be more that one) that minimizes the sum of the weights. A Minimum weight Spanning Tree (MST) of a given network points out the strongest relationships between the nodes. Searching for a spanning subtree is also preferred compared with other spanning subgraphs since both the availability of efficient algorithms for finding a spanning tree even on huge networks, and the ease of interpreting its geometry. Thus, along with evaluating the nodes dynamics with the classical B&S model, we are able to evaluate the changes in the relationships between firms over times by observing the changes in the topology of a MST. Actually, MST topologies and particular kind of spanning subtrees may represent structures relevant for Economics, like, for instance, the raise of oligopolies. In the next subsections we show the conditions that give rise to very different shapes of a MST, leaving the most proper definition of weights to econometric work. In order to evaluate this dynamics we considered different probability distributions for assigning weights to the nodes and edges of the underlying network, and again we refer to the uniform, Normal, lognormal, Pareto, log-Pareto and Weibull distributions.

4.1 Weights not Correlated with the Degrees of the Nodes

Numerical results on MST shapes when random edge weights uncorrelated with nodes weights are considered, appear in Dobrin et al. [19] for square and cubic lattices. For scale-free networks, in Szabo et al. [47] is shown that if the weights of the links incident to a given node are independent from the node's degree (i.e., the degree of connectivity of the node) then the geometry of a MST depends only on the topology of the network. Therefore, the probability distribution of the nodes values (fitnesses) is not relevant for the MST topology, and the B&S dynamics changes the MST, but not its topology. As an example, Fig. 1 reports on the result in the case of a Weibull distribution on a two-dimensional square lattice. At time t, the edge connecting nodes i and j has weight $w_{ij} = \mid f_i^d(t) - f_j^d(t) \mid$. This is a first raw measure of distance between i and j, and it is independent of the degree of connectivity of both i and j. The degree distribution of the nodes is stable under the B&S dynamics and for the different probability distributions considered.

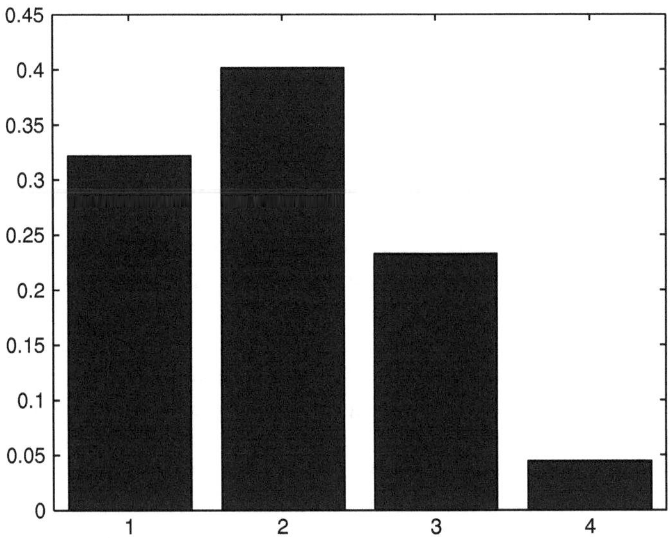

Fig. 1 Distribution of the degrees of the nodes in the MSTs on a lattice network and with a Weibull probability distribution

4.2 Weights Correlated with the Degrees of the Nodes

In order to observe other MST shapes, it is necessary to assign weights to the links depending on the degrees of the nodes. We leave the definition and interpretation of these weights, and their exact meaning problem-oriented, to empirical econometric papers addressing real-world problems.

In real-world applications, weights should be correlated to both the fitnesses and the degree of the nodes. Although the values of the fitnesses may change during the dynamics, only the nodes degree may change the topology of the MST. Referring to SF networks, due to their practical relevance, an immediate implication of the above remark is that any rewiring procedure mapping the network into another network having the same SF property will lead to the same MST topology. Therefore, it is worth examining the effects of the correlation function between edge weights and connectivity degree for the general class of SF networks, without adding any rewiring dynamics.

In [36] it is shown that solely by changing the nature of the correlations between wights and network topology, the structure of a MST may change from SF to exponential. In particular, they explore the MST behavior considering weights w_{ij} from node i to node j defined as being directly proportional to $(k_i k_j)\vartheta$, $\vartheta > 0$, $max(k_i,k_j)$, $min(k_i,k_j)$, $1/(k_i k_j)$, $1/min(k_i,k_j)$, $1/max(k_i,k_j)$.

Their numerical results indicate that in the presence of correlations, two classes of MSTs exist for scale-free networks, having either a power law or an exponential degree distribution. Correlated weights choices $w_{ij} \propto k_i k_j$, or $w_{ij} \propto max(k_i,k_j)$ give rise to MSTs with exponential degree distributions, while the other choices result in MSTs with power law distributions.

The exponential nature of the first two weights choices is due to the tendency of a MST algorithm to avoid links with large weights, so that, for this weight selection, a MST algorithm effectively shuns the highly connected nodes by using, when possible, links connecting low degree nodes. The remaining weights choices give rise to power law degree distribution of a MST [28, 36, 47].

Of course, intermediate degree of correlation between edge weights and nodes degree may give rise to very different shapes of a MST.

4.3 Numerical Results

We examined the B&S dynamics under different probability distributions. We considered two-dimensional square lattices with the usual boundary conditions and scale-free networks as underlying networks. Since the B&S dynamics shows a transient phase, we considered 10^9 iterations of the B&S dynamics before starting our analysis, and statistics were drawn on the next 10^7 iterations. On a two-dimensional square lattice we considered $L^d = 10,000$ evolving species. In order to provide statistics about the MST evolution, at each time $t = 0,\dots,T$, we used the distance measure between nodes i and j defined by $w_{ij} = \mid f_i^d(t) - f_j^d(t) \mid$. The more the values of the nodes are close to each other, the lower is the distance w_{ij}. Mutations following the B&S dynamics randomly change the $f_i^d(t)$, $i = 1,\dots,L^d$, and therefore the edge weights w_{ij}, for each edge (i, j). The sampling of the fitnesses $f_i^d(t)$, $i = 1,\dots,L^d$, was performed according to the uniform, Normal, lognormal, Pareto, log-Pareto and Weibull distributions. Figure 1 shows the histogram representing the distribution of the degrees of the nodes in the MSTs in the case of a Weibull distribution.

The analysis on the edge weights in a MST, presented in [19, 36], does not consider any evolution w.r.t. the time. On the other hand, we provide a synthetical analysis that is obtained by considering the behavior of the weights of the MSTs obtained during the dynamics. We notice that the $f_i^d(t)$, $i = 1,\dots,L^d$, and w_{ij} are not co-monotone, thus allowing the social interpretation of the possibility of disagreement among agents as they change their opinion. In the stationary state we have that all the $f_i^d(t) \in (f_c, 1)$, $i = 1,\dots,L^d$ apart from avalanches. Thus, most of the weights w_{ij} belong to $[0, 1 - f_c)$, and exhibit avalanches as a consequence of fitnesses avalanches. Following Dobrin et al. [19], we also provide the probability that a link with a given weight lies on the MST (see Fig. 2).

We also note that for $d = 2$ and for lattice networks, a changing of the fitness of a node only implies a change of the weights of its $2d = 4$ adjacent nodes. This corresponds to modify the weights of 16 edges at a time, thus resulting in a fast updating procedure for the nodes and the edges values. The number of links changed at each step of the dynamics is centered at 5 over the 16 links changed. The MST constructing algorithm is based on the Kruskal procedure.

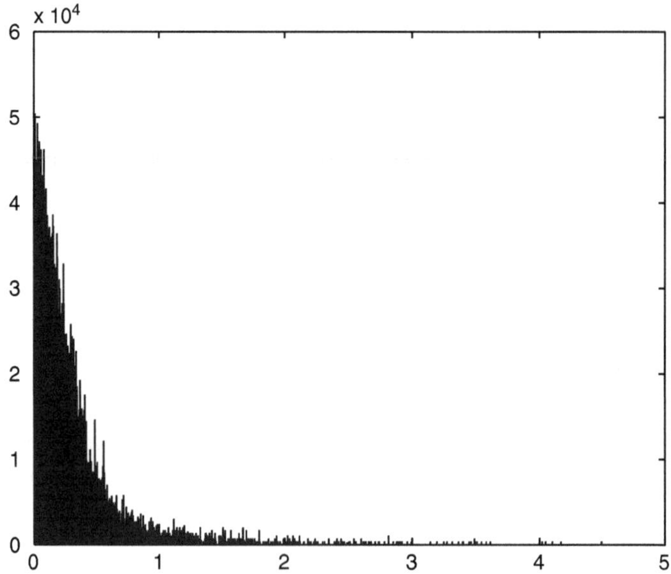

Fig. 2 Probability that a link with a given weight lies on the MST in a lattice and with a log-Pareto distribution

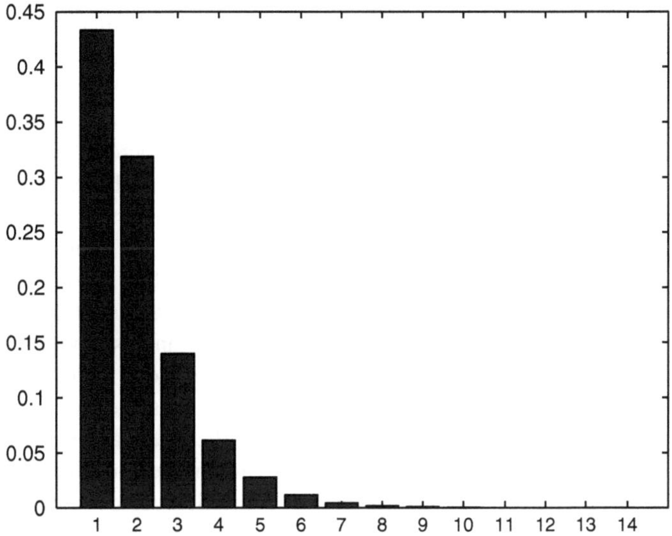

Fig. 3 Power law property for the distribution of the degrees of the nodes in a MST by using the Weibull distribution on scale-free networks

As introduced before, we also examined the results of the co-evolutive model when applied on scale-free networks, that are supposed to be the most spread form of contacts organizations. We used the free downloadable Barabasi–Albert software for scale-free networks generation. We run simulations on 1,005 nodes, and 20,000 edges. Then we run the B&S dynamics again considering the uniform, Normal, lognormal, Pareto, log-Pareto, and the Weibull distribution. As shown in Fig. 3, the distribution of the degrees of the nodes of the MSTs follows a power law, and this holds also for all the other probability distributions considered. All the programs implementing the B&S dynamics as well as the Kruskal algorithms for finding the MSTs are written in C code.

5 Conclusions

The paper goes beyond the B&S model features introducing weights and analyzing the co-evolution of nodes and their relationship with the MST evolution of a given underlying network. Actually, given a network representing the relationships between firms (nodes of the networks), a Minimum weight Spanning Tree points out the strongest relationships between the firms. In this paper, we are interested in evaluating the evolution of an MST and, in particular, the evolution of the geometry of an MST since different MST shapes may reveal structures relevant, for instance, in an economic system, like the raise of oligopolies. We notice that this paper differs from other papers in the literature (see e.g., [4]), where the values assigned to each node and edge of a network are constant. In fact, while their dynamics concerns activation/disactivation of links, thus considering the so called *rewiring problem*, we are interested in studying the evolution of a spanning subgraph of the network that summarizes the relationships between the nodes of the network. Indeed, in social models, a MST may represent the most strong social interactions between agents, that can be transferred to management links when considering the evolution of firms.

MST dynamics and its evolution strictly depends on the underlying network as well as on the sampling distribution of the values assigned to its nodes and edges. Our toy model provides a dynamical co-evolutive model that considers functional properties observed in empirical studies, and can constitute a starting point for the construction of more realistic models. Finer econometric analysis evidenced fluctuations of the parameters of probability distribution of the size of the firms, depending on the selection of the time windows [13]. The models presented here can be easily adapted to accomplish the arise and the change of the shape and scale parameters of probability distribution during cycles of recession and expansion of economies. The introduction of edge weights correlated to the degrees of nodes in place of weights that are proportional to them, and the rewiring mechanism for changing network topologies are the main tools for calibrating the models. We leave this for future research as well as further investigations introducing asymmetry in the network's links.

References

1. Albert R, Barabasi AL (2002) Statistical mechanics of complex networks. Rev Mod Phys 74:47–97
2. Amaral LAN, Buldyrev SV, Havlin S, Leschhorn H, Maass P, Salinger MA, Stanley HE, Stanley MHR (1997) Scaling behavior in economics: I. Empirical results for company growth. J Phys I 7:621–633
3. Amaral LAN, Buldyrev SV, Havlin S, Leschhorn H, Maass P, Salinger MA, Stanley HE, Stanley MHR (1997) Scaling behavior in economics: II. Modeling of company growth. J Phys I 7:635–650
4. Andergassen R, Nardini F, Ricottilli M (2007) The emergence of paradigm setters through firms interaction and network formation (to appear)
5. Ausloos M, Pekalski A (2007) Model of wealth and goods dynamics in a closed market. Physica A 373:560–568
6. Ausloos M, Petroni F (2007) Statistical dynamics of religions and adherents. Europhys Lett 77:38002
7. Ausloos M, Clippe P, Pekalski A (2004) Evolution of economic entities under heterogeneous political/environmental conditions within a Bak–Sneppen-like dynamics. Physica A 332:394–402
8. Axtell RL (2001) Zipf distribution of u.s. firm sizes. Science 293:1818–1820
9. Bak P (1996) How nature works. Springer, New York
10. Bak P, Sneppen K (1993) Punctuated equilibrium and criticality in a simple model of evolution. Phys Rev Lett 71:4083–4086
11. Cerqueti R, Rotundo G (2007) Processi di rinnovamento nei cluster di imprese. In Garofalo G, Capitalismo distrettuale, localismi d'impresa, globalizzazione. Firenze University Press: 129–143

12. Cuniberti G, Valleriani A, Vega JL (2001) Effects of regulation on a self-organized market. Quantitative Finance 1:332–338
13. Delli Gatti D, Di Guilmi C, Gaffeo E, Giulioni G, Gallegati M, Palestrini A (2004) Business cycle fluctuations and firms' size distribution dynamics. Adv Complex Syst 7(2):1–18
14. Delli Gatti D, Di Guilmi C, Gaffeo E, Giulioni G, Gallegati M, Palestrini A (2005) A new approach to business fluctuations: heterogeneous interacting agents, scaling laws and financial fragility. J Econ Behav Organ 56:489–512
15. Di Guilmi C, Gaffeo E, Gallegati M (2003) Power law scaling in the world income distribution. Econ Bull 15(6):1–7
16. Di Guilmi C, Gaffeo E, Gallegati M (2004) Empirical results on the size distribution of business cycle phases. Physica A 333:325–334
17. Di Guilmi C, Gallegati M, Ormerod P (2004) Scaling invariant distributions of firms exit in OECD countries. Physica A 334:267–273
18. Dickman R, Muñoz MA, Vespignani A, Zapperi S (2000) Paths to self-organized criticality. Bras J Phys 30:27–41
19. Dobrin R, Duxbury PM (2001) Minimum spanning trees on random networks. Phys Rev Lett 86(22):5076–5079
20. Föllmer H, Horst U, Kirman A (2005) Equilibria in financial markets with etherogeneous agents: a prbobabilistic perspective. J Math Econ 41:123–125
21. Fujiwara Y, Di Guilmi C, Aoyama H, Gallegati M, Souma W (2000) Do Pareto–Zipf and Gibrat laws hold true? An analysis with European firms. Physica A Stat Mech Appl 335 (1–2):197–216
22. Gaffeo E, Gallegati M, Palestrini A (2003) On the size distribution of firms: additional evidence from the G7 countries. Physica A Stat Mech Appl 324(1–2):117–123
23. Grassberger P (1995) The Bak–Sneppen model for punctuated evolution. Phys Lett A 200:277–282
24. Hart PE, Oulton N (1996) Growth and size of firms. Econ J 106:1242–1252
25. Hart PE, Oulton N (1997) Zipf and the size distribution of firms. Appl Econ Lett 4:205–206
26. Jensen HJ (1998) Self-organized criticality. Cambridge Lecture Notes in Physics
27. Kertesz J, Kullmann L, Zawadowski AG, Karadi R, Kaski K (2003) Correlations and response: absence of detailed balance on the stock market. Physica A 324(1):74–80
28. Kim DH, Noh JD, Jeong H (2004) Scale-free trees: skeleton of complex networks. Phys Rev E 70:046126
29. Kirman A, Teyssiere G (2002) Microeconomic models for long-memory in the volatility of financial time series. Studies in Noninear Dynamics and Econometrics 5:281–302
30. Kulkarni RV, Almaas E, Stroud D (1999) Evolutionary dynamics in the Bak–Sneppen model on small-world networks. arXiv:cond-mat/9905066 v1 6
31. Lambiotte R, Ausloos M (2005) Uncovering collective listening habits and music genres in bipartite networks. Phys Rev E 72:066107
32. Lambiotte R, Ausloos M (2006) Modeling the evolution of coupled networks. In: First World congress on Social Simulation (WCSS 2006), e-Proceedings 1:375–381
33. Lambiotte R, Ausloos M, Holyst JA (2007) Majority model on a network with communities. Phys Rev E 75
34. Lee S, Kim Y (2005) Coevolutionary dynamics on scale-free networks. Phys Rev E 71:057102
35. Lee KE, Hong BH, Lee JW (2005) Universality class of Bak–Sneppen model on scale-free network. arXiv:cond-mat/0510067
36. Macdonald PJ, Almaas E, Barabási AL (2005) Minimum spanning trees of weighted scale-free networks. Europhys Lett. 72:308314
37. Maslov S (1996) Infinite hierarchy of exact equations in the Bak–Sneppen model. Phys Rev Lett 77:1182
38. Meester R, Znamenski D (2004) Critical thresholds and the limit distribution in the Bak–Sneppen model. Commun Math Phys 246
39. Paczuski M, Maslov S, Bak P (1996) Avalanche dynamics in evolution, growth, and depinning models. Phys Rev E 53:414

40. Palestrini A (2007) Analysis of industrial dynamics: a note on the relationship between firms' size and growth rate. Econ Lett 94(3):367–371
41. Rotundo G, Ausloos M (2007) Microeconomic co-evolution model for financial technical analysis signals. Physica A 373:569–585
42. Salvemini MT, Simeone B, Succi R (1995) Analisi del possesso integrato nei gruppi di imprese mediante grafi. L'Industria XVI:641–662
43. Sornette D (2002) Why the stock markets crash. Princeton University Press, Princeton
44. Stanley MHR, Amaral LAN, Buldyrev SV, Havlin S, Leschhorn H, Maass P, Salinger MA, Stanley HE (1996) Scaling behaviour in the growth of companies. Lett Nature 379:804–806
45. Stauffer D, Sornette D (1999) Self-organized percolation model for stock market fluctuations. Physica A 271:496
46. Sutton J (2003) The variance of corporate growth rates. Physica A 324(1):45–48
47. Szabó GJ, Alava M, Kertesz J (2003) Geometry of minimum spanning trees on scale-free networks. Physica A 330:31–36
48. Vergeles M (1995) Self-prganization at nonzero temperatures. Phys Rev Lett 75
49. Watts DJ, Strogatz SH (1998) Collective dynamics of 'small-world' networks. Nature 393:4
50. Yamano T (2001) Regulation effects on market with Bak–Sneppen model in high dimenzions. Int J Mod Phys C 12:13291333

Part III
Markov Chains and Topology

Betweenness Centrality: Extremal Values and Structural Properties

R. Grassi, R. Scapellato, S. Stefani, and A. Torriero

Abstract In this paper we investigate the structural properties of betweenness centrality and determine some cases in which betweenness reaches its extremal values. Special attention is paid to $Star(G)$, the set of vertices adjacent to all other vertices in a graph and we prove several results about the betweenness of the elements of this set. We introduce the new concept of total betweenness and relate it to group betweenness. We prove a necessary and sufficient condition for the two measures to coincide. Next we consider cutsets and cutvertices and we find a lower bound for their betweenness; in particular for a cutvertex this lower bound is the cutting number. Finally we apply the previous results to trees, proving an alternative formula for betweenness based on cutvertex properties.

1 Introduction

Centrality is a major issue in network theory. When referred to a vertex, centrality helps to quantify the role that the vertex plays in the overall structure of the network. There are several definitions of vertex centrality, each one related to the network topological properties and applicable to specific contexts. Among

R. Grassi and S. Stefani
Dipartimento di Metodi Quantitativi per le Scienze Economiche ed Aziendali
Università degli Studi di Milano – Bicocca, Milan, Italy
rosanna.grassi@unimib.it, silvana.stefani@unimib.it

R. Scapellato
Dipartimento di Matematica, Politecnico di Milano, Milan, Italy
rafsca@mate.polimi.it

A. Torriero
Dipartimento di Discipline matematiche, Finanza matematica e Econometria
Università Cattolica del Sacro Cuore di Milano, Milan, Italy
anna.torriero@unicatt.it

A.K. Naimzada et al. (eds.), *Networks, Topology and Dynamics*,
Lecture Notes in Economics and Mathematical Systems 613,

the vertex centrality measures we may recall betweenness centrality, degree centrality, eigencentrality and closeness centrality. For a thorough description of vertex centrality see for example [1] and [8]. The degree and the eigenvector centrality got much attention recently and have been deeply investigated (see [4] and [12]). For a unifying approach to the vertex centrality measures from a topological point of view see [3]. In this paper we investigate the betweenness centrality, a measure for the *intermediary* role of a vertex in a graph; this gives an idea of the influence of a vertex over the spread of information through the network and measures how often a vertex is located on a shortest path between other vertices in the graph. Betweenness centrality has a wide range of applications. For instance we recall that in the Internet it provides a measure of the traffic load that a vertex is supposed to handle. Another application regards social networks, as betweenness measures the influence that an individual vertex has in the spread of information within the network, assuming that the information, to be spread most rapidly, follows the shortest path. For a discussion of betweenness in scale-free networks see [10].

What seems natural in investigating betweenness is to find under which conditions betweenness attains its extremal values, i.e. in the least and most central vertices. First we give a necessary and sufficient condition for a vertex to have zero betweenness. Then we consider $Star(G)$, the star set of a graph, i.e. the set of vertices that are connected to all other vertices in the graph. The betweenness of the vertices in $Star(G)$ is maximum. We show that their betweenness decreases as long as edges are added to the graph. The interpretation is very clear: a vertex adjacent to all other vertices in a network plays an essential role in spreading information but, if we increase the number of links, the vertices in $Star(G)$ are less crucial since other geodesics can be created, bypassing them. The other extreme, zero betweenness, occurs when the graph is complete, i.e. $Star(G) = V$: no vertex is necessary to pass information since all vertices are adjacent to each other.

A natural extension of the concept of vertex centrality is the centrality of a subset of vertices. To recall an example, the centrality of a subset makes it possible to evaluate in a social context the importance of a department within an organization or of a holding and its associates within a financial participation network. Subset centrality for trees was considered in [11], focusing on connected subsets (i.e. subtrees). We introduce the new concept of total betweenness and relate it to group betweenness, recently introduced in [5]. Our total betweenness, defined as the sum of the betweenness values of the vertices in the subset, is easy to compute and measures the whole extent of the centrality of the subset, including in the count the geodesics between vertices belonging to the subset. Group betweenness centrality, on the other hand, gives the betweenness of the subset only referring to the paths leading to the *external* vertices, i.e. the vertices not belonging to the subset. We prove a necessary and sufficient condition for the two measures to coincide, thereby characterizing both concepts in a formal way. We use here the new concept of pure set.

Further, we clarify the role of cutvertices and cutsets, i.e. sets of vertices whose removal can disconnect the graph. We find lower bounds for the betweenness of cutvertices based on the number of connected components that the removal of the vertices generates and we show that their betweenness is increasing with the number of components; this generalizes a result of [7]. Again, the interpretation is easy: cutvertices are as more important in spreading information through geodesics as more components are generated by their removal. The more components the removal generates, the more central in betweenness the vertex is. Finally, we consider the particular case of cutvertices in trees and are able to provide an alternative formula for betweenness.

The paper is structured as follows: in Sect. 2 we recall the fundamental definitions and we define the pure subsets of a graph, characterizing the connected ones; in Sect. 3 the extremal values of the vertex betweenness centrality are investigated. Next we introduce the definition of total betweenness, analyze its properties (Sect. 4) and related it to cutsets (Sect. 5). Conclusions are in Sect. 6.

2 Preliminaries

We recall here some graph-theoretical definitions and notations. Let $G(V,E)$ be a simple undirected graph, with n vertices and m edges.

For every $v \in V$, $N(v) = \{w \in V : w \sim v\}$ is the set of vertices adjacent to v. Its cardinality is called the degree of v and denoted by $d(v)$. $W \subseteq V$ is an independent set of a graph G if, for all pairs $\{u,v\}$ of vertices in W, we have $u \notin N(v)$.

We will write $H \subseteq G$ if H is a subgraph of a graph G; the induced subgraph of a set $W \subseteq V$ is the maximal subgraph of G with vertex set W; we denote by $\langle W \rangle$ the induced subgraph of W.

Two subgraphs of G are said to be disjoint if they have no common vertices. If $H_1, H_2, \dots, H_m$ are pairwise disjoint subgraphs of G such that $G = H_1 \cup \dots \cup H_m$, we say that G is their disjoint union. When $m = 1$ the term will be used to simply mean that $G = H_1$.

A star graph $K_{1,n-1}$ is a graph with one vertex having degree $n-1$ and the other $n-1$ having degree 1.

A path is a sequence of distinct adjacent vertices; let us denote by $v_0 v_1 \dots v_k$ a path passing through vertices $v_0, v_1, \dots, v_k$. The relation $x \equiv y$, *if and only if there is a path joining x and y*, is an equivalence on the set V. Its equivalence classes $G_1, G_2, \dots, G_m$ are called connected components of G. Clearly, G is a disjoint union of $G_1, G_2, \dots, G_m$. If $m = 1$, then G is said to be connected. In the sequel we always suppose that G is connected.

A $v-w$ geodesic is a shortest path between vertices w and v. Its length is called the distance between v and w and denoted by $d(v,w)$. The betweenness centrality of a vertex $v \in V$ is

$$\beta_G(v) = \sum_{u,w} \frac{g_{uw}(v)}{g_{uw}}, \qquad u,w \neq v\,,$$

where g_{uw} is the number of geodesics from u to w, and $g_{uw}(v)$ is the number of geodesics between u and w passing through v. In the definition of betweenness we always suppose that the pairs u,w appear only once in the sum.

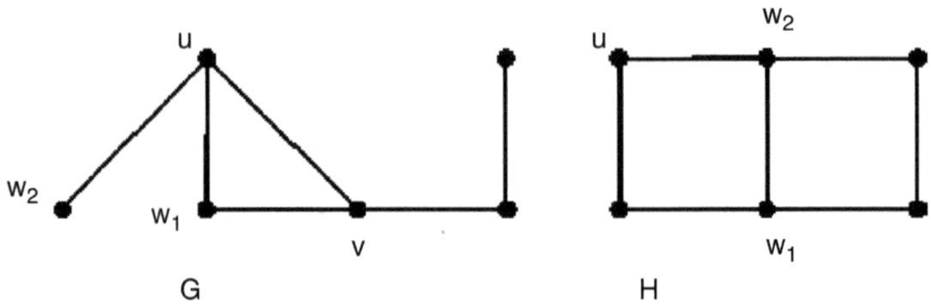

Fig. 1 Pure and non-pure sets

A vertex $v \in V$ is a cutvertex if the graph $G \setminus v$ is not connected; $W \subseteq V$ is a cutset if the graph $G \setminus W$ obtained by removing the set of vertices W is not connected.

For other graph-theoretical definitions see for example [2].

Definition 1. A pure set $W \subseteq V$ is a set such that, for each $v \in V$ and $w \in W$, no geodesic between v and w meets W in vertices other than v and w.

In Fig. 1 the graph G contains the pure set $W = \{w_1, w_2\}$; every geodesic from w_1 to another vertex of V meets W only in w_1 (the same for vertex w_2). In the graph H, the set $W = \{w_1, w_2\}$ is not pure, because one of the two geodesics from u to w_1 meets W in w_2.

Clearly, all subsets of a pure set are pure. The subgraphs induced by pure subsets are characterized as follows. Note that here by *clique* we mean any non-empty set X of vertices such that $\langle X \rangle$ is complete, including singletons. Thus a pure set W can be independent, which corresponds to the case where all involved cliques are singletons.

Moreover, if a pure set W is connected, it must consist of a unique clique. A full characterization of the latter case will be offered by Corollary 1.

Lemma 1. *A pure set W of G is a disjoint union of cliques.*

Proof. It suffices to show that any two distinct maximal cliques X and Y of W are disjoint. By contradiction, assume that $X \cap Y$ is non-empty; let $z \in X \cap Y$, $x \in X \setminus Y$ and $y \in Y \setminus X$. As x, y, z are elements of the pure set W and xzy is a path, x must be adjacent to y. Therefore, every two elements of $X \cup Y$ are adjacent and so $X \cup Y$ is a clique. As X and Y are distinct, X is a proper subset of $X \cup Y$. This contradicts the hypothesis of maximality. □

Theorem 1. *Let G be a graph and $W \subseteq V$ be pure. Let $W_1, W_2, \ldots, W_m$ be the connected components of $\langle W \rangle$. Then for all $v \in V$, and $w, w' \in W_i \setminus v$, $i = 1, \ldots, m$ we have $d(v,w) = d(v,w')$.*

Proof. Note that, by Lemma 1, each W_i is a clique. Choose $v \in V, w, w' \in W_i \setminus v$, $i = 1, \ldots, m$. If $w = w'$ there is nothing to prove. Otherwise, w and w' are adjacent as W_i is a clique. Assume, by contradiction, that $d(v,w')$ and $d(v,w)$ are different, say $d(v,w') < d(v,w)$. Let $v = v_0 v_1 \ldots v_m = w'$ be a geodesic joining v and w'. The path $v_0v_1 \dots v_m\, w$ joins v and w and has length less or equal to $d(v,w)$. Then it is necessarily a geodesic and passes through w'. This contradicts the fact that W is pure. □

Corollary 1. *Let G be a graph. A connected subset W of V is pure if and only if for all $v \in V$, and $w, w' \in W \setminus v$ we have $d(v,w) = d(v,w')$.*

Proof. The "only if" part follows from Theorem 1. We assume now that the condition holds and prove that W is pure. Let $v \in V$ and $w \in W$ and consider a geodesic joining v and w. Suppose, by contradiction, that it meets W in another vertex, say w'; so it contains a geodesic from v to w', hence it is longer than $d(v,w')$. As $d(v,w') = d(v,w)$ in view of our assumption, this is a contradiction. This shows that W is pure. □

3 Extremal Values

In this section we investigate the extremal values that the betweenness of a vertex v can take. It is known that, for a graph G and for each $v \in V$:

$$0 \leq \beta_G(v) \leq \binom{n-1}{2}. \tag{1}$$

Clearly, if G is complete the minimum value is attained by all vertices of G. Likewise, all pendant vertices have betweenness zero. The minimum value is completely characterized by the following proposition:

Proposition 1. *Let $v \in V$. Then $\beta_G(v) = 0$ if and only if $\langle N(v) \rangle$ is a complete subgraph of G.*

Proof. Assume that $\langle N(v) \rangle$ is complete. Letting x, y be two vertices of G, different from v, consider any path from x to y passing through v, say: $x = x_1 \dots x_{k-1}x_k = vx_{k+1} \dots x_m = y$.

By the assumption on v, the vertices x_{k-1} and x_{k+1}, both adjacent to v, are adjacent to each other. So we can remove the vertex v, and we obtain again a path from x to y but it does not pass through v and its length is less than the previous one. It follows that a path from x to y passing through v cannot be of minimum length. Then $\beta_G(v) = 0$.

On the other hand, if the neighborhood of v is not complete, it contains two non-adjacent vertices x and y. Now x, v, y is a path joining x and y, whose length is necessarily minimum and whose contribution to the betweenness value is positive. Then $\beta_G(v) > 0$. □

Remark 1. This was essentially noticed in [6, Theorem 2], with a quite different terminology.

Let us now deal with the maximum value. Let G be of order n; then define the following set:

$$Star(G) = \{v \in V : d(v) = n-1\}. \tag{2}$$

The elements of $Star(G)$ are called *star vertices*.

Note that $Star(G)$, along with all its subsets, is always pure. This is a straightforward consequence of Corollary 1.

In a graph all star vertices have the same maximum betweenness (see [3, Theorem 3.2]); in particular for $K_{1,n-1}$, $Star(K_{1,n-1}) = \{v\}$, $\beta_{K_{1,n-1}}(v) = \binom{n-1}{2}$, and $\beta_{K_{1,n-1}}(w) = 0\ \forall w \in V \setminus v$.

The above results suggest that the betweenness of the vertices in $Star(G)$ decreases as its cardinality increases. This is true, as we show in the following:

Theorem 2. *Let G be a graph, and let H be a graph such that $V(G) = V(H)$; suppose that $E(G) \subseteq E(H)$, i.e. G is a spanning subgraph of H; if $Star(G) \neq \varnothing$ and $E(G) \subset E(H)$, then $\beta_G(v) > \beta_H(v)$, $v \in Star(G)$.*

Proof. First observe that $Star(G) \subseteq Star(H)$: this is obvious if $Star(G) = \varnothing$; when $Star(G) \neq \varnothing$, let $v \in Star(G)$; then the degree of v is $n-1$ in the graph H, so $v \in Star(H)$ as well.

Suppose now that $Star(G) \neq \varnothing$, then $\forall v \in Star(G)$, $\forall w \in V(G)$, $\beta_G(v) \geq \beta_G(w)$; also $Star(H) \neq \varnothing$ being G a spanning subgraph of H. The graph H is obtained from G by adding one (or more) edges. Being $v \in Star(G)$, the contribution of x, y in G is equal to zero only if $x \sim y$. So, letting xy be the new edge, its contribution to the betweenness of v, which in G is different from zero, vanishes in H, because x and y are adjacent in H.

Finally, let x, y two vertices that contribute to $\beta_G(v)$; these vertices contribute to $\beta_H(v)$; by adding new edges, it is possible that new geodesics from x to y appear, so that $\beta_G(v) > \beta_H(v)$. □

In Fig. 2, G is a graph with $n = 8$; $v_1 \in Star(G)$ and $\beta_G(v_1) = 3.087$. G is a spanning subgraph of H where $d(v_1) = d(v_2) = n-1$ and $\beta_H(v_1) = 1.827$.

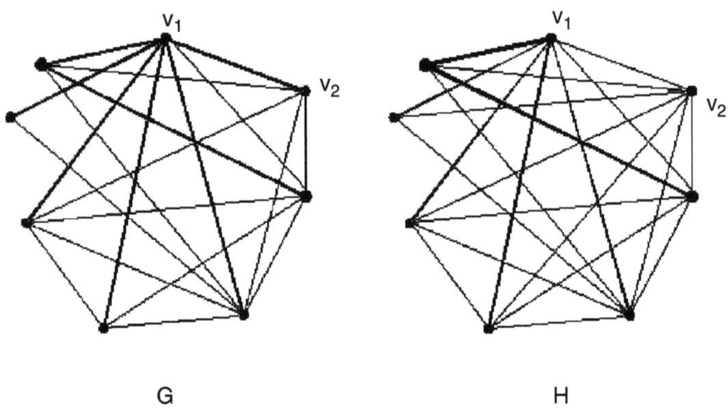

Fig. 2 The betweenness of $v_1 \in Star(G)$ decreases by adding edges to the node v_2

Remark 2. The previous result does not hold when $v \notin Star(G)$; in general, the betweenness does not necessarily decrease as the number of edges increases. For a graph G of order $n > 3$, let us consider for instance a pendant vertex v; $v \notin Star(G)$ and $\beta_G(v) = 0$, and by adding an edge to v its betweenness does not decrease, because $\beta_G(v) \geq 0$.

4 Total Betweenness Centrality

In this section we analyze the betweenness of subsets of vertices of V. The underlying idea is to study a natural extension of the concept of single-vertex betweenness centrality to a set of vertices. Letting $W \subseteq V$, we define the betweenness centrality of the set W *(total betweenness centrality)* as follows:

Definition 2. The total betweenness centrality of W is the sum of the betweenness values of its vertices, i.e. $\beta_G(W) = \sum_{w \in W} \beta_G(w)$.

The concept of group centrality was introduced in a sociological framework for various centrality measures earlier in [5]; we recall here the definition of group betweenness centrality. For all $u, v \in V$, let $g_{uv}(W)$ be the number of $u - v$ geodesics passing through at least one vertex of W.

Definition 3. The group betweenness centrality of W is

$$\alpha_G(W) = \sum_{u,v} \frac{g_{uv}(W)}{g_{uv}} \qquad u, v \in V \setminus W \tag{3}$$

where all pairs u, w appear only once in the sum.

Note that the word "group" is not used in the standard algebraic sense. The total betweenness is easy to compute and measures the whole extent of the centrality of the subset, including in the count the geodesics between vertices in the set. On the other hand, the group betweenness gives the betweenness of subset only referring to the paths leading to the *external vertices*, i.e. vertices not belonging to the subset.

The two concepts are related as follows:

Theorem 3. *For all $W \subseteq V$, we have $\alpha_G(W) \leq \beta_G(W)$. Moreover, $\alpha_G(W) = \beta_G(W)$ if and only if W is pure.*

Proof. Let u, v be vertices not in W. Every geodesic from u to v passing through W contributes to the betweenness of all its vertices in W, then each term $\frac{g_{uv}(W)}{g_{uv}}$ appears in $\beta_G(W)$. Hence the sum of all $\beta_G(w)$, $w \in W$, is larger or equal than $\alpha_G(W)$ and this proves the first part of the statement.

Assume now that W is pure. Consider a pair u, v of vertices outside W. Each geodesic u, v that meets W in a vertex w gives contribution 1 to the numerator of the corresponding summand in $\alpha_G(W)$, 1 to the numerator of $\beta_G(w)$ and 0 to all the $\beta_G(w')$ with w' different than w. Overall, the contribution of u,v to $\alpha_G(W)$ and $\beta_G(W)$ is the same for each w. Hence $\alpha_G(W) = \beta_G(W)$.

Finally, assume that $\alpha_G(W) = \beta_G(W)$; if W is not pure, consider a $w-v$ geodesic with $w \in W$ that meets W in a vertex w'; it gives a positive contribution to $\beta_G(W)$, but not to $\alpha_G(W)$. Then $\alpha_G(W) < \beta_G(W)$, and this is a contradiction. Hence W is pure. □

From Theorem 3 we see that the two measures coincide when the concerned subset is pure. In fact, in a pure set the contribution in the total betweenness of pairs of vertices in the set is zero being the connected components of a pure set, cliques (Lemma 1), hence with betweenness zero.

Proposition 2. *Let G be a graph and let W be a pure set in G. Let U be the set of all vertices of $v \in V \setminus W$ such that $d(v,w) = 1$ for some $w \in W$. If $\langle U \rangle$ is a clique, then the total betweenness of W is zero.*

Proof. Let $w \in W$. Consider a pair x,y of vertices of V that might give a positive contribution to $\beta(w)$. Since W is pure, we can assume that neither x nor y belongs to W. Let $x = x_1 \dots x_h w y_k \dots y_1 = y$ be a path joining x and y and passing through w. Since $x \notin W$, there exists i such that $x_i \notin W$ and $x_{i+1} \in W$. As x_{i+1} and w lie in the same connected component of W, from $d(x_i, x_{i+1}) = 1$ it follows that $d(x_i, w) = 1$ (Theorem 1). Hence $x_i \in U$. Likewise, letting j be such that $y_i \notin W$ and $y_{j+1} \in W$, we get $y_j \in U$. Since $\langle U \rangle$ is a clique, x_i and y_j are adjacent. Consequently, $x_1 \dots x_h w y_k \dots y_1$ is a path joining x and y. This proves that the original path was not a geodesic. Therefore $\beta(w) = 0$. □

Corollary 2. *Let G be a graph and let W be a pure set in G. If $\langle V \setminus W \rangle$ is a clique then the total betweenness of W is zero.*

Proof. Let U be as in Proposition 2; U is a subset of $V \setminus W$, hence a clique. Now Proposition 2 applies. □

5 Total Betweenness and Cutsets

Now we study the total betweenness of a cutset $W \subseteq V$, in order to obtain some useful lower bounds.

In Fig. 3 some graphs with pure and non-pure cutsets are depicted.

Remark 3. Let W be a pure cutset of G and let $A_1, A_2, \dots, A_k$ be the connected components of $G \setminus W$. If x,y are in different connected components, all geodesics joining x and y meet W in (only) one vertex; if x,y, both adjacent to $w \in W$, are in the same connected component but $x \not\sim y$, then the path xwy is a geodesic.

Observe that if a vertex w lies in a cutset W, w is not necessarily a cutvertex; if $W = \{w\}$, then w is a cutvertex, and in this case the set W is always a pure cutset. The following result extends [7, Lemma 2]:

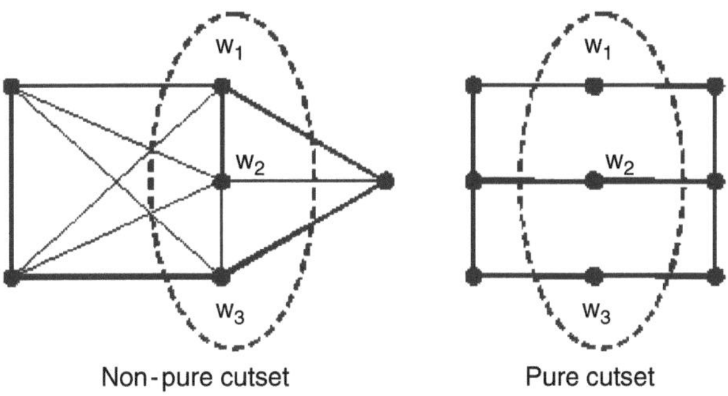

Fig. 3 Pure and non-pure cutsets

Theorem 4. *Let W be a cutset of G and let $A_1, A_2, \ldots, A_k$ be the connected components of $G \setminus W$, with $a_i = |V(A_i)|$. Then*

$$\beta_G(W) \geq \sum_{i<j} a_i a_j . \tag{4}$$

Equality holds if and only if W is a pure cutset, and $\forall w \in W$ and $\forall i = 1, \ldots, k$, the vertices in A_i adjacent to w are pairwise adjacent.

Proof. Let $x \in A_i$, $y \in A_j$ with $i \neq j$. Being W a cutset, all geodesics joining x and y meet W in a vertex w, so the pair x, y gives to $\beta_G(w)$ a contribution at least equal to 1. As x, y run over the above sets, the total amount of their contribution to $\beta_G(W)$ is at least $a_i a_j$. Summing up over all pairs i, j, we have that $\beta_G(W)$ is at least $\sum_{i<j} a_i a_j$, then inequality (4) follows.

We prove the second part of the statement by contradiction; let us fix w and i such that the neighborhood of w $\langle A_i \cup w \rangle$ is not a complete graph; there are x and y adjacent to w but not to each other. The pair x, y gives a positive contribution to $\beta_G(w)$, since xwy is a geodesic. By the above counting argument, this contribution to $\beta_G(W)$ was not included in the sum in the right-hand side of (4), therefore, inequality (4) holds with sign $>$. Likewise, if $\forall w \in W$ and $\forall i = 1, \ldots, k$, the vertices in A_i adjacent to w are pairwise adjacent, no further contributions to $\beta_G(W)$ can be found other than those already counted in the sum at the right-hand side of (4), because W is a pure cutset. This completes the proof. □

A particular case occurs when $W = \{v\}$ and v is a cutvertex such that $d(v) = k$.

From Theorem 4, in a pure set, being zero the contribution in the total betweenness of the pairs of vertices in the set, the total betweenness reaches its lower bound and equals the cutting number.

Corollary 3. *Let G be a graph, let v be a cutvertex of G and let $A_1, \ldots, A_k$ be the connected components of $G \setminus v$. If $k = d(v)$ then $\beta_G(v) = \sum_{i<j} a_i a_j$.*

Proof. Being v a cutvertex, $W = \{v\}$ is a pure cutset of G; since $d(v) = k$, in each $A_i \cup \{v\}$ the vertex v has degree 1, hence its neighborhood is a complete graph, so the inequality (4) of Theorem 4 is satisfied as equality. □

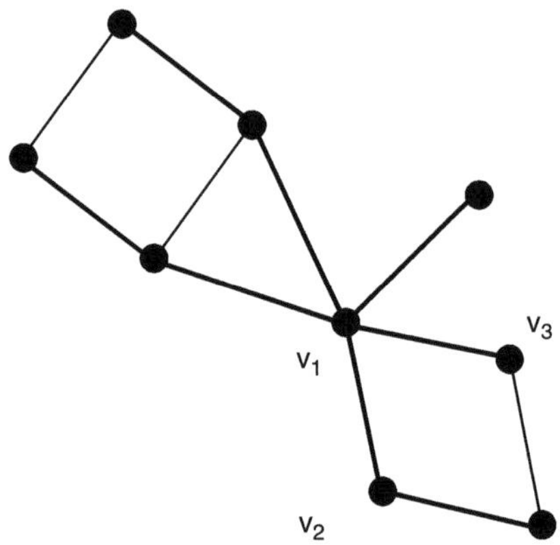

Fig. 4 v_1 is a cutvertex and $\beta(v_1) > \gamma(v_1)$

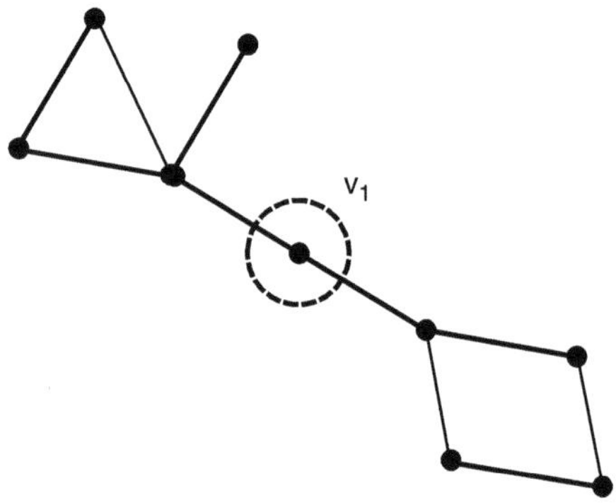

Fig. 5 v_1 is a cutvertex and $\beta(v_1) > \gamma(v_1)$

Remark 4. Observe that in this case $\sum_{i<j} a_i a_j$ is the cutting number $\gamma(v)$ of the vertex v, i.e. the number of pairs $\{u,w\}$ such that u and w are in different connected components of $G \setminus v$ (see [9]); from Corollary 3 we deduce that the cutting number is a lower bound for the betweenness of v.

Remark 5. Every cutvertex such that $d(v) = k$ is adjacent to a pendant vertex, or to a cutvertex.

Example 1. Let us consider the graph G in Fig. 4; in this case $W = \{v_1\}$ and $k = 3$; there is a pair of vertices (v_2 and v_3) belonging to the same connected component of $G \setminus W$, that contributes to the betweenness of v_1; so $\beta_G(v_1) = 19.488 > \sum_{i<j} a_i a_j = 19$.

Example 2. Let us consider the graph in Fig. 5; the vertex v_1 is a cutvertex, $k = d(v_1) = 2$. In this case $\beta_G(v_1) = 16 = \sum_{i<j} a_i a_j$.

Theorem 5. *Let G be a graph of order n, let $v \in V$ be a cutvertex of G and let $A_1, A_2, \ldots, A_k$ be the connected components of $G \setminus v$ of cardinalities $a_1, a_2, \ldots, a_k$, respectively, where $k = 2, 3, \ldots, n-1$. Let r be the remainder of the division of $n-1$ by k. If $\beta_G(v) = \sum_{\substack{i<j \\ i,j=1,\ldots,k}} a_i a_j$ then:*

$$\beta_G(v) \le \binom{k-r}{2}\left(\frac{(n-1)-r}{k}\right)^2 + \binom{r}{2}\left(\frac{(n-1)-r}{k}+1\right)^2 \tag{5}$$

$$+r(k-r)\frac{(n-1)-r}{k}\left(\frac{(n-1)-r}{k}+1\right),$$

where the equality holds if and only if all the cardinalities a_i $(i=2,\dots,k)$ differ by 1 at most. In particular, there are $k-r$ $(k>r)$ components of the same cardinality and r components that differ by 1 at most.

Proof. Let $\sum_{i=1}^{k} a_i = n-1$; observe that $n-1 = kp+r$, $0 \le r < k$. From the following equality:

$$\sum_{\substack{i<j \\ i,j=1,\dots,k}} a_i a_j = \sum_{\substack{i<j \\ i,j=1,\dots,k}} \left[\left(\frac{a_i+a_j}{2}\right)^2 - \left(\frac{a_i-a_j}{2}\right)^2\right] \tag{6}$$

$$= \sum_{\substack{i<j \\ i,j=1,\dots,k}} \left(\frac{a_i+a_j}{2}\right)^2 - \sum_{\substack{i<j \\ i,j=1,\dots,k}} \left(\frac{a_i-a_j}{2}\right)^2$$

$\sum_{\substack{i<j \\ i,j=1,\dots,k}} a_i a_j$ attains its maximum value whenever $\sum_{\substack{i<j \\ i,j=1,\dots,k}} \left(\frac{a_i-a_j}{2}\right)^2$ attains its minimum. We distinguish two cases:

1. The a_i's are all equal for every $i=1,\dots,k$. In this case $a_1 = a_2 = \dots = a_k = p$, and $n = kp+1$, so we obtain that $p = \frac{n-1}{k}$; furthermore $\sum_{\substack{i<j \\ i,j=1,\dots,k}} \left(\frac{a_i-a_j}{2}\right)^2 = 0$ and:

$$\sum_{\substack{i<j \\ i,j=1,\dots,k}} a_i a_j = \sum_{\substack{i<j \\ i,j=1,\dots,k}} \left(\frac{a_i+a_j}{2}\right)^2 = \sum_{i<j}\left(\frac{2p}{2}\right)^2 \tag{7}$$

$$= \binom{k}{2}\left(\frac{n-1}{k}\right)^2 = \frac{k(k-1)}{2}\left(\frac{n-1}{k}\right)^2$$

$$= \frac{(n-1)^2}{2}\cdot\frac{k-1}{k}$$

that is formula (5) with $r=0$.

2. The a_i's are not all equal, then at least one pair $\{a_i, a_j\}$ exists such that $a_i = a_j + h$, where h is an integer ≥ 1; in this case the addendum

$$\left(\frac{a_i-a_j}{2}\right)^2 = \left(\frac{a_j+h-a_j}{2}\right)^2 = \frac{h^2}{4}$$

is minimum only for $h = 1$, so a_i differs from a_j by 1. In particular, if r is the number of components with cardinalities all equal to $p+1$, then $k-r$ are the components with cardinalities p, with $p = \frac{n-1-r}{k}$.
From (6) we have that

$$\sum_{\substack{i<j\\ i,j=1,\dots,k}} a_i a_j = \sum_{i<j}\left(\frac{2p}{2}\right)^2 + \sum_{i<j}\left(\frac{2p+2}{2}\right)^2 + \sum_{i<j}\left(\frac{2p+1}{2}\right)^2 - \sum_{i<j}\left(\frac{p-p-1}{2}\right)^2$$
$$= \binom{k-r}{2} p^2 + \binom{r}{2}(p+1)^2 + (k-r)r\,(p^2+p)$$
$$= \binom{k-r}{2}\left(\frac{n-1-r}{k}\right)^2 + \binom{r}{2}\left(\frac{n-1-r}{k}+1\right)^2$$
$$+(k-r)r\left(\frac{n-1-r}{k}\right)\left(\frac{n-1-r}{k}+1\right)$$

□

As a consequence of Theorem 5 the betweenness $\beta_G(v)$ attains its maximum value $\frac{(n-1)^2}{2}\cdot\frac{k-1}{k}$ only when the a_i's are all equal.

Corollary 4. *Under the hypothesis of Theorem 5, for equal a_i the betweenness $\beta_G(v)$ increases as k increases and the maximum value $\beta_G(v) = \binom{n-1}{2}$ is attained being v the center of the star graph.*

Proof. From Theorem 5, for a_i all equal $\beta_G(v) = \frac{(n-1)^2}{2}\cdot\frac{k-1}{k}$. The statement follows immediately by observing that the sequence $\frac{k-1}{k}, k = 2,\dots,n-1$ is strictly increasing; the maximum value is attained when $k = n-1$, i.e. when v is the center of the star graph. □

Example 3. Let us consider the graph G in Fig. 6; v is a cutvertex, $k = 4$ and $r = 2$; $\gamma(v) = 73 = \beta_G(v)$. According to Theorem 5 we have

$$\binom{k-r}{2}\left(\frac{(n-1)-r}{k}\right)^2 + \binom{r}{2}\left(\frac{(n-1)-r}{k}+1\right)^2$$
$$+r(k-r)\frac{(n-1)-r}{k}\left(\frac{(n-1)-r}{k}+1\right) = 73\,. \tag{8}$$

When G is a tree every vertex v is either a pendant vertex or a cutvertex, and $d(v) = k$, so for a tree we can establish that $\beta_G(v) = \sum_{i<j} a_i a_j$ and inequality (5) of Theorem 5 holds.

Moreover, this result implies:

Corollary 5. *Let G be a tree and $v \in V$ a vertex adjacent only to pendant vertices (except at most one) then:*

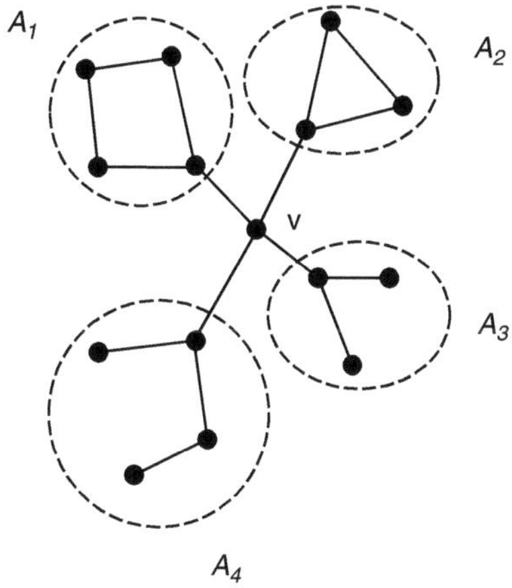

Fig. 6 Removing v the graph has four connected components

$$\beta_G(v) = \begin{cases} (n-d(v))(d(v)-1) + \binom{d(v)-1}{2} & \text{if } d(v) > 2 \\ (n-d(v))(d(v)-1) & \text{if } d(v) = 2 \end{cases} . \quad (9)$$

Proof. If v is adjacent only to pendant vertices, then $a_1 = a_2 = \cdots = a_k = 1$, where $k = d(v) = n-1$ (i.e. $G = K_{1,n-1}$) and from the formula:

$$\begin{aligned}(n-d(v))(d(v)-1) + \binom{d(v)-1}{2} &= (n-n+1)(n-2) + \binom{n-2}{2} \\ &= (n-2) + \frac{(n-2)!}{2(n-4)!} = (n-2) + \frac{(n-2)(n-3)}{2} \\ &= \frac{(n-2)(n-1)}{2} = \beta_G(v)\end{aligned}$$

that is exactly the betweenness value of the vertex v.

Let's now consider the case when v is again adjacent only to pendant vertices, and also to a unique vertex w, of degree $d(w) > 1$, let A_1 be the component where w lies. If G has n vertices we have $a_1 = n - d(v)$ and $a_2 = a_3 = \cdots = a_k = 1$ (here k is equal to $d(v)$) Then the pairs of vertices $w_i, w_j \neq v$, with both $w_i, w_j \notin A_1$ give contribution 1 to $\beta_G(v)$; there are $\binom{d(v)-1}{2}$ such pairs. Moreover, for every vertex $u_i \in A_1$, the pairs u_i, $w_i \neq v$ each give contribution $n-d(v)$ to $\beta_G(v)$; there are $d(v)-1$ such pairs. It follows that

$$\beta_G(v) = (n-d(v))(d(v)-1) + \binom{d(v)-1}{2}.$$

□

Example 4. Let us consider the graph G in Fig. 7. The betweenness values, computed via the formula of Corollary 5 are shown in Table 1.

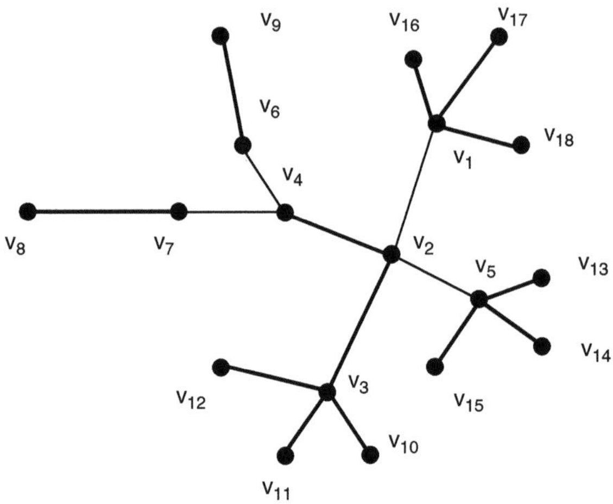

Fig. 7 A tree whose betweenness values are shown in Table 1

Table 1 Total betweenness values computed via the formula of Corollary 5

v	$d(v)$	$\beta_G(v)$
v_1	4	45
v_2	4	108
v_3	4	45
v_4	3	56
v_5	4	45
v_6	2	16
v_7	2	16
$v_8, \ldots, v_{18}$	1	0

6 Conclusions

In this paper we concentrate on betweenness centrality, referring to both vertex and subset centrality. We believe that the new concept we introduced, the total betweenness centrality, can help in clarifying the role of subsets in a graph. This can be interesting from both a practical and a theoretical point of view. Further research can be done on the cutsets and more in general on the relationship among total betweenness, group betweenness and the cutting number.

Acknowledgements Thanks to Giovanni Zambruno for useful comments and careful reading. We would like to thank the referee for his helpful comments. All errors are our responsibility.

References

1. Borgatti S (2005) Centrality and network flow. Social Network 27:55–71
2. Buckley F, Harary F (1990) Distance in graphs. Addison-Wesley, Reading, MA
3. Grassi R, Stefani S, Torriero A (2006) Central vertices in networks: a unified approach. Rapporto di Ricerca del Dipartimento di Metodi Quantitativi per le Scienze Economiche ed Aziendali n. 105
4. Grassi R, Stefani S, Torriero A (2007) Some new results for eigenvector centrality. J Math Sociol 31:237–248
5. Everett MG, Borgatti SP (1999) The centrality of groups and classes. J Math Sociol 23: 181–201
6. Everett MG, Seidman SB (1985) The Hull Number of a graph. Discrete Math 97:217–223
7. Everett MG, Sinclair P, Dankelmann P (2004) Some centrality results new and old. J Math Sociol 28:215–227
8. Freeman LC (1979) Centrality in social networks: conceptual clarifications. Social networks 1:215–239
9. Harary F, Ostrand P (1971) The cutting center theorem for trees. Discrete Math 1:7–18
10. Li L, Alderson D, Tanaka R, Doyle JC, Willinger W (2005) Toward a theory of scale-free graphs: definition, properties and implications (extended version). arXiv:cond-mat/0501169v2 18 Oct 2005
11. Nieminen J, Peltola M (1999) The subtree center of a tree. Networks 34:272–278
12. Papendieck B, Recht P (2000) On maximal entries in the principal eigenvector of graph. Linear Algebra and its Applications 310:129–138

How to Reduce Unnecessary Noise in Targeted Networks

Giacomo Aletti and Diane Saada

Abstract This work is a review of previous works on the stopping laws in networks. Among other results, we show a non combinatorial method to compute the stopping law, the existence of a minimal Markov chain without oversized information, the existence of a polynomial algorithm which projects the Markov chain onto the minimal Markov chain. Several applied examples are presented.

1 Introduction

Consider a finite state Markov chain, with state space E. The process is stopped when it reaches a sub-class T of E. It turns out that one does not need the whole information carried by its transition matrix in order to compute the law of reaching this class. The following paper, which is a compilation of several papers, deals with how to reduce this unnecessary information, first in real time and then for large times. An extension is also given for R-networks in [3]. The problem of finding general closed-forms for different kinds of waiting problems is widely studied, following various approaches. See, for example, [9] in the case of Bernoulli trials, [6], [10], [1] and [12] for its extensions to Markov-dependent trials, and [13] for another methodology. A new approach was given in [2–5], where it was proved that there exists an optimal projection for any given Markov problem which leaves the probability of reaching the target set unchanged. A simple ε-approximation of this projection exists, provided the system has evolved for a sufficient amount of time

G. Aletti
ADAMSS and Dipartimento di Matematica, Università di Milano, 20133 Milan, Italy
giacomo.aletti@unimi.it

D. Saada
Department of Statistics, Hebrew University, Jerusalem, Israel
dianes@mscc.huji.ac.il

A.K. Naimzada et al. (eds.), *Networks, Topology and Dynamics*,
Lecture Notes in Economics and Mathematical Systems 613,

and some conditions on the Markov chain are satisfied. In the framework of [2–4], a compatible projection is an equivalence relationship S on the indexing set E s.t.:

- $\forall e_i \in T,\ e_i S e_j \iff e_j \in T$.
- For any $\{e_i, e_j, e_k\} \subseteq E$: $e_i S e_j$, we have

$$\sum_{e_l S e_k} P(e_i, e_l) = \sum_{e_l S e_k} P(e_j, e_l),$$

  where T (the absorbing target class) and $P : E \times E \to R$ are given (P is the Markov matrix of the network and R is a semiring, see [3]).

In [2, 3] it was proved the existence of a polynomial-time algorithm which reaches the minimum Markov network. In [5], the question of a further reduction is posed, when time tends to infinity. An asymptotic conditional law of exit will exist, according to the shape of the transition submatrix which corresponds to the states leading to the target class. The methodology is based on spectral theory for non-negative matrices and in particular on the Perron–Frobenius theorem. The framework in [2–5] regards a huge class of problems which occur in many real situations. We recall here how this class of problem may appear:

1. In *finance* the filter rule for trading is a special case of the Markov chain stopping rule suggested in [4] (see, e.g., [11]).
2. "When enough is enough!" for example, an insured has an accident only occasionally in a while. How many accidents in a specified number of years should be used as a stopping time for the insured (in other words, when it should be discontinued the insurance contract).
3. *State dependent Markov chains.* Namely, the transition probabilities are given in terms of the history. In many situations, the matrix of the embedded problem may be reduced.
4. *Medical sciences.* Given that the length of a menstrual cycle has a known distribution, what is the probability that the length of a woman's menstrual cycle is the same three consecutive times?
5. *Small-world Networks.* Given one of the networks as in Fig. 1 (either as Markov network or as a graph), is it possible to reduce it in polynomial time and to preserve the law of reaching a given absorbing state?

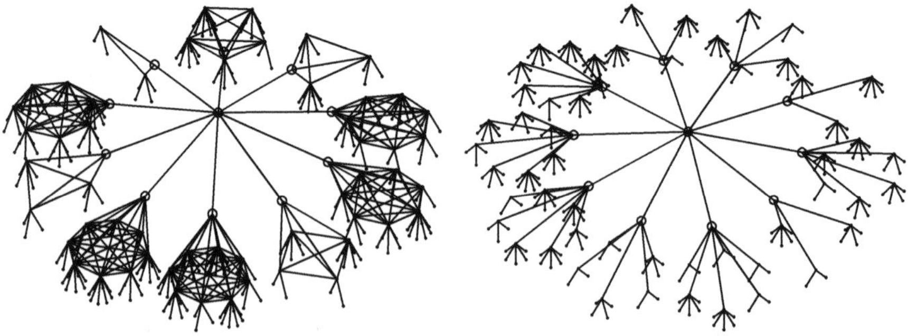

Fig. 1 Networks that may be compressed (see [4])

There are of course many other such examples (e.g., records: Arnold et al. [7] and optimization: Cairoli and Dalang [8]).

1.1 A Combinatorial Problem

Let $X = \{X\tau, \tau \in \mathbb{N}\}$ be a Markov chain on a finite state space $E = \{e_1, \ldots, e_n\}$:

$$\begin{array}{c|ccc} & e_1 & \ldots & e_n \\ \hline e_1 & p_{1,1} & \cdots & p_{1,n} \\ \vdots & \vdots & \ddots & \vdots \\ e_1 & p_{n,1} & \cdots & p_{n,n} \end{array} \left.\vphantom{\begin{array}{c} e_1 \\ \vdots \\ e_1 \end{array}}\right\} =: P\,.$$

The process is stopped when it reaches one of some given states $T := (e_{n_i})_{i=1}^{k} \subseteq E$. For sake of clarity, we suppose $F = \{e_1, \ldots, e_k\}$ and $E \setminus T = \{e_{k+1}, \ldots, e_n\}$. To compute the law of stopping, we may consider a new Markov chain $X' = \{X'\tau, \tau \in \mathbb{N}\}$ on $T \cup (E \setminus T)$:

$$\begin{array}{cc|ccccccc} & & & T & & & E\setminus T & \\ & & e_1 & \ldots & e_k & e_{k+1} & \ldots & e_n \\ \hline & e_1 & \mathbf{1} & \ldots & \mathbf{0} & 0 & \ldots & 0 \\ T & \vdots & \vdots & \ddots & \vdots & \vdots & \ddots & \vdots \\ & e_k & \mathbf{0} & \ldots & \mathbf{1} & 0 & \ldots & 0 \\ & e_{k+1} & p_{k+1,1} & \cdots & p_{k+1,k} & p_{k+1,k+1} & \cdots & p_{k+1,n} \\ E\setminus T & \vdots & \vdots & \ddots & \vdots & \vdots & \ddots & \vdots \\ & e_n & p_{n,1} & \cdots & p_{n,k} & p_{n,k+1} & \cdots & p_{n,n} \end{array} \Bigg\} =: \widehat{P}\,.$$

Thus, the probability of reaching T by time τ is reduced to the computation of the τ-th power of $\widehat{P}$:

$$\mathbb{P}(\cup_{i=1}\tau\{X_i \in T\}) = \mathbb{P}(\{X'\tau \in T\}) = \overbrace{(p_1^0, \ldots, p_n^0)}^{\mathbf{p_0}} \quad (\widehat{P})^{\tau} \quad (\overbrace{1, \ldots, 1}^{k \text{ terms}}, \overbrace{0, \ldots, 0}^{n-k \text{ terms}})'\,. \tag{1}$$

There exists a trivial reduction which preserves the above calculation for any $\tau \in \mathbb{N}$ and initial distribution $\mathbf{p_0}$:

$$\mathbb{P}(\cup_{i=1}\tau\{X_i \in T\}) =$$

$$\left(\sum_{i=1}^{k} p_i^0, p_{k+1}^0, \ldots, p_n^0\right) \begin{pmatrix} \mathbf{1} & 0 & \ldots & 0 \\ \sum_{i=1}^{k} p_{k+1,i} & p_{k+1,k+1} & \cdots & p_{k+1,n} \\ \vdots & \vdots & \ddots & \vdots \\ \sum_{i=1}^{k} p_{n,i} & p_{n,k+1} & \cdots & p_{n,n} \end{pmatrix}^{\tau} \begin{pmatrix} 1 \\ 0 \\ \vdots \\ 0 \end{pmatrix}. \tag{2}$$

2 Target Problems

We begin the mathematical framework in this section with an example. Suppose (2) is written in the following way

$$\left(p_1^0, p_2^0, p_3^0, p_4^0, p_5^0\right) \begin{pmatrix} \mathbf{1} & 0 & 0 & 0 & 0 \\ \frac{1}{2} & \frac{1}{6} & \frac{1}{6} & \frac{1}{12} & \frac{1}{12} \\ \frac{1}{2} & \frac{1}{4} & \frac{1}{12} & 0 & \frac{1}{6} \\ 0 & \frac{1}{2} & \frac{1}{4} & \frac{1}{8} & \frac{1}{8} \\ 0 & \frac{3}{8} & \frac{3}{8} & \frac{1}{4} & 0 \end{pmatrix}^{\tau} \begin{pmatrix} 1 \\ 0 \\ 0 \\ 0 \\ 0 \end{pmatrix} . \tag{3}$$

We define a problem of compression of redundant information, in terms of equivalence relationships on E (see [2–4]). First, we extend P to $\mathbb{P}_P$ in the classical way

$$\mathbb{P}_P : E \times \mathfrak{P}(E) \to \mathbb{R}_+ \cup \{0\},$$

where $\mathfrak{P}(E)$ is the set of subsets of E and $\mathbb{P}_P(e,A) = \sum_{e_i \in A} P(e,e_i)$. Obviously, for each $e \in E$, $\mathbb{P}_P(e,\cdot)$ is a probability on $(E,\mathfrak{P}(E))$ that gives the conditional probability of reaching $\cdot$ given that we are in the state e.

As stated in the introduction, a compatible projection is an equivalence relationship S on the indexing set E s.t.:

1. $\forall e_i \in T,\, e_i S e_j \iff e_j \in T$.
2. For any $\{e_i, e_j, e_k\} \subseteq E$: $e_i S e_j$, we have

$$\sum_{e_l S e_k} P(e_i, e_l) = \sum_{e_l S e_k} P(e_j, e_l).$$

1 and 2 are satisfied if and only if the matrix $P^* : E/S \times E/S \to \mathbb{R}_+ \cup \{0\}$ such that the following diagram commutes, is well-defined:[1]

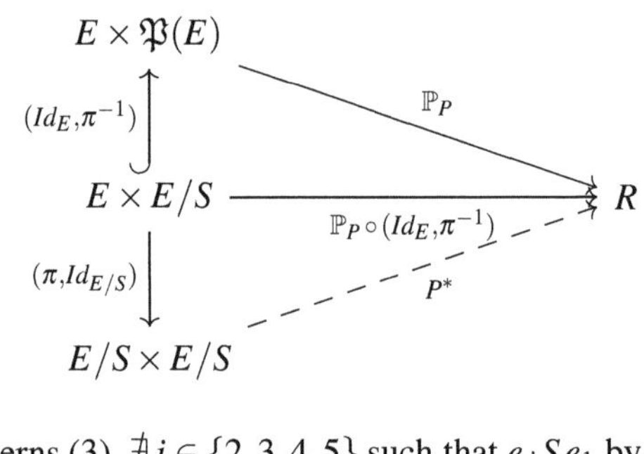

(4)

For what concerns (3), $\nexists j \in \{2,3,4,5\}$ such that $e_j S e_1$ by 1. Note that finding a nontrivial projection is not a local search. For example, we have $P(e_4,e_2) \neq P(e_5,e_2)$ but $P(e_4,e_2) + P(e_4,e_3) = P(e_5,e_2) + P(e_5,e_3)$, which means that $e_4 S e_5$ may be found if *we know* that $e_2 S e_3$.

[1] Here, $\pi : E \to E/S$ is the canonical projection.

Now, a nontrivial projection is given by

$$E/S = \{\underbrace{T = \{e_1\}}_{f_1}, \underbrace{\{e_2, e_3\}}_{f_2}, \underbrace{\{e_4, e_5\}}_{f_3}\}.$$

Accordingly, the new matrix, associated with the projected states $E/S = \{f_1, f_2, f_3\}$, is given by

$$\left.\begin{array}{c|ccc} & f_1 & f_2 & f_3 \\ \hline f_1 & 1 & 0 & 0 \\ f_2 & \frac{1}{2} & \frac{1}{3} & \frac{1}{6} \\ f_2 & 0 & \frac{3}{4} & \frac{1}{4} \end{array}\right\} =: P^* .$$

The new Markov problem $(\{f_1, f_2, f_3\}, P^*)$ carries all the necessary information for the target problem. In fact, the states in f_i play all together with respect to the target, like if they were the same point. For example, problem (3) becomes

$$\left(p_1^0, p_2^0 + p_3^0, p_4^0 + p_5^0\right) \begin{pmatrix} 1 & 0 & 0 \\ \frac{1}{2} & \frac{1}{3} & \frac{1}{6} \\ 0 & \frac{3}{4} & \frac{1}{4} \end{pmatrix}^{\tau} \begin{pmatrix} 1 \\ 0 \\ 0 \end{pmatrix} . \tag{5}$$

In the general case, we deal with a (at most) countable indexing set E (in (3), $E = \{e_1, \ldots, e_5\}$). We then take a suitable space $\mathcal{M}_E$ of matrices on E (again, for what concerns (3), $\mathcal{M}_E$ is the space of 5×5 stochastic matrices); in general it will be a monoid $(\mathcal{M}_E, \cdot)$ where $\cdot$ can be seen as the matrix multiplication. The function $\mathbb{P}$, as defined above, is assumed to, and plays the role of the conditional probability of reaching any family of states starting from a given state. A target set $T \subseteq E$ is fixed, and it is assumed to be an absorbing class. A target problem is therefore a triple (E, P, T) where:

- E is an, at most countable, indexing set.
- $P \in \mathcal{M}_E$ is a given matrix and $\mathbb{P}_P$ is well-defined.
- T is an absorbing class: $\mathbb{P}_P(t, T) = 1$ and $\mathbb{P}_P(t, E \setminus T) = 0$, for any $t \in T$.

We are dealing here with Markov matrices only, and we leave more general extensions to [3]. In [3], an extended framework includes graph connection, as well. Note that one may always change the matrix P as in (2) so that T is an absorbing class, by defining

$$\widehat{P}(e_1, e_2) = \begin{cases} P(e_1, e_2), & \text{if } e_1 \notin T; \\ 1, & \text{if } e_1 \in T \text{ and } e_1 = e_2; \\ 0, & \text{if } e_1 \in T \text{ and } e_1 \neq e_2. \end{cases}$$

Let $\widetilde{E}$ be the set of all equivalence relations on E. Let $V, S \in \widetilde{E}$. We say that $V \preceq S$ if $a_1 V a_2$ implies $a_1 S a_2$ (if you think E as the set of all men and V is "belonging to the same state" while S is "belonging to the same continent", then $V \preceq S$). An equivalence relationship $S \in \widetilde{E}$ is called *compatible projection with respect to the target problem* (E, P, T) if:

1. $\forall e \in T, eSe_j \iff e_j \in T$ (i.e., the target set defines an equivalence class).
2. There exists $P^* \in \mathcal{M}_{E/S}$ such that (4) commutes.

We call $\mathcal{S} = \mathcal{S}(E,P,T)$ the set of all compatible projections.

The previous definition of compatible projection states when it is possible to project our target problem (E,P,T) into the smaller one $(E/S,P^*,t = \pi(T))$, without loosing necessary information (see [4]). In this framework, we can state the following general result.

Theorem 1 ([3, 4]). *For any target problem (E,P,T), there exists the optimal projection: $\exists S \in \mathcal{S}$ s.t. $V \preceq S$, $\forall V \in \mathcal{S}$.*

For example, note that the compatible projection $S \in \mathcal{S}$ which projects (3) into (5) is optimal. In fact, suppose there exists $S^* \in \widetilde{E}$ such that (a) $\{f_1\} \in E/S^*$, by 1 above; (b) $S \precneqq S^*$. (a) and (b) imply $E/S^* = \{\{e_1\},\{e_2,e_3,e_4,e_5\}\}$, which is not a compatible projection, since $1/2 = \mathbb{P}_P(e_2,\{e_1\}) \neq \mathbb{P}_P(e_4,\{e_1\}) = 0$.

2.1 Target Algorithm

The proof for the existence of the optimal solution was based in [3, 4] on the fact that the set of compatible projections $\mathcal{S}$ has its $\preceq$-join in $\widetilde{S}$.

This proof is useless in practice when the Markov chain is so big that a search in $\widetilde{E}$ can be impracticable. In fact, as stated in the previous section, searching for a compressing map is not a local search and it appears as a non-polynomial search, in the sense that we have to look at the whole set of equivalent relations on E. In [2, 3] it was proved the existence of a polynomial algorithm which reaches the optimal projection. We give here the algorithm and we state this result in Theorem 2. The idea is to reach the optimal projection E/S – unknown – starting from a trivial and known relation $M_E \in \widetilde{E}$, given by the problem. $M_E \in \widetilde{E}$ is defined by the relation "being or not a member of T": for any $(e_i,e_j) \in E \times E$,

$$e_i M_E e_j \iff \{e_i,e_j\} \subseteq T \text{ or } \{e_i,e_j\} \subseteq (E \setminus T).$$

It is clear that M_E is not in general a compatible projection (see, for example, S^* at the end of the previous section). By definition, it is obvious that $S \preceq M_E$, $\forall S \in \mathcal{S}$ and hence, if M_E is compatible, then it is optimal.

We denote here by $F\pi$ the optimal equivalence unknown map, and we build a monotone operator $\mathcal{F}$ on $\widetilde{E}$ which will reach $E/F\pi$ starting from E/M_E. $\mathcal{F} : \widetilde{E} \to \widetilde{E}$ is defined as follows: for any $S \in \widetilde{E}$, let $s_1, s_2, \ldots$ be the classes of equivalence of E induced by S. For any $(e_l,e_k) \in E \times E$, define

$$e_l \mathcal{F}_{s_i} e_k \iff \mathbb{P}_P(e_l,s_i) = \mathbb{P}_P(e_k,s_i)$$
$$\mathcal{F}(S) = \bigcap_{i=1,2,\ldots} \mathcal{F}_{s_i} \cap S.$$

$\mathcal{F}(S)$ is a new equivalence relation, that defines, in consequence, new classes of equivalence. Two states belong to the same new class if they have the same behavior towards the classes of S. If this step does not define the compatible projection, then we go a step further applying $\mathcal{F}$ to the classes of $\mathcal{F}(S)$. For example, take $E/M_E = \{\{1\},\{2,3,4,5\}\}$ as in (3). For $s_1 = \{1\}$ we have

$$\mathbb{P}(1,\{1\}) = 1\,, \quad \mathbb{P}(2,\{1\}) = \mathbb{P}(3,\{1\}) = 1/2\,, \quad \mathbb{P}(4,\{1\}) = \mathbb{P}(5,\{1\}) = 0$$

and hence $E/\mathcal{F}(s_1) = \{\{1\},\{2,3\},\{4,5\}\}$. Note that $E/\mathcal{F}(s_1) = E/\mathcal{F}(s_2) \subseteq E/M_E$ and hence $E/\mathcal{F}(M_E) = \{\{1\},\{2,3\},\{4,5\}\}$. The new relationship $\mathcal{F}(M_E)$ is a fixed point for $\mathcal{F}$ and it is also the optimal relationship.

This leads to the following theorem.

Theorem 2 ([2, 3]). *For any target problem (E,T,P), there exists $m = m(E,T,P)$ s.t. $m \leq N-2$ and*

$$F\pi = \underbrace{\mathcal{F} \circ \mathcal{F} \circ \cdots \circ \mathcal{F}}_{m \text{ times}}(M_E)\,,$$

where N is the cardinality of $E/F\pi$.

For what concerns the problem (3), we already knew that $|E/F\pi| = 3$ and hence $m \leq 1$ by Theorem 2. We have indeed noted that $\mathcal{F}(\mathcal{F}(M_E)) = \mathcal{F}(M_E)$. In fact, in the proof (see [2, 3]), it is also shown that $F\pi$ is the unique fixed point for a suitable restriction of $\mathcal{F}$ and this algorithm "works" on this restriction.

Remark 1. Note that the operator $\mathcal{F}$ may be computed in a $|E|$-polynomial time. Theorem 2 ensures that

$$\underbrace{\mathcal{F} \circ \mathcal{F} \circ \cdots \circ \mathcal{F}}_{\text{at most } |E/F\pi| - 2 \text{ times } (\leq |E|)}$$

will reach F, given any triple (E,T,P). A MATLAB version of such an algorithm for multitarget $\mathbf{T}$ may be downloaded at http://www.mat.unimi.it/ aletti.

3 Large Time Projections

Suppose now that the Markov chain $\{X\tau, \tau \in \mathbb{N}\}$ is stationary on a finite set E and denote by P its transition matrix and by μ_0, the initial probability measure. Let A be the class of transient states which lead to T (the target class), and let $T_\infty = \{i \in E; P\tau_{ij} = 0, \forall j \in T, \forall \tau\}$ the class of the remaining states. As before, the transition matrix can be decomposed as follows:

$$P = \left(\begin{array}{cc|c} 1 & 0 & \\ 0 & 1 & 0 \\ \hline & & \\ v_\infty & v & A \end{array} \right),$$

v_∞ (resp. v) is the vector of probabilities of hitting T_∞ (resp. T) from A, and A is the sub-matrix of the states lying in A. We suppose that $A \geq 0$ -that is, if $A = (a_{ij})_{(i,j)\in\{1,\cdots,n\}^2}$ is a square matrix, then every a_{ij} is nonnegative- and that $A \neq 0$. This matrix also satisfies lim $\tau A \tau = 0$. Here, $\underline{1}$ denotes the vector whose components are all equal to one. In order to know whether an asymptotic reduction can be done or not, we had to check whether the following limit exists

$$\lim_{\tau\to\infty} \mathbb{P}(X_{\tau+1} \in T | X\tau \in A) \quad \text{for any } \mu_0$$

and under which conditions this limit is independent of μ_0. It is easy to show that this limit is in fact equal to

$$\lim_{\tau\to\infty} \frac{\mu_{0|A}^T A \tau v}{\mu_{0|A}^T A \tau \underline{1}}.$$

$\mu_{0|A}$ is the trace on A of the initial probability μ_0. We suppose in the following that $\mu_{0|A} > 0$. We can decompose A into disjoint classes of communication, where i communicates with j if $i = j$ or if i leads to j and vice versa, and we obtain $A = \cup_{i=1}^N C_i^*$. We denote by $p(i)$ the period of state i. We recall that $p(i)$ is defined as the greatest common divisor of all integers $n \geq 1$ for which $A_{ii}^n > 0$, when it exists, otherwise we set $p(i) = \infty$. All the elements of a same class have the same period. The whole discussion in the following on the existence (and uniqueness) of a solution will depend upon the number of classes and their periodicity.

3.1 The Irreducible Case

This is the case when all the states communicate, and so $N = 1$. The existence of a limit depends on the three different cases for the common period of the states; either (1) $p(1) = 1$, or (2) $p(1) = k > 1$, or (3) $p(1) = \infty$, where 1 denotes the first state in the matrix A.The third case is not to be taken into account, see [5].

Theorem 3. *Suppose A is irreducible and aperiodic, then*

$$\lim_{\tau\to\infty} \frac{\mu_{0|A}^T A \tau v}{\mu_{0|A}^T A \tau \underline{1}} \tag{6}$$

exists for any initial probability, is independent of μ_0 and is equal to $\frac{(f_0)^T v}{(f_0)^T \underline{1}}$, where f_0 is the first left eigenvector of A. If A is irreducible and periodic, then (6) exists if and only if the asymptotic probabilities of exit from each class of periodicity are equal.

As a consequence, one obtains at infinity the behavior outlined in Fig. 2.

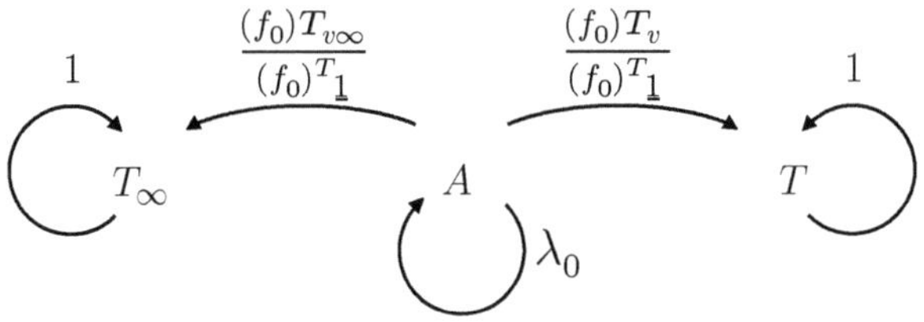

Fig. 2 Behavior when A is irreducible and aperiodic

3.2 The Reducible Case

In the reducible case, A is decomposed into $N > 1$ disjoint classes C_i^*, $i = 1, \ldots, N$. As the classes are disjoint, when the chain exits one class, it does not go back to it. It follows that we can reorder the matrix A so that it will be equal to an upper triangular (non-negative) matrix with block square irreducible matrices on its diagonal, each corresponding to a class. The existence of the asymptotic probability of reaching T from A will therefore depend on the percent of mass that is distributed on each class. If the more important mass is associated with the final classes, this means that the mass in each of the remaining classes will decrease with time and we will only have to take into account this family of final classes. In this case, a limit will exist. A class is called basic if the sub-matrix of A associated with it, admits as spectral radius the spectral radius of A.

Theorem 4. *Suppose the matrix A is reducible with spectral radius $\lambda_0 > 0$, and suppose the final classes are the only basic classes. Then there exists a unique asymptotic probability (depending on μ_0) of exit from A to T if and only if the asymptotic probability of exit from each final class also exists. Moreover, if the final classes have dominant spectral radius and if some of them are the only basic classes, then there exists a unique asymptotic probability of exit from A to T if μ_0 charges these classes.*

Finally, a particular case,

Theorem 5. *Suppose A admits a Jordan decomposition of the form $D+N$, where D is a diagonal block primitive sub-matrices with the same spectral radius λ_0 and N is an upper triangular non-negative matrix. Then the limiting conditional distribution exists.*

4 Extension to Multiple Targets and Examples of Markov Networks

The previous results may be extended to multiple targets problems. More precisely, let $\mathbf{T} = \{T_1, T_2, \ldots\}$ be target disjoint sets on the same R-network (E, P) over $\mathcal{M}_E(R)$. We are interested in the optimal $\{T_1, T_2, \ldots\}$-compatible relationship S such that (4) holds.

The answer is trivial, since each target class T_i defines its equivalence relationship S_i. It is not difficult to show that the required set S is just $S = \cap S_i$, see [2] and [5]. We start here by showing some classical Network problems that cannot be projected in smaller ones.

Example 1 (Negative Binomial Distribution). Repeat independently a game with probability p of winning until you win n games.

Let $S\tau = \sum_{i=1} \tau Y_i$, where $\{Y_i, i \in \mathbb{N}\}$ is a sequence of i.i.d. bernoulli random variable with $Prob(\{Y_i = 1\}) = 1 - Prob(\{Y_i = 0\}) = p$. Our interest is engaged by the computation of the probability of reaching n starting from 0. Let $E = \{0, 1, \ldots, n\}$ be the set of levels we have reached. We have

$$
\begin{array}{c|cccccc}
 & 0 & 1 & 2 & \ldots & n-1 & n=T \\
\hline
0 & (1-p) & p & 0 & \ldots & 0 & 0 \\
1 & 0 & (1-p) & p & \ddots & 0 & 0 \\
2 & 0 & 0 & (1-p) & \ddots & 0 & 0 \\
\vdots & \vdots & \vdots & \vdots & \ddots & \ddots & \vdots \\
n-1 & 0 & 0 & 0 & \ldots & (1-p) & p \\
n=T & 0 & 0 & 0 & \ldots & 0 & 1
\end{array}
\left.\vphantom{\begin{array}{c}0\\0\\0\\0\\0\\0\\0\end{array}}\right\} =: P
$$

Since the length of the minimum path for reaching the target state n from different states is different, the problem cannot be projected on a smaller one by [4, Proposition 31].

Example 2 (Consecutive winning). Repeat independently a game with probability p of winning until you win n consecutive games. The problem is similar to the previous one, where

$$
\begin{array}{c|cccccc}
 & 0 & 1 & 2 & \ldots & n-1 & n=T \\
\hline
0 & (1\quad p) & p & 0 & \ldots & 0 & 0 \\
1 & (1-p) & 0 & p & \ddots & 0 & 0 \\
2 & (1-p) & 0 & 0 & \ddots & 0 & 0 \\
\vdots & \vdots & \vdots & \vdots & \ddots & \ddots & \vdots \\
n-1 & (1-p) & 0 & 0 & \ldots & 0 & p \\
n=T & 0 & 0 & 0 & \ldots & 0 & 1
\end{array}
\left.\vphantom{\begin{array}{c}0\\0\\0\\0\\0\\0\\0\end{array}}\right\} =: P
$$

The problem is again not projectable by [4, Proposition 31].

Example 3 (Gambler's ruin). Let two players each have a finite number of pennies (say, n_1 for player one and n_2 for player two). Now, flip one of the pennies (from either player), with the first player having p probability of winning, and transfer a penny from the loser to the winner. Now repeat the process until one player has all the pennies.

Let $S\tau = \sum_{i=1} \tau(2Y_i - 1)$, where $\{Y_i, i \in \mathbb{N}\}$ is a sequence of i.i.d. bernoulli random variable with $Prob(\{Y_i = 1\}) = 1 - Prob(\{Y_i = 0\}) = p$. Our interest is engaged by the computation of the probability of reaching $T_1 = n_2$ or $T_2 = -n_1$ (multiple target) starting from 0. Let $E = \{-n_2, \dots, -1, 0, 1, \dots, n_1\}$ be the set of levels we have reached. We have

	$-n_1 = T_2$	$-n_1+1$	$\dots$	-1	0	1	$\dots$	n_2-1	$n_2 = T_1$
$-n_1 = T_2$	1	0	$\dots$	0	0	0	$\dots$	0	0
$-n_1+1$	$(1-p)$	0	$\ddots$	0	0	0	$\dots$	0	0
$\vdots$	$\vdots$	$\ddots$	$\ddots$	$\ddots$	$\vdots$	$\vdots$	$\vdots$	$\vdots$	$\vdots$
-1	0	0	$\ddots$	0	p	0	$\dots$	0	0
0	0	0	$\vdots$	$(1-p)$	0	p	$\dots$	0	0
1	0	0	$\vdots$	0	$(1-p)$	0	$\ddots$	0	0
$\vdots$	$\vdots$	$\vdots$	$\vdots$	$\vdots$	$\vdots$	$\ddots$	$\ddots$	$\ddots$	$\vdots$
n_2-1	0	0	$\vdots$	0	0	0	$\dots$	0	p
$n_2 = T_1$	0	0	$\dots$	0	0	0	$\dots$	0	1

This problem is clearly not projectable on a smaller one, since it is for T_1 (for example). The problem may be reduced if and only if we are interested in the time of stopping (without knowing who wins, i.e. $\mathbf{T} = T_1 \cup T_2$) and $p = 1/2$. In this case, the relevant information is the distance from the nearest border and hence the problem may be half-reduced.

The following classical problem may be reduced.

Example 4 (Random walk on a cube). A particle performs a symmetric random walk on the vertices of a unit cube, i.e., the eight possible positions of the particle are $(0,0,0)$, $(1,0,0)$, $(0,1,0)$, $(0,0,1)$, $(1,1,0), \dots, (1,1,1)$, and from its current position, the particle has a probability of $1/3$ of moving to each of the three neighboring vertices. This process ends when the particle reaches $(0,0,0)$ or $(1,1,1)$.

Let $T_1 = (0,0,0)$, $T_2 = (1,1,1)$. The following transition matrix

	(0,0,0)	(1,0,0)	(0,1,0)	(0,0,1)	(1,1,0)	(1,0,1)	(0,1,1)	(1,1,1)
(0,0,0)	1	0	0	0	0	0	0	0
(1,0,0)	1/3	0	0	0	1/3	1/3	0	0
(0,1,0)	1/3	0	0	0	1/3	0	1/3	0
(0,0,1)	1/3	0	0	0	0	1/3	1/3	0
(1,1,0)	0	1/3	1/3	0	0	0	0	1/3
(1,0,1)	0	1/3	0	1/3	0	0	0	1/3
(0,1,1)	0	0	1/3	1/3	0	0	0	1/3
(1,1,1)	0	0	0	0	0	0	0	1

can be easily reduced on

	t_1	f_1	f_2	t_2
t_1	1	0	0	0
f_1	1/3	0	2/3	0
f_2	0	2/3	0	1/3
t_2	0	0	0	1

where $t_i = T_i$ and $f_i = \{e = (e_1, e_2, e_3) : \sum e_j = i\}$. If we are only interested in the time of stopping (i.e. $\mathbf{T} = T_1 \cup T_2$), the previous problem may be reduced to a geometrical one. Clearly, this results hold also for random walk on a d-dimensional cube.

We give, in the following, several examples to the different results we obtained before.

Example 5 (Medical science). We intend to find the probability that the length of a woman's menstrual cycle can be the same three consecutive times. If the length of a menstrual cycle is uniformly distributed between 26 and 35 days (and the length of menstrual cycles being independent from one another), then the process may be seen as a Markov chain on $E = \{26, \ldots, 35\}$, where

$$P = \begin{pmatrix} 1/10 & \ldots & 1/10 \\ \vdots & \ddots & \vdots \\ 1/10 & \ldots & 1/10 \end{pmatrix}.$$

The problem can be solved by introducing the stopping time defined by

$$S = \inf\{\tau \in \mathbb{N} \colon X_{\tau-2} = X_{\tau-1} = X\tau\}.$$

This problem can naturally be embedded in a 21 states Markov problem whose transition matrix is defined in (7), see [4].

	T	26.26	27.27	…	35.35	26	27	…	35
T	1	0	0	…	0	0	0	…	0
26.26	1/10	0	0	…	0	0	1/10	…	1/10
27.27	1/10	0	0	…	0	1/10	0	…	1/10
⋮	⋮	⋮	⋮	⋮	⋮	⋮	⋮	⋮	⋮
35.35	1/10	0	0	…	0	1/10	1/10	…	0
26	0	1/10	0	…	0	0	1/10	…	1/10
27	0	0	1/10	…	0	1/10	0	…	1/10
⋮	⋮	⋮	⋮	⋮	⋮	⋮	⋮	⋮	⋮
35	0	0	0	…	1/10	1/10	1/10	…	0

(7)

It is simplified by considering the process $Z\tau$ with the following three states:

$$\widehat{E} = \begin{cases} & \text{there is no } N \leq \tau \colon X_{N-2} = X_{N-1} = X_N : \\ 1, & \text{if } X_{\tau-1} \neq X\tau \\ 2, & \text{if } X_{\tau-2} \neq X_{\tau-1} = X\tau \\ 3 = T, & \text{if there exists } N \leq \tau \colon X_{N-2} = X_{N-1} = X_N \end{cases}$$

with initial distribution $\mu_0^T = (1,0,0)$ and matrix

$$\left.\begin{array}{r|lll} & 1 & 2 & 3=T \\ \hline 1 & p & (1-p) & 0 \\ 2 & p & 0 & (1-p) \\ 3=T & 0 & 0 & 1 \end{array}\right\} =: \widehat{P},$$

where $p = 9/10$. Simple calculations (see [4]) give the corresponding Hazard rate:

$$H_S(\tau) = 1 - \underbrace{\frac{1}{20}\frac{(9+\sqrt{117})^{\tau+1} - (9-\sqrt{117})^{\tau+1}}{(9+\sqrt{117})^{\tau} - (9-\sqrt{117})^{\tau}}}_{\underset{\tau\to\infty}{\longrightarrow} \hat{\lambda}_0} = \mathbb{P}(X_{\tau+1} \in T | X\tau \in A).$$

The matrix A of the transient states is

$$A = \begin{pmatrix} p & 1-p \\ p & 0 \end{pmatrix}.$$

Therefore, the first eigenvalue is

$$\lambda_0 = \frac{p+\sqrt{-3p^2+4p}}{2} \underset{(p=9/10)}{=} \frac{9+\sqrt{117}}{20},$$

while the first left eigenvector f_0 associated to λ_0 is

$$(f_0)^T = \left(\tfrac{p+\sqrt{-3p^2+4p}}{2(1-p)}, 1\right),$$

and the limit conditional distribution is

$$\frac{(f^0)^T}{(f^0)^T\underline{1}} = \left(\frac{p+\sqrt{-3p^2+4p}}{2-p+\sqrt{-3p^2+4p}}, \frac{2(1-p)}{2-p+\sqrt{-3p^2+4p}}\right).$$

We have

$$\frac{(f^0)^T v}{(f^0)^T\underline{1}} = \frac{(f^0)^T}{(f^0)^T\underline{1}} v = \left(\frac{p+\sqrt{-3p^2+4p}}{2-p+\sqrt{-3p^2+4p}}, \frac{2(1-p)}{2-p+\sqrt{-3p^2+4p}}\right)\begin{pmatrix} 0 \\ 1-p \end{pmatrix} = 1-\lambda_0.$$

As expected, we note that $\lambda_0 = \hat{\lambda}_0$.

Example 6 (The gambler ruin). A gambler A plays against a gambler B a sequence of heads or tails independent games. The total sum of their wealth is a\$. At each game, A wins one dollar or loses it with probability p and $q = 1-p$ respectively. The game stops when one of the gamblers is ruined. Denote by $X\tau$ the wealth of A at the end of the τ-th game. $X\tau$ is a Markov chain with set of states $E = \{0,\ldots,a\}$. Its transition matrix is given by

$$P = \begin{pmatrix} 1 & 0 & 0 & \cdots & 0 & 0 & 0 \\ q & 0 & p & \cdots & 0 & 0 & 0 \\ \vdots & \vdots & \vdots & \ddots & \vdots & \vdots & \vdots \\ 0 & 0 & 0 & \cdots & q & 0 & p \\ 0 & 0 & 0 & \cdots & 0 & 0 & 1 \end{pmatrix}.$$

If we take the position of A, then $T = \{a\}$ and $T_\infty = \{0\}$.

Then the previous matrix can be rearranged:

$$P = \begin{pmatrix} 1 & 0 & & & \mathbf{0} & & \\ 0 & 1 & & & & & \\ q & 0 & 0 & p & \cdots & 0 & 0 \\ 0 & 0 & q & 0 & p & \cdots & 0 \\ \vdots & \vdots & \vdots & \ddots & 0 & \ddots & \vdots \\ 0 & 0 & 0 & \cdots & q & 0 & p \\ 0 & p & 0 & \cdots & \cdots & q & 0 \end{pmatrix}.$$

In this case,

$$A = \begin{pmatrix} 0 & p & \cdots & 0 & 0 \\ q & 0 & p & \cdots & 0 \\ \vdots & \ddots & 0 & \ddots & \vdots \\ 0 & \cdots & q & 0 & p \\ 0 & \cdots & \cdots & q & 0 \end{pmatrix}, \qquad v = \begin{pmatrix} 0 \\ 0 \\ \vdots \\ 0 \\ p \end{pmatrix}, \qquad v_\infty = \begin{pmatrix} q \\ 0 \\ \vdots \\ 0 \\ 0 \end{pmatrix}.$$

A is an irreducible matrix with period 2. Suppose first that $a = 2k+1$ ($2k$ transient states $A = \{1,\ldots,2k\}$ plus $T = \{a\}$ and $T_\infty = \{0\}$). Then, if we permute the order of the states,

$$\hat{A} = \begin{pmatrix} \mathbf{0} & \begin{matrix} p & 0 & \cdots & \cdots & 0 \\ q & p & \cdots & \cdots & 0 \\ 0 & q & p & \cdots & 0 \\ & & \ddots & \ddots & \\ 0 & \cdots & \cdots & q & p \end{matrix} \\ \begin{matrix} q & p & \cdots & \cdots & 0 \\ 0 & q & p & \cdots & 0 \\ & & \ddots & \ddots & \\ 0 & \cdots & \cdots & q & p \\ 0 & \cdots & \cdots & 0 & q \end{matrix} & \mathbf{0} \end{pmatrix}, \qquad \hat{v} = v, \qquad \hat{v}_\infty = v_\infty.$$

Suppose now that $a = 2k$ ($2k-1$ transient states $A = \{1,\ldots,2k-1\}$ plus $T = \{a\}$ and $T_\infty = \{0\}$). Then

$$\hat{A} = \begin{pmatrix} \mathbf{0} & \begin{matrix} p & 0 & \cdots & \cdots & 0 \\ q & p & \cdots & \cdots & 0 \\ 0 & q & p & \cdots & 0 \\ & & \ddots & \ddots & \\ 0 & \cdots & \cdots & q & p \\ 0 & \cdots & \cdots & 0 & q \end{matrix} \\ \begin{matrix} q & p & \cdots & \cdots & 0 \\ 0 & q & p & \cdots & 0 \\ & & \ddots & \ddots & \\ 0 & \cdots & \cdots & q & p \end{matrix} & \mathbf{0} \end{pmatrix}, \qquad \hat{v} = \begin{pmatrix} 0 \\ 0 \\ 0 \\ \vdots \\ 0 \\ p \\ 0 \\ 0 \\ \vdots \\ 0 \end{pmatrix}, \qquad \hat{v}_\infty = v_\infty.$$

In both cases, $a = 2k+1$ and $a = 2k$, $\hat{A}^2$ takes the form

$$\hat{A}^2 = \begin{pmatrix} C_1 & 0 \\ 0 & C_2 \end{pmatrix}$$

with $(C_i)^m \gg 0$ for all $m \geq k-1$. Moreover,

$$v_{2-\text{mod}(a,2)} = (0,0,\ldots,0)^T, \qquad v_{1+\text{mod}(a,2)} = (0,0,\ldots,p)^T$$

In this case it is obvious that the conditional limit does not exist (see [5]); however, we can compute the asymptotic conditional law given each class of periodicity. For example, if $a = 10$ and $p = 1/2$, we obtain

$$C_1 = \begin{pmatrix} 1/4 & 1/4 & 0 & 0 & 0 \\ 1/4 & 1/2 & 1/4 & 0 & 0 \\ 0 & 1/4 & 1/2 & 1/4 & 0 \\ 0 & 0 & 1/4 & 1/2 & 1/4 \\ 0 & 0 & 0 & 1/4 & 1/4 \end{pmatrix} \quad \text{and} \quad C_2 = \begin{pmatrix} 1/2 & 1/4 & 0 & 0 \\ 1/4 & 1/2 & 1/4 & 0 \\ 0 & 1/4 & 1/2 & 1/4 \\ 0 & 0 & 1/4 & 1/2 \end{pmatrix}.$$

We have

$$\lambda_0^{(1)} = \lambda_0^{(2)} = \frac{5+\sqrt{5}}{8},$$

while the corresponding first eigenvectors are

$$\frac{(f^0)^{T(1)}}{(f^0)^T \underline{1}} = \left(\frac{1}{2(3+\sqrt{5})}, \ 1/4, \ \frac{\sqrt{5}+1}{2(3+\sqrt{5})}, \ 1/4, \ \frac{1}{2(3+\sqrt{5})} \right)$$

$$\frac{(f^0)^{T(2)}}{(f^0)^T \underline{1}} = \left(\frac{1}{3+\sqrt{5}}, \ \frac{\sqrt{5}+1}{2(3+\sqrt{5})}, \ \frac{\sqrt{5}+1}{2(3+\sqrt{5})}, \ \frac{1}{3+\sqrt{5}} \right).$$

Example 7 (Random walk on a polygon). A particle can move on a regular polygon with r sides. Its vertices are numbered from 0 to $r-1$. If at some time the particle is on the vertex i $(0 \leq i \leq r-1)$ then, right afterwards, it will be in state $i+1$ (mod r) with probability p and in the state $i-1$ (mod r) with probability q. We assume that $r = 2N$. We also suppose that the particle can exit the polygon to a target set from each vertex 0 to $N-1$ and to T_∞ from each other vertex, both with probability s. We denote by $X\tau$ the vertex visited by the particle at time τ, its transition matrix is given hereafter

$$P = \begin{pmatrix} 1 & 0 & 0 & 0 & \cdots & 0 & 0 & 0 & 0 \\ 0 & 1 & 0 & 0 & \cdots & 0 & 0 & 0 & 0 \\ s & 0 & 0 & p & 0 & \cdots & 0 & 0 & q \\ s & 0 & q & 0 & p & \cdots & 0 & 0 & 0 \\ s & 0 & 0 & q & 0 & \cdots & 0 & 0 & 0 \\ \vdots & \vdots & \vdots & \vdots & \vdots & \ddots & \vdots & \vdots & \vdots \\ 0 & s & 0 & 0 & 0 & \cdots & 0 & p & 0 \\ 0 & s & 0 & 0 & 0 & \cdots & q & 0 & p \\ 0 & s & p & 0 & 0 & \cdots & 0 & q & 0 \end{pmatrix}.$$

Consider now the sub-matrix A of transient states leading to T. This matrix is irreducible, periodic with period 2 and $A/(p+q)$ bi-stochastic. The largest eigenvalue is $\lambda_0 = p+q$, see [5]. The left eigenvector associated with λ_0 is $v^0 = \underline{1}$. It follows that the limiting conditional probability law is equal to $\frac{(f^0)^T}{(f^0)^T \underline{1}} = (1/r, \ldots, 1/r)$, which is the result one would expect for a matrix that is bi-stochastic (up to a multiplicative constant), and irreducible. It, indeed, admits a unique stationary probability given by $u = (u_1, \ldots, u_r)$ such that $u_i = 1/r$, for each i.

Then, if we permute the order of the states,

$$\hat{A} = \begin{pmatrix} 0 & \begin{matrix} p & 0 & \cdots & \cdots & q \\ q & p & \cdots & \cdots & 0 \\ 0 & q & p & \cdots & 0 \\ & & \ddots & \ddots & \\ 0 & \cdots & \cdots & q & p \end{matrix} \\ \begin{matrix} q & p & \cdots & \cdots & 0 \\ 0 & q & p & \cdots & 0 \\ & \ddots & \ddots & & \\ 0 & \cdots & \cdots & q & p \\ p & \cdots & \cdots & 0 & q \end{matrix} & 0 \end{pmatrix}, \qquad \hat{v} = v, \qquad \hat{v}_\infty = v_\infty.$$

Notice that the first sub-matrix A_1 is composed by the even states (we start from 0) and the second sub-matrix by the odd ones.

It follows that $\hat{v} = (s, s, \ldots, 0, 0, s, s, \ldots, 0)^T \qquad \hat{v}_\infty = v_\infty$. In fact, if N is itself an even number, $\hat{v}_1 = \hat{v}_2 = \begin{pmatrix} s \\ s \\ \vdots \\ 0 \\ 0 \end{pmatrix}$ and the conditional limit of exit is $\frac{f_0^{(1)} \hat{v}_1}{f_0^{(1)} \underline{1}} = \frac{f_0^{(2)} \hat{v}_2}{f_0^{(2)} \underline{1}} = \frac{sN/2}{N} = \frac{s}{2}$. The result will be the same if we consider a random walk on the line with reflecting barriers together with a jump to the other side.

Example 8 (The bonus and malus model). An insurance company orders the bonus–malus levels of its clients according to integers $0, 1, 2, \ldots$. The level 0 is the most advantageous for the client. Let $0 \leq i \leq j$. If the bonus-malus level of an insured is i at time τ, it will be j at time $\tau + 1$ if, between times τ and $\tau + 1$, he had $j - i$ accidents. We denote by $(X\tau)$ the sequence of the bonus-malus levels for this insured. Time unit is a year and we suppose that the number of accidents during a year is a Poisson random variable with parameter λ. The probability that the insured moves from level i to level j, $j \geq i$ is equal to $\pi_{\lambda, j-i} = e^{-\lambda} \lambda^{j-i}/(j-i)!$. Furthermore, we suppose that the contract is canceled once the insured has had N accidents. $(X\tau)$ is a Markov chain with set of states $\{0, \ldots, N\}$ and transition matrix

$$P = \begin{pmatrix} \pi_{\lambda,0} & \pi_{\lambda,1} & \pi_{\lambda,3} & \cdots & \pi_{\lambda,N} \\ 0 & \pi_{\lambda,0} & \pi_{\lambda,1} & \cdots & \pi_{\lambda,N-1} \\ 0 & 0 & \pi_{\lambda,0} & \cdots & \pi_{\lambda,N-2} \\ \vdots & \vdots & \vdots & \cdots & \vdots \\ 0 & 0 & 0 & \cdots & 1 \end{pmatrix}.$$

N can be considered to be an absorbing state and to represent the target set T.

This is the Jordan case with $D+N$, where D is a diagonal matrix, and N is an upper triangular matrix. According to Theorem 5, the conditional limit probability of exit from the last class exists.

References

1. Aki S, Hirano K (1999) Sooner and later waiting time problems for runs in Markov dependent bivariate trials. Ann Inst Statist Math 51(1):17–29
2. Aletti G (2006) Compressing redundant information in Markov chains. http://arxiv.org/math.PR/0605099
3. Aletti G (2008) Compressing redundant information in networks (submitted)
4. Aletti G, Merzbach E (2006) Stopping markov processes and first path on graphs. J Eur Math Soc (JEMS) 8(1):49–75
5. Aletti G, Saada D (2008) Asymptotic reduction of Markov chains (submitted)
6. Antzoulakos DL, Philippou AN (1997) Probability distribution functions of succession quotas in the case of Markov dependent trials. Ann Inst Statist Math 49(3):531–539
7. Arnold BC, Balakrishnan N, Nagaraja HN (1998) Records. Wiley series in probability and statistics: probability and statistics. Wiley, New York
8. Cairoli R, Dalang RC (1996) Sequential stochastic optimization. Wiley series in probability and statistics: probability and statistics. Wiley, New York
9. Ebneshahrashoob M, Sobel M (1990) Sooner and later waiting time problems for Bernoulli trials: frequency and run quotas. Statist Probab Lett 9(1):5–11
10. Koutras MV, Alexandrou VA (1997) Sooner waiting time problems in a sequence of trinary trials. J Appl Probab 34(3):593–609
11. Lam K, Yam HC (1997) Cusum techniques for technical trading in financial markets. Financial Engineering & the Japanese Markets 4(3):257–274
12. Stefanov VT (2000) On some waiting time problems. J Appl Probab 37(3):756–764
13. Stefanov VT, Pakes AG (1997) Explicit distributional results in pattern formation. Ann Appl Probab 7(3):666–678

The Dynamic Behaviour of Non-Homogeneous Single-Unireducible Markov and Semi-Markov Chains

Guglielmo D'Amico, Jacques Janssen, and Raimondo Manca

Abstract In this paper single-unireducible Markov and semi-Markov chains are defined and their dynamic behaviour is analysed. The main results concern the asymptotic study of these processes. In fact it is proved that the topological structure of single-unireducibility represents a sufficient condition that guarantees the absorption after a sufficiently long period in the absorbing class for both Markov and semi-Markov chains. The probabilistic results are based on graph theory using relations between the graphs and transition matrices.

1 Introduction

The study of asymptotic behaviour of Markov chain in homogeneous environment is totally explored, see for example [2, 9, 10]. In non-homogeneous case this topic was studied and many results were obtained see for example [3, 4, 12, 13, 15]. But, at authors' knowing, an explicit study of the asymptotic behaviour of non-homogenous Markov chains formed by unireducible transition matrices was never done.

Indeed, in multi-state insurance and migration credit risk models the transition matrices that rule the system evolution usually are unireducible matrices (see for example [10]) with the unique recurrent class constituted by a unique absorbing state and in this paper we would face this particular case.

G. D'Amico
Dipartimento di Scienze del Farmaco Università "G.D'Annunzio", Chieti, Italy
g.damico@unich.it

J. Janssen
Jacan EURIA & Université de Bretagne Occidentale, Brest, France
jacques.janssenwets@skynet.be

R. Manca
Dipartimento di Matematica per le Decisioni Economiche, Finanziarie ed Assicurative Università di Roma "La Sapienza", Rome, Italy
raimondo.manca@uniroma1.it

A.K. Naimzada et al. (eds.), *Networks, Topology and Dynamics*,
Lecture Notes in Economics and Mathematical Systems 613,

In some previous papers [6, 8] some results on the asymptotic behaviour of these kind of matrices were found. More precisely in the first paper it was shown the behaviour of these models in the case of a homogeneous semi-Markov environment. In the second paper the study of asymptotic behaviour of a non-homogeneous Markov and semi-Markov models was presented. But this study was limited at the case with only one class of transient states. In this framework the transition matrices and consequently the model were called mono-unireducible.

In this paper we present a general unireducible approach to the problem generalizing the results obtained in the previous papers. The paper develops the theoretical aspects of these special kinds of transition models. The applicative aspects will be faced in some future works.

The study of these models is strictly connected with some graph theory notions. In the next section the relations between the graphs and the transition matrices are highlighted. In the section three some particular unireducible transition matrices are defined generalizing the results given in [7]. The obtained results are extended in a non-homogeneous discrete time semi-Markov environment.

2 Graphs and Matrices

In this paper the matrices are non negative. As it is well known, a directed weighted graph corresponds to each matrix and vice-versa.

Definition 2.1. A Hamiltonian matrix is a matrix in which the corresponding digraph contains a circuit (Hamiltonian cycle).

Proposition 2.1. *A matrix is Hamiltonian iff it is an irreducible matrix.*

Proof. Trivial. □

Definition 2.2. ([1]) Given a matrix **B** the adjacency matrix associated to the matrix **A** is defined in the following way:

$$a_{ij} = \begin{cases} 1 \text{ if } b_{ij} \neq 0 \\ 0 \text{ if } b_{ij} = 0. \end{cases}$$

Remark 1. The topological structure of a matrix corresponds to its adjacency matrix.

Definition 2.3. Given two matrices $\mathbf{B}_1 = \left[b^1_{ij}\right]$ and $\mathbf{B}_2 = \left[b^2_{ij}\right]$ and let $\mathbf{A}_1$, $\mathbf{A}_2$ be the two corresponding adjacency matrices we say that

$$\mathbf{A}_1 \subseteq \mathbf{A}_2 \overset{\text{def}}{\Leftrightarrow} \left(a^1_{ij} = 1 \Rightarrow a^2_{ij} = 1\right).$$

Definition 2.4. Given the permutation

$$\begin{pmatrix} 1 & 2 & \cdots & m \\ r_1 & r_2 & \cdots & r_m \end{pmatrix} \in \Pi_m, \tag{1}$$

where Π_m is the set of permutations of $\{1,2,\ldots,\boldsymbol{m}\}$. The permutation matrix $\boldsymbol{\Pi}$ corresponding to (1) is defined in the following way

$$\pi_{ij} = \begin{cases} 1 \; if \; j = r_i \\ 0 \; if \; j \neq r_i. \end{cases}$$

Definition 2.5. A circular permutation matrix is a permutation matrix that corresponds to a circular permutation, where a circular permutation can be denoted in the following way:

$$(r_1, r_2, \ldots, r_{m-1}).$$

It means that 1 goes to r_1, r_1 goes to $r_2, \ldots, r_{m-1}$ goes to 1.

Remark 2. A circular permutation matrix $\boldsymbol{\Pi}$ is a Hamiltonian matrix and it results that $\pi_{ii} = 0 \; \forall i$.

Definition 2.6. A circuit matrix $\mathbf{C}$ is a primitive Hamiltonian matrix in which $c_{ii} \neq 0 \; \forall i$.

Definition 2.7. Let $\mathbf{C}$ a circuit matrix it is a simple circuit matrix if its adjacency matrix $\mathbf{A}$ is given by

$\mathbf{A} = \mathbf{I} + \boldsymbol{\Pi}$, where $\boldsymbol{\Pi}$ is a circular permutation matrix and $\mathbf{I}$ is the unitary matrix.

Remark 3. Each simple circuit matrix has two elements not equal to 0 for each row and each column.

Remark 4. Each adjacency matrix of a circuit matrix will contain an adjacency matrix of a simple circuit matrix.

Proposition 2.2. *Let* $(\mathbf{C}_n)_{n\in\mathbb{N}}$ *a sequence of simple circuit matrices and* $(\mathbf{A}_n)_{n\in\mathbb{N}}$ *the corresponding sequence of their adjacency matrices. Given*

$$\mathbf{S}^{(n)} = \mathbf{A}_1 * \mathbf{A}_2 * \cdots * \mathbf{A}_n,$$

where $*$ *represents the usual row column product, then*

$$\sum_{k=1}^{m} s_{ik}^{(n)} = 2^n.$$

Proof. The proposition is true for n = 1. It is supposed that it is true for n − 1. Then

$$\mathbf{S}^{(n)} = \mathbf{S}^{(n-1)} * \mathbf{A}_n. \tag{2}$$

A row of (2) is given by

$$\begin{aligned} \mathbf{s}_{i*}^{(n)} = \sum_{k=1}^{m} s_{ik}^{(n-1)} \mathbf{a}_{k*}^{n} &= \left(s_{i1}^{(n-1)} a_{11}^{n}, 0, \ldots, 0, s_{ir_1}^{(n-1)} a_{r_1 1}^{n}, 0, \ldots, 0\right) + \cdots \\ &\cdots + \left(0, \ldots, 0, s_{ir_m}^{(n-1)} a_{r_m m}^{n}, 0, \ldots, 0, s_{im}^{(n-1)} a_{mm}^{n}\right), \end{aligned} \tag{3}$$

where $(r_1, \ldots, r_m)$ is a circular permutation of $\{1,2,\ldots,m\}$.

From (3), it results

$$\sum_{k=1}^{m} s_{ik}^{(n)} = \sum_{k=1}^{m} s_{ik}^{(n-1)} + \sum_{k=1}^{m} s_{ir_k}^{(n-1)} = 2^n.$$

□

Remark 5. $s_{ij}^{(n)}$ represents the number of different paths of length n that start from i and arrive in j.

Remark 6. 2^n represents the number of different paths of length n that start from i.

Remark 7. Similar results can be proved for the columns.

The next result can be useful for a better understanding of the main theorem.

Proposition 2.3. *Let **B** and **C** two simple circuit matrices if*

$$\mathbf{D} = \mathbf{B} * \mathbf{C}$$

then each row $\mathbf{d}_{i*}$ will have three or four non-zero elements.

Proof. From Definition 2.7 it results that $\mathbf{b}_{i*}$ has only the two elements $b_{ii}, b_{ir_i} \neq 0$, $i \neq r_i$, in the same way also $c_{ii}, c_{is_i} \neq 0$, $i \neq s_i$ will be the only two elements not equal to zero in the i-th row, where respectively

$$\pi_1 = \begin{pmatrix} 1 & 2 & \cdots & m \\ r_1 & r_2 & \cdots & r_m \end{pmatrix} \quad \text{and} \quad \pi_2 = \begin{pmatrix} 1 & 2 & \cdots & m \\ s_1 & s_2 & \cdots & s_m \end{pmatrix}$$

are two circular permutations. This time the notation given in (1) is used also for circular permutations.

Given $\mathbf{D} = \mathbf{B} * \mathbf{C}$ from the hypotheses it results that $\mathbf{d}_{i*}$ can have only

$$d_{ii} = b_{ii}c_{ii}, d_{ir_i} = b_{ir_i}c_{r_ir_i}, d_{is_i} = b_{ii}c_{is_i}, d_{is_{r_i}} = b_{ir_i}c_{r_is_{r_i}}$$

not equal to zero. If i, r_i, s_i, s_{r_i} are all each other different then there will be four elements not equal to zero in the i-th row of the matrix **D**.

It is possible that $r_i = s_i$ then $s_{r_i} \neq i$. In fact if $s_{r_i} = i$ then it results that $c_{is_i} = c_{ir_i} \neq 0$ and $c_{r_is_{r_i}} = c_{r_ii} = c_{s_ii} \neq 0$ and π_2 is no more a circular permutation and matrix **C** is not a simple circuit matrix.

It is also possible that $s_{r_i} = i$ then $r_i \neq s_i$. In fact if $r_i = s_i$ then it results $c_{is_i} = c_{ir_i} \neq 0$ and $c_{s_is_{s_i}} = c_{r_is_{r_i}} = c_{r_ii} \neq 0$ and π_2 is no longer a circular permutation and matrix **C** is not a simple circuit matrix. These two last results implies that there are at least three elements not equal to zero for each row of **D** □

Theorem 2.1. *Let $\{\mathbf{C}_1, \mathbf{C}_2, \ldots, \mathbf{C}_n\}$ be circuit matrices with $n < m$ then if*

$$\mathbf{S}^{(n)} = \mathbf{C}_1 * \mathbf{C}_2 * \cdots * \mathbf{C}_n \tag{4}$$

then each row (column) of (4) will have at least $n+1$ elements greater then 0.

Proof. Each circuit matrix contains a simple circuit matrix, so it is enough to prove the proposition for simple circuit matrices.

Let **B** be a matrix and **C** a simple circuit matrix where, as before,

$$c_{ii}, c_{is_i} \neq 0,\ i \neq s_i \tag{5}$$

are the only two non-zero elements for each row.

Let $\mathbf{b}_{i*}$ the i-th row of matrix **B** in which $b_{ir_1}, b_{ir_2}, \ldots, b_{ir_h} \neq 0$, $h < m$ are the only non zero elements of the i-th row.

Let $\mathbf{D} = \mathbf{B} * \mathbf{C}$ then it results

$$d_{ir_1}, d_{ir_2}, \ldots, d_{ir_h} \neq 0, \quad d_{ir_k} = b_{ir_k} c_{r_k r_k} \neq 0 k = 1, \ldots, h.$$

Taking into account (5) the other elements of $\mathbf{d}_{i*}$ that are not zero are the following:

$$\begin{aligned}
d_{is_{r_1}} &= b_{ir_1} c_{r_1 s_{r_1}}, \quad r_1 \neq s_{r_1} \\
d_{is_{r_2}} &= b_{ir_2} c_{r_2 s_{r_2}}, \quad r_2 \neq s_{r_2} \\
d_{is_{r_3}} &= b_{ir_3} c_{r_3 s_{r_3}}, \quad r_3 \neq s_{r_3} \\
&\vdots \\
d_{is_{r_{h-1}}} &= b_{ir_{h-1}} c_{r_{h-1} s_{r_{h-1}}}, \quad r_{h-1} \neq s_{r_{h-1}} \\
d_{is_{r_h}} &= b_{ir_h} c_{r_h s_{r_h}}, \quad r_h \neq s_{r_h}.
\end{aligned} \tag{6}$$

From the properties of **C** the following can happen

1.1) $s_{r_1} \notin \{r_2, r_3, \ldots, r_h\}$,
1.2) $s_{r_1} \in \{r_2, r_3, \ldots, r_h\}, s_{r_1} = r_{k_1}, k_1 \in \{2, 3, \ldots, h\}$.

If *ii*) holds then there will be $h+1$ non-zero elements in the row $\mathbf{d}_{i*}$ otherwise the following can happen:

$$\begin{aligned}
&2.1)\ s_{r_{k_1}} \notin \{r_1, r_2, r_3, \ldots, r_h\}, \\
&2.2)\ s_{r_{k_1}} \in \{r_1, r_2, r_3, \ldots, r_h\} - \{r_1, r_{k_1}\},\ s_{r_{k_1}} = r_{k_2},\ k_2 \in \{1, 2, 3, \ldots, h\} - \{1, k_1\}.
\end{aligned} \tag{7}$$

If 2.1) holds then there will be $h+1$ non-zero elements in the row $\mathbf{d}_{i*}$. Now it will be explained because if 2.1) does not hold then it must be $s_{r_{k_1}} \neq r_1, r_{k_1}$. In fact if

$$s_{r_{k_1}} = r_{k_1} \Leftrightarrow d_{is_{r_{k_1}}} = b_{ir_{k_1}} c_{r_{k_1} r_{k_1}},$$

that contradicts (6) and if

$$s_{r_{k_1}} = r_1 \Leftrightarrow d_{is_{r_{k_1}}} = b_{ir_{k_1}} c_{r_{k_1} r_1} \neq 0 \text{ and } d_{is_{r_1}} = b_{ir_1} c_{r_1 r_{k_1}} \neq 0 \tag{8}$$

and **C** is not a simple circuit matrix (r_1 goes to r_{k_1} and r_{k_1} goes to r_1).

If 1.2) and 2.2) hold then the following can happen:

3.1) $s_{r_{k_2}} \notin \{r_1, r_2, r_3, \ldots, r_h\}$,
3.2) $s_{r_{k_2}} \in \{r_1, r_2, r_3, \ldots, r_h\} - \{r_1, r_{k_1}, r_{k_2}\}, s_{r_{k_2}} = r_{k_3}, k_3 \in \{1, 2, 3, \ldots, h\} - \{1, k_1, k_2\}$.

If 3.1) holds then there will be $h+1$ non-zero elements in the row $\mathbf{d}_{i*}$. Now it will be explained because if 3.1) does not hold then it must be $s_{r_{k_2}} \neq r_1, r_{k_1}, r_{k_2}$. In fact $s_{r_{k_2}} = r_{k_2}$ contradicts (6), $s_{r_{k_2}} = r_{k_1}$ corresponds to (8), and, finally, if $s_{r_{k_2}} = r_1 \Leftrightarrow d_{is_{r_{k_2}}} = b_{ir_{k_2}} c_{r_{k_2} r_1} \neq 0$, $d_{ir_{k_2}} = b_{ir_{k_1}} c_{r_{k_1} r_{k_2}} \neq 0$ and $d_{ir_{k_1}} = b_{ir_1} c_{r_1 r_{k_1}} \neq 0$ and $\mathbf{C}$ is not a simple circuit matrix (r_1 goes to r_{k_1}, r_{k_1} goes to r_{k_2} and r_{k_2} goes to r_1). Continuing with this construction the following can happen

$$s_{r_{k_j}} \notin \{r_1, r_2, r_3, \ldots, r_h\},\ j = 1, 2, \ldots, h-2 \tag{9}$$

and, in this case, there will be $h+1$ no-zero elements in the row $\mathbf{d}_{i*}$. If (9) does not hold it results that

$$s_{r_{k_{h-1}}} \neq r_{k_{h-1}}, r_{k_{h-2}}, \ldots, r_{k_2}, r_{k_1}, r_1$$

and all the r_j will be different from each other, otherwise a cycle will be closed and $\mathbf{C}$ will not be a simple circuit matrix. This implies that

$$s_{r_{k_{h-1}}} \notin \{r_1, r_2, r_3, \ldots, r_h\}.$$

So if a row of a matrix $\mathbf{B}$ has $h < m$ elements greater than 0 then after the multiplication with a simple circuit matrix it will have at least $h+1$ elements greater than 0.

Now let $\{\mathbf{C}_1, \mathbf{C}_2, \ldots, \mathbf{C}_n\}$ be simple circuit matrices with $n < m$ then $\mathbf{C}_1 * \mathbf{C}_2$ shall have at least three elements greater than 0 because $\mathbf{C}_1$ has two elements greater than 0 in each row and so on. Finally, the following results:

$$\mathbf{S}^{(n)} = \mathbf{C}_1 * \mathbf{C}_2 * \cdots * \mathbf{C}_n \tag{10}$$

has at least $n+1$ elements in each row

Remark 8. In (10) if $n = m-1$ the matrix $\mathbf{S}^{(m-1)}$ will be a full matrix.

Remark 9. Given a non-negative matrix $\mathbf{B}$ of orders $p \times n$ that has only the element $b_{ij} > 0$ and the matrix $\mathbf{S}^{(n)}$ then the matrix $\mathbf{A} = \mathbf{B} * \mathbf{S}^{(n)}$ ($\mathbf{A} = \mathbf{S}^{(n)} * \mathbf{B}$) has at least $n+1$ elements greater than 0 in the row $\mathbf{a}_{i*}$ (column $\mathbf{a}_{*j}$).

3 Single-Unireducible Non-Homogeneous Markov Chains

As it is well known, see [10], the states of a Markov chain can be partitioned in equivalence classes. A class of states can be transient or recurrent (absorbing).

If the Markov chain has only one class then the Markov chain is irreducible if there is more than one class then the process can be unireducible if there is only one recurrent class and reducible if there are two or more recurrent classes. One particular case of unireducible matrix is the one with the absorbing class formed by only one state. Matrices of this kind are present in many finance and insurance models.

In homogeneous Markov chains, the behaviour of these models is well known (see books [9, 10, 14, 16] or the paper [8]).

In this paper, we will show how these matrices can behave in particular way in a non-homogeneous discrete time process environment.

Among the classes of the Markov chain states an order relation can be defined (see [8]).

In this paper we suppose that, in unireducible and reducible cases, the states that belong to the same class are close. Furthermore we suppose that the classes are already ordered following the partial order as defined in [8]. This ordering means that in the first transition matrix row there are the maximal classes, that are a particular case of transient classes from which it is not possible to entry from other classes but it is possible to get out, after there are the non-maximal transient classes in which it is possible to arrive and to get out and in the last rows there are the recurrent (absorbing) classes that are also named minimal classes because it is only possible to arrive inside them but not to get out.

By means of this hypothesis we do not loose generality because once the classification of classes is done (see [8]) then with a permutation of rows and columns it is possible to get the transition matrix ordered in the way described before.

In this case, there is only one absorbing (recurrent) class A, in the following the absorbing class will be formed by only one state. The maximal classes are $C^1, C^2, \dots, C^k$. The remaining classes can be related as follows:

$$\begin{aligned} &C^1 \geq C_2^1 \geq C_3^1 \geq \dots \geq C_{l_1}^1 \geq A, \\ &C^2 \geq C_2^2 \geq C_3^2 \geq \dots \geq C_{l_2}^2 \geq A, \\ &\dots\dots\dots\dots\dots\dots\dots\dots \\ &C^k \geq C_2^k \geq C_3^k \geq \dots \geq C_{l_s}^k \geq A, \end{aligned} \tag{11}$$

where C_j^i is a non-maximal transient class and $C_j^i \geq C_{j+1}^i$ means that it is possible to migrate from class C_j^i to class C_{j+1}^i. We define the elements of (11) chains of classes.

If $C^h \geq C_2^h \geq C_3^h \geq \dots \geq C_{l_h}^h \geq A$ is one of the class chains of (11) then it results

$$\begin{aligned} C^h &= \left\{i_0^h, i_0^h+1, i_0^h+2, \dots, i_0^h+h_0-1\right\} \\ C_1^h &= \left\{i_1^h, i_1^h+1, i_1^h+2, \dots, i_1^h+h_1-1\right\} \\ &\vdots \\ C_{l_h}^h &= \left\{i_{l_h}^h, i_{l_h}^h+1, i_{l_h}^h+2, \dots, i_{l_h}^h+h_{l_h}-1\right\} \\ A &= \{m\}. \end{aligned} \tag{12}$$

Remark 10. Let $(\mathbf{P}(s))_{s\in \mathrm{N}}$ be the sequence of a discrete time non-homogeneous m-states Markov chain. If $\mathbf{P}(s)$, $s \in \mathrm{N}$ are circuit matrices then from Theorem 2.1, $\forall t \geq m-1$ their row column product, $\mathbf{P}^{(t)}(s) = \mathbf{P}(s+1) * \mathbf{P}(s+2) * \cdots * \mathbf{P}(s+t)$, that represents the stochastic process t-step evolution equation, is a full matrix.

Some definitions and results given in [7] are reported.

Definition 3.1. Let

$$\mathbf{A} = \begin{bmatrix} \mathbf{A}_{1,1} & \mathbf{A}_{1,2} \\ \mathbf{0}^{\mathrm{T}} & \mathbf{A}_{2,2} \end{bmatrix}$$

be a $m \times m$ matrix, where $\mathbf{A}_{1,1}$ is a $m-1 \times m-1$ circuit matrix, $\mathbf{A}_{1,2}$ is a $m-1$ non negative column vector in which at least 1 element is positive, $\mathbf{A}_{2,2} = [a_{mm}]$, $a_{mm} > 0$ and $\mathbf{0}^{\mathrm{T}}$ is a $m-1$ null row vector; then $\mathbf{A}$ is a mono-unireducible matrix.

Definition 3.2. A homogeneous Markov chain is mono-unireducible if its transition probability matrix $\mathbf{P}$ is a mono-unireducible matrix.

Consequently the chain has:

1) Only two classes of states. The first is a transient class and the second is an absorbing class.
2) The absorbing class is constituted by only one state.
3) All the elements of the main diagonal of the transition matrix are always greater than 0.

Theorem 3.1. ([7]) *Let* $(\mathbf{P}(s))_{s\in\mathrm{N}}$ *be the sequence of a discrete time non-homogeneous m-states Markov chain. If each* $\mathbf{P}(s)$, $s \in \mathrm{N}$, *is a mono-unireducible matrix.*

Fixed $\mathbf{P}^{(t)}(s) = \mathbf{P}(s+1) * \mathbf{P}(s+2) * \cdots * \mathbf{P}(s+t), \forall t \geq m-1$ *then we have the following results:*

$$1)\ \mathbf{P}^{(t)}(s) = \begin{bmatrix} p^{(t)}_{1,1}(s) & p^{(t)}_{1,2}(s) & \cdots & p^{(t)}_{1,m-1}(s) & p^{(t)}_{1,m}(s) \\ p^{(t)}_{2,1}(s) & p^{(t)}_{2,2}(s) & \cdots & p^{(t)}_{2,m-1}(s) & p^{(t)}_{2,m}(s) \\ \vdots & \vdots & \ddots & \vdots & \vdots \\ p^{(t)}_{m-1,1}(s) & p^{(t)}_{m-1,2}(s) & \cdots & p^{(t)}_{m-1,m-1}(s) & p^{(t)}_{m-1,m}(s) \\ 0 & 0 & & 0 & 1 \end{bmatrix}, \tag{13}$$

where $\forall t \geq m-1 \quad p^{(t)}_{ij}(s) > 0$;

$$2)\ \lim_{t\to\infty} \mathbf{P}^{(t)}(s) = \begin{bmatrix} 0 & 0 & \cdots & 0 & 1 \\ 0 & 0 & \cdots & 0 & 1 \\ \vdots & \vdots & \ddots & \vdots & \vdots \\ 0 & 0 & \cdots & 0 & 1 \\ 0 & 0 & \cdots & 0 & 1 \end{bmatrix} \quad \forall s \in \mathrm{N}. \tag{14}$$

The result of this theorem is really useful in insurance and credit risk models because usually the transition matrices in these environment are mono-unireducible.

In some particular credit risk models it can be interesting the generalization of the Theorem 3.1 in this line.

Definition 3.3. Let

$$\mathbf{A} = \begin{bmatrix} \mathbf{A}_{11} & \mathbf{A}_{12} & \mathbf{A}_{13} & \cdots & \mathbf{A}_{1s} & \mathbf{A}_{1s+1} \\ \mathbf{0} & \mathbf{A}_{22} & \mathbf{A}_{23} & \cdots & \mathbf{A}_{2s} & \mathbf{A}_{2s+1} \\ \mathbf{0} & \mathbf{0} & \mathbf{A}_{33} & \cdots & \mathbf{A}_{3s} & \mathbf{A}_{3s+1} \\ \vdots & \vdots & \vdots & \ddots & \vdots & \vdots \\ \mathbf{0} & \mathbf{0} & \mathbf{0} & \cdots & \mathbf{A}_{ss} & \mathbf{A}_{ss+1} \\ \mathbf{0} & \mathbf{0} & \mathbf{0} & \cdots & \mathbf{0} & \mathbf{A}_{s+1s+1} \end{bmatrix} \tag{15}$$

be a $m \times m$non negative matrix, where

$\mathbf{A}_{i,i}$ are square circuit matrices,

$$\forall i \, \exists j > i : \mathbf{A}_{i,j} \geq \mathbf{0}, \mathbf{A}_{i,j} \neq \mathbf{0},$$

$\mathbf{A}_{s+1s+1} = [a_{mm}]$, $a_{mm} > 0$, then $\mathbf{A}$ is a single-unireducible matrix.

Definition 3.4. An homogeneous Markov chain with a single-unireducible transition probability matrix $\mathbf{P}$ is said to be a single-unireducible Markov chain.

Consequently a single-unireducible homogeneous Markov chain has:
- The absorbing class with only one state.
- Each sub-matrix that represents a class constitutes a circuit matrix.

A mono-unireducible Markov chain is a particular case of single-unireducible Markov chain. An example of single-unireducible Markov chain is given in Fig. 1. The sub-matrices represent, if they are in the main diagonal ($\mathbf{A}_{i,i}$), the relation inside a class. These sub-matrices are circuit matrices. A sub-matrix outside the main diagonal represents the connections between two classes; it has to be a non-zero matrix. This fact expresses the possibility to go from one class to the other. All the

$$\begin{bmatrix} \begin{bmatrix}X&X\\X&X\end{bmatrix} & & & \begin{bmatrix}X&X\\X&X\end{bmatrix} & & & \\ & \begin{bmatrix}X&X\\X&X\end{bmatrix} & \begin{bmatrix}X&X\\X&X\end{bmatrix} & & & & \\ & & \begin{bmatrix}X&X\\X&X\end{bmatrix} & & \begin{bmatrix}X&X\\X&X\end{bmatrix} & \begin{bmatrix}X&X\\X&X\end{bmatrix} & \\ & & & \begin{bmatrix}X&X\\X&X\end{bmatrix} & \begin{bmatrix}X&X\\X&X\end{bmatrix} & & \\ & & & & \begin{bmatrix}X&X\\X&X\end{bmatrix} & & \begin{bmatrix}X\\X\end{bmatrix} \\ & & & & & \begin{bmatrix}X&X\\X&X\end{bmatrix} & \begin{bmatrix}X\\X\end{bmatrix} \\ & & & & & & [X] \end{bmatrix}$$

Fig. 1 Seven classes of states single unireducible Markov matrix

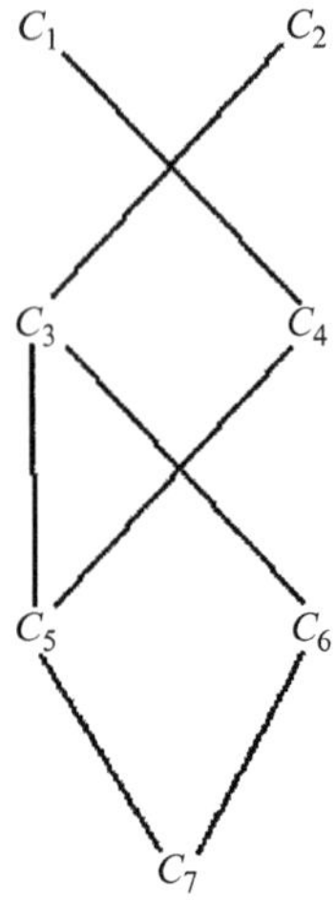

Fig. 2 Graph of the partial order among the classes

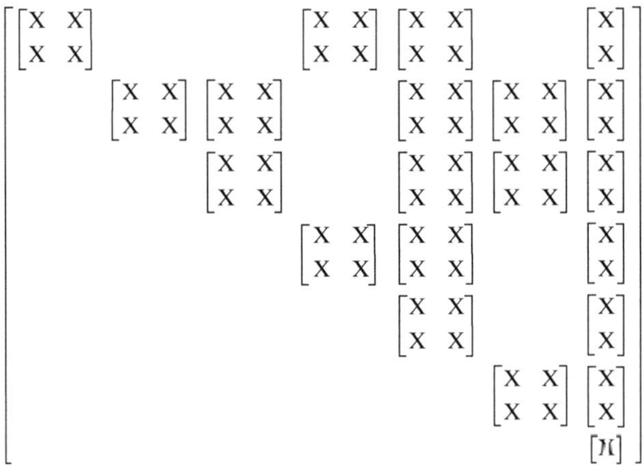

Fig. 3 Reachability matrix of Markov matrix given in Fig. 1

sub-matrices related to the transient classes do not have the same order; the matrix in Fig. 1 is constructed in this way only to distinguish the transient classes by the unique absorbing class.

Figure 2 represents the partial relation order defined on the classes of states of the matrix depicted in Fig. 1. In this example it results

$$C_1 \geq C_4 \geq C_5 \geq C_7 = A;$$
$$C_2 \geq C_3 \geq C_5 \geq C_7 = A;$$
$$C_2 \geq C_3 \geq C_6 \geq C_7 = A.$$

In the Fig. 3 it is given the reachability matrix (see [1]) of the matrix shown in Fig. 1.

Definition 3.5. Two block matrices have the same topological block structure if the non-zero blocks are in the same positions.

Definition 3.6. A non-homogeneous Markov chain with transition probability matrices $(\mathbf{P}(s))_{s\in \mathrm{N}}$ is said to be single-unireducible if $\forall s$ $\mathbf{P}(s)$ is a single-unireducible matrix and they have the same topological block structure.

Theorem 3.2. *Let* $(X(s))_{s\in \mathrm{N}}$ *a m-states single-unireducible non-homogeneous Markov chain with transition probability matrices* $(\mathbf{P}(s))_{s\in \mathrm{N}}$. *Then*

$$\lim_{t\to\infty}\mathbf{P}^{(t)}(s)=\begin{bmatrix}0 & 0 & \cdots & 1\\ 0 & 0 & \cdots & 1\\ \vdots & \vdots & \ddots & \vdots\\ 0 & 0 & \cdots & 1\end{bmatrix}.$$

Proof. $\forall i\neq m\exists$ a class C_j^h to which $i\in C_j^h$. Then since $C_j^h\geq C_{l_h}^h$ it exists a sequence of times $\{s,s+t_1,s+t_2,\dots,s+t_h\}$ and states $\{i,x_{s+t_1},x_{s+t_2},\dots,x_{s+t_h}\}$ such that

$$P(X_{t_h}=x_{t_h}\in C_{l_h}^h,X_{t_{h-1}}=x_{t_{h-1}},\dots,X_s=i\in C_j^h)>0$$

consequently $p_{i,x_{t_h}}^{(t_h)}(s)>0$.

Let consider whatever time $t>t_h>s$. It results that

$$p_{i,m}^{(t)}(s)=\sum_{k\neq m}p_{i,k}^{(t-1)}(s)p_{k,m}(t+s)+p_{i,m}^{(t-1)}(s)$$

from which $p_{i,m}^{(t)}(s)-p_{i,m}^{(t-1)}(s)=\sum\limits_{k\neq m}p_{i,k}^{(t-1)}(s)p_{k,m}(t+s)$.

So we can see that the sequences $\left(p_{i,m}^{(t)}(s)\right)_{\substack{t\in \mathrm{N}\\ t\geq s\in \mathrm{N}}}$ is an increasing bounded sequence then its limit exists. Let suppose that $\lim\limits_{t\to\infty}p_{i,m}^{(t)}(s)=1-\varepsilon$, $0<\varepsilon<1$ then $\forall\delta>0\,\exists\bar{n}\in \mathrm{N}:\forall n\geq\bar{n}\; 1-\varepsilon-\delta<p_{i,m}^{(t)}(s)<1-\varepsilon$ which implies that $\forall t>\bar{n}\; p_{i,m}^{(t)}(s)-p_{i,m}^{(t-1)}(s)<\delta$ if and only if $\sum\limits_{k\neq m}\sum\limits_{\alpha\in E}p_{i,\alpha}^{(t_h)}(s)\cdot p_{\alpha,k}^{(t-1-t_h)}(t_h+s)\cdot p_{k,m}(t+s)<\delta$ which implies

$$\sum_{\alpha\in C_{l_h}^h}\sum_{k\in C_{l_h}^h}p_{i,\alpha}^{(t_h)}(s)\cdot p_{\alpha,k}^{(t-1-t_h)}(t_h+s)\cdot p_{k,m}(t+s)<\delta. \tag{16}$$

Now since $C_{l_h}^h$ is a minimal transient class it exists $\exists k\in C_{l_h}^h:p_{k,m}(t+s)>0$; moreover for $i\in C_j^h\;\exists\alpha\in C_{l_h}^h:p_{i,\alpha}^{(t_h)}(s)>0$; furthermore $\forall\alpha,k\in C_{l_h}^h$ *and* $(t-1-t_h)>m-1\; p_{\alpha,h}^{(t-1-t_h)}(t_h+s)>0$ (see Theorem 2.1).

Now call $b=\min\limits_{\alpha,k\in C_{l_h}^h}\left\{p_{\alpha,k}^{(t-1-t_h)}(s+t_h)\right\}>0$ then (16) implies that

$\sum_{\alpha\in C^h_{l_h}} \sum_{k\in C^h_{l_h}} p^{(t_h)}_{i,\alpha}(s)\cdot b\cdot p_{k,m}(t+s)<\delta$ and setting $B=b\sum_{\alpha\in C^h_{l_h}} p^{(t_h)}_{i,\alpha}(s)>0$ it should be

$$\sum_{k\in C^h_{l_h}} p_{k,m}(t+s)<\frac{\delta}{B} \tag{17}$$

But nothing can ensure (17) for single-unireducible Markov chains, consequently $\lim_{t\to\infty} p^{(t)}_{im}(s)=1-\varepsilon,\ 0<\varepsilon<1$ can not be true for all ε and $\lim_{t\to\infty} p^{(t)}_{im}(s)=1, \forall i=1,\ldots,m-1$. □

4 Single Unireducible Semi-Markov Chains

First of all we define the DTNHSMP and the most used variables.

In SMP environment, two random variables (r.v.) run together. J_n $n\in \mathrm{N}$ with state space $E=1,2,\ldots,m$ represents the state at the n-th transition. $T_n, n\in \mathrm{N}$ with state space equal to N represents the time of the n-th transition,

$$J_n:\Omega\to E \quad T_n:\Omega\to \mathrm{N}.$$

We suppose that the process (J_n,T_n) is a non-homogeneous Markov renewal process.

The kernel $\mathbf{B}=[\mathrm{b}_{ij}(s,t)]$ associated to the process is defined in the following way:

$$b_{ij}(s,t)=P[J_{n+1}=j,T_{n+1}=t| \quad J_n=i,T_n=s]$$

and it results

$$p_{ij}(s)=\lim_{t\to\infty}\sum_{\tau=s+1}^{t} b_{ij}(s,\tau); \quad i,j\in E,\ s\in \mathrm{N},\ s\le t,$$

where $\mathbf{P}(s)=[p_{ij}(s)]$ is the transition matrix of the embedded non-homogeneous Markov chain in the process. Furthermore, it is necessary to introduce the probability that process will leave state i from time s up to time t

$$H_i(s,t)=\mathrm{P}\left[T_{n+1}\le t|J_n=i,\ T_n=s\right]=\sum_{j=1}^{m}\sum_{\tau=s+1}^{t} b_{ij}(s,\tau).$$

Now it is possible to define the distribution function of the waiting time in each state i, given that the state successively occupied is known

$$\begin{aligned}\mathrm{G_{ij}}(\mathrm{s,t})&=\mathrm{P}\left[\mathrm{T_{n+1}}\le t|J_n=i,J_{n+1}=j,T_n=s\right]\\&=\begin{cases}\sum_{\tau=s+1}^{t} b_{i,j}(s,\tau)/p_{i,j}(s) & \text{if } p_{i,j}(s)\neq 0\\ 1 & \text{if } p_{i,j}(s)=0.\end{cases}\end{aligned}$$

Now let $N(t) = \sup\{n \in \mathrm{N} | T_n \le t\}$ then the NHSMP $Z(t) = J_{N(t)}$, $t \in \mathrm{N}$ can be defined.

It represents, for each waiting time, the state occupied by the process.

The semi-Markov transition probabilities are defined by

$$\phi_{i,j}(s,t) = \mathrm{P}\left[Z(t) = j | Z(s) = i\right].$$

The evolution equation of DTNHSMP is

$$\phi_{i,j}(s,t) = (1 - H_i(s,t))\,\delta_{ij} + \sum_{k=1}^{m} \sum_{\vartheta=s+1}^{t} b_{i,k}(s,\vartheta)\phi_{k,j}(\vartheta,t).$$

Definition 4.1. Let $B(t) = t - T_{N(t)}$ be the backward recurrence time process (see [10, 11]). We denote

$$^{b}\phi_{i,j}(l,s;t) = \mathrm{P}\left[Z(t) = j | Z(s) = i, B(s) = l\right], \tag{18}$$

$$\phi^{b}_{i,j}(s;l,t) = \mathrm{P}\left[Z(t) = j, B(t) = l | Z(s) = i\right]. \tag{19}$$

Definition 4.2. A discrete time non-homogeneous semi-Markov process (DTNH SMP) is mono-unireducible if:

(1) $\mathbf{B}(s,s+1)$, $\forall s \in \mathrm{N}$ is mono-unireducible.
(2) $b_{i,i}(s,t) > 0\ \forall i \in E, \forall s < t;\ s,t \in \mathrm{N}$.

Definition 4.3. It is possible to define the following probability:

$$b^{(n)}_{i,j}(s,t) = P\left[J_n = j, T_n = t | J_0 = i, T_0 = s\right],$$

where

$$\mathbf{B}^{(1)}(s,t) = \mathbf{B}(s,t),$$

$$\mathbf{B}^{(2)}(s,t) = \mathbf{B}(s,t) \bullet \mathbf{B}^{(1)}(s,t);\ b^{(2)}_{i,j}(s,t) = \sum_{k \in E} \sum_{\vartheta=s+1}^{t} b_{i,k}(s,\vartheta)\, b^{(1)}_{k,j}(\vartheta,t),$$

$$\vdots$$

$$\mathbf{B}^{(n)}(s,t) = \mathbf{B}(s,t) \bullet \mathbf{B}^{(n-1)}(s,t);\ b^{(n)}_{i,j}(s,t) = \sum_{k \in E} \sum_{\vartheta=s+1}^{t} b_{i,k}(s,\vartheta)\, b^{(n-1)}_{k,j}(\vartheta,t).$$

Lemma 4.1. ([7]) *Let $(Z_t | t \in \mathrm{N})$ be a mono-unireducible DTNHSMC and fixed*

$$\mathbf{B}^{(n)}(s,t) = \begin{bmatrix} \mathbf{B}^{(n)}_{1,1}(s,t) & \mathbf{B}^{(n)}_{1,2}(s,t) \\ \mathbf{0}^{\mathrm{T}} & 1 \end{bmatrix};\ n \ge m;\ s < t;\ s,t \in \mathrm{N}$$

then it results

$$\mathbf{B}^{(n)}_{1,1}(s,t) \gg \mathrm{O}, \mathbf{B}^{(n)}_{1,2}(s,t) \gg \mathbf{0}, \tag{20}$$

where $\mathbf{O}$ is an $m-1 \times m-1$ zero matrix.

Theorem 4.1. ([7]) *Let* $(Z(t)|t \in \mathrm{N})$ *be a semi-Markov chain and* $\phi_{i,j}(s,t) = P[Z(t) = j|Z(s) = i]$. *If the DTNHSMP is mono-unireducible then*

$$\lim_{t\to\infty} \mathbf{\Phi}(s,t) = \begin{bmatrix} 0\,0 \cdots 0\,1 \\ 0\,0 \cdots 0\,1 \\ \vdots\,\vdots\,\ddots\,\vdots\,\vdots \\ 0\,0 \cdots 0\,1 \\ 0\,0 \cdots 0\,1 \end{bmatrix}, \forall s \in \mathrm{N}.$$

Remark 11. From the previous theorem it results that

1) the sequence $(\phi_{i,m}(s,t))_{t\in\mathrm{N}}$, $\forall s \in \mathrm{N}, \forall i \in E$ is increasing.
2) $\lim_{t\to\infty} \phi_{i,m}(s,t) = 1$.

This implies that $(\phi_{i,m}(s,t))_{t\in\mathrm{N}}$, $\forall s \in \mathrm{N}, \forall i \in E$ is the d.f. of going in the absorbing state starting from state $i$at time s.

In credit risk problems it can be interesting the generalization of the Theorem 4.1 in this line.

Definition 4.4. A unireducible non-homogeneous semi-Markov chain such that:

(1) $\mathbf{B}(s,s+1)$, $\forall s \in \mathrm{N}$ is single-unireducible and they have the same topological structure.
(2) $\mathbf{B}(s,t) \forall s < t; s,t \in$ Nis a matrix with the same diagonal block structure as $\mathbf{B}(s,s+1)$.

is said to be a single-unireducible semi-Markov chain.

Theorem 4.2. *Let* $(Z(t)|t \in \mathrm{N})$ *be a semi-Markov chain and* $\phi_{i,j}(s,t) = P[Z(t) = j|Z(s) = i]$. *If the DTNHSMP is single-unireducible then*

$$\lim_{t\to\infty} \mathbf{\Phi}(s,t) = \begin{bmatrix} 0\,0 \cdots 0\,1 \\ 0\,0 \cdots 0\,1 \\ \vdots\,\vdots\,\ddots\,\vdots\,\vdots \\ 0\,0 \cdots 0\,1 \\ 0\,0 \cdots 0\,1 \end{bmatrix}, \forall s \in \mathrm{N}.$$

Proof. It results

$$\phi_{i,m}(s,t) = (1 - H_i(s,t))\,\delta_{im} + \sum_{k=1}^{m} \sum_{\vartheta=s+1}^{t} b_{i,k}(s,\vartheta)\phi_{k,m}(\vartheta,t). \tag{21}$$

If $i \in E - \{m\} \; \exists\, h, j : i \in C_j^h$ and (21) becomes

$$\phi_{i,m}(s,t) = \sum_{k=1}^{m} \sum_{\vartheta=s+1}^{t} b_{i,k}(s,\vartheta)\phi_{k,m}(\vartheta,t).$$

Taking into account all the possible states and backward values (see [6]) it results

$$\begin{aligned}\phi_{i,m}(s,t) &= \mathrm{P}[Z(t)=m|Z(s)=i]\\ &= \sum_{l=0}^{t-1-s}\sum_{k\in E}\mathrm{P}[Z(t)=m,\, Z(t-1)=k,\, B(t-1)=l|Z(s)=i]\\ &= \sum_{l=0}^{t-1-s}\sum_{k\in E}\mathrm{P}[Z(t)=m|\, Z(t-1)=k,\, B(t-1)=l, Z(s)=i]\\ &\quad\times\mathrm{P}[Z(t-1)=k,\, B(t-1)=l|Z(s)=i].\end{aligned}$$

From markovianity of backward processes (see [11]) it results

$$\begin{aligned}\phi_{i,m}(s,t) = &\sum_{l=0}^{t-1-s}\sum_{k\in E}\mathrm{P}[Z(t)=m|\, Z(t-1)=k,\, B(t-1)=l]\\ &\times\mathrm{P}[Z(t-1)=k,\, B(t-1)=l|Z(s)=i],\end{aligned}$$

that is

$$\phi_{i,m}(s,t)=\sum_{k=1}^{m}\sum_{l=0}^{t-s-1}\,{}^{b}\phi_{i,k}^{b}(s;l,t-1)^{b}\phi_{k,m}(l,t-1;t), \tag{22}$$

where, as already stated, the b apex on the left stands for initial backward and on the right for final backward

$$\begin{aligned}\phi_{i,m}(s,t) = &\sum_{l=0}^{t-s-1}\,{}^{b}\phi_{i,m}^{b}(s;l,t-1)^{b}\phi_{m,m}(l,t-1;t)\\ &+\sum_{k=1}^{m-1}\sum_{l=0}^{t-s-1}\,{}^{b}\phi_{i,k}^{b}(s;l,t-1)^{b}\phi_{k,m}(l,t-1;t),\end{aligned}$$

that is

$$\phi_{i,m}(s,t)=\phi_{i,m}(s,t-1)+\sum_{k=1}^{m-1}\sum_{l=0}^{t-s-1}\,{}^{b}\phi_{i,k}^{b}(s;l,t-1)^{b}\phi_{k,m}(l,t-1;t). \tag{23}$$

From (23) it results that the sequences $(\phi_{im}(s,t))_{t>s}$, $i=1,\ldots,m-1;\, s,t\in\mathrm{N}$ are increasing and $\phi_{im}(s,t)\leq 1$that means $\exists\lim_{t\to\infty}\phi_{i,m}(s,t),\ \forall i.$

Now we suppose that

$$\lim_{t\to\infty}\phi_{i,m}(s,t)=1-\varepsilon,\ 0<\varepsilon<1. \tag{24}$$

From (23) and (24) it results

$$\forall\delta>0\exists\bar{n}\in\mathrm{N}:\forall t>\bar{n}\Rightarrow\sum_{k=1}^{m-1}\sum_{l=0}^{t-s-1}\,{}^{b}\phi_{i,k}^{b}(s;l,t-1)^{b}\phi_{k,m}(l,t-1;t)<\delta. \tag{25}$$

Such condition implies that

$$\sum_{k=1}^{m-1} \phi^{b}_{i,k}(s;0,t-1)^{b}\phi_{k,m}(0,t-1;t) < \delta. \tag{26}$$

Furthermore:

1) ${}^{b}\phi_{k,m}(0,t-1;t) = \phi_{k,m}(t-1,t) = b_{k,m}(t-1,t)$ because it denotes the absence of backward at the starting time.

2) $\phi^{b}_{i,k}(s;0,t-1) \geq b^{(n)}_{i,k}(s,t-1)\ \forall n \leq (t-1)-s$ because the left part denotes the probability of entering in state k at time $t-1$, starting from state i at times, in any possible number of transitions. The right part denotes the same event considering just n transitions.

From (26) it results

$$\begin{aligned}
&\sum_{k=1}^{m-1} b^{(n)}_{i,k}(s,t-1) b_{k,m}(t-1,t) < \delta \Rightarrow \\
&\sum_{k=1}^{m-1} \sum_{\alpha \in E} b^{(r)}_{i,\alpha}(s,t_h) \cdot b^{(n-r)}_{\alpha,k}(t_h,t-1) \cdot b_{k,m}(t-1,t) < \delta \Rightarrow \\
&\sum_{k \in C^h_{l_h}} \sum_{\alpha \in C^h_{l_h}} b^{(r)}_{i,\alpha}(s,t_h) \cdot b^{(n-r)}_{\alpha,k}(t_h,t-1) \cdot b_{k,m}(t-1,t) < \delta.
\end{aligned} \tag{27}$$

Without loss of generality we assume that $m-1 < r < n-(m-1)$.

Since $C^h_{l_h}$ is a minimal transient class it$\exists\, k \in C^h_{l_h} : b_{k,m}(t-1,t) > 0$.

Setting $\theta = \min_{\alpha,k \in C^h_{l_h}} \left\{ b^{(n-r)}_{\alpha,k}(t_r,t-1) \right\}$it results that $\theta > 0$ in fact $\alpha \in C^h_{l_h}$, $k \in C^h_{l_h}$ and $C^h_{l_h}$ corresponds to a circuit matrix then from Lemma 4.1 $\mathbf{B}^{(n-r)}(t_r,t) \gg \mathbf{0}$.

It should be $\theta \sum_{k \in C^h_{l_h}} \sum_{\alpha \in C^h_{l_h}} b^{(r)}_{i,\alpha}(s,t_r) \cdot b_{k,m}(t-1,t) < \delta$ and setting $\Theta = \theta \cdot \sum_{\alpha \in C^h_{l_h}} b^{(r)}_{i,\alpha}(s,t_r), \Theta > 0$ because $i \in C^h_j$ and $\alpha \in C^h_{l_h}$ with $C^h_j \geq C^h_{l_h}$ then it exists a sequence of times $\{s,t_1,t_2,\ldots,t_r\}$ and a sequence of states $\{i,x_{t_1},x_{t_2},\ldots,x_{t_r}\}$ such that

$$P(X_r = x_{t_r} \in C^h_{l_h}, T_r = t_r | X_{N(s)} = i \in C^h_j, T_{N(s)} = s) = b^{(r)}_{i,x_{t_r}}(s,t_r) > 0.$$

Then we should have

$$\sum_{k \in C^h_{l_h}} b_{k,m}(t-1,t) < \frac{\delta}{\Theta} \tag{28}$$

but nothing can ensure (28) for single-unireducible semi-Markov chains, consequently, (28) is false. This means that $\lim_{t\to\infty} \phi_{i,m}(s,t) = 1-\varepsilon$, $0 < \varepsilon < 1$ can not be true $\forall \varepsilon$ and

$$\lim_{t\to\infty} \phi_{im}(s,t) = 1, \quad \forall i = 1,\ldots,m-1.$$

□

References

1. Christofides N (1975) Graph theory. An algorithmic approach. Academic, New York
2. Cinlar E (1975) Introduction to stochastic processes. Prentice Hall, New York
3. Cohn H (1975) Finite non-homogeneous Markov chains: asymptotic behaviour. Adv Appl Probab 8:502–516
4. Cohn H (1989) Products of stochastic matrices and applications. Int J Math Math Sci 12:209–233
5. D'Amico G, Janssen J, Manca R (2005) Homogeneous discrete time semi-Markov reliability models for credit risk Management. Decis Econ Finance 28:79–93
6. D'Amico G, Janssen J, Manca R (2005) Starting and ending backward and forward non-homogeneous semi-Markov recurrence time processes. In: Proceedings of the XXIX AMASES, Palermo
7. D'Amico G, Janssen J, Manca R (2006) Discrete time non-homogeneous semi-Markov reliability transition credit risk models and the default distribution functions (submitted)
8. De Dominicis R, Manca R (1984) A computational procedure for the asymptotic analysis of a homogeneous semimarkov process. Stat Probab Lett 2:249–253
9. Isaacson DL, Madsen RW (1976) Markov chains. Theory and applications. Wiley Series in Probability and Mathematical Statistics. Wiley, New York
10. Janssen J, Manca R (2006) Applied semi-Markov processes. Springer, New York
11. Limnios N, Oprian G (2001) Semi-Markov processes and reliability modeling. Birkhauser, Boston
12. Madsen RW, Isaacson DL (1973) Strongly ergodig behaviour for non-stationary Markov processes. Ann Probab 1:329–335
13. Paz A (1970) Ergodic theorems for infinite probabilistic tables. Ann Math Stat 41:539–550
14. Seneta E (1981) Non-negative matrices and Markov chains. Springer, New York
15. Sonin I (1990) The asymptotic behaviour of a general finite nonhomogeneous Markov chain (the decomposition-separation theorem). Statistics, Probability and Game Theory. IMS Lecture Notes – Monograph Series, vol 30, 337–346
16. Tewarson RP (1973) Sparse matrices. Academic, London

Part IV
Applications

Shareholding Networks and Centrality: An Application to the Italian Financial Market

M. D'Errico, R. Grassi, and S. Stefani, and A. Torriero

Abstract In this paper we studied the Shareholding Network (SN) embedded in the Italian Stock Market (MIB). We identified the central companies both in the role of transferring information flows and controlling companies. To this end we used betweenness and flow betweenness centrality measures, together with in and out degree. We tested the scale-free property on in and out degree, betweenness and flow betweenness centrality. The effect of external shocks to SN and the different extent on which companies react to them are measured relating asset volatility and betweenness.

1 Introduction

As it is well known from firm growth theory, companies can substantially reduce their risk by diversifying production (even in presence of important institutional constraints), create scale economies thereby reducing their average costs and coping with the necessary dimensional growth due to the market structure, globalization, etc., increase their market power by implementing strategies aimed at controlling other companies.

For these reasons companies tend to acquire other companies and exercise their shareholding power. Due to the need to survive in a market becoming large and integrated, mergers and control acquisition over companies are phenomena quite

M. D'Errico, R. Grassi, and S. Stefani
Dipartimento di Metodi Quantitativi per le Scienze Economiche ed Aziendali
Università degli Studi di Milano – Bicocca, Milan, Italy
ma.derrico@gmail.com, rosanna.grassi@unimib.it, silvana.stefani@unimib.it

A. Torriero
Dipartimento di Discipline matematiche, Finanza matematica e Econometria
Università Cattolica del Sacro Cuore di Milano, Milan, Italy
anna.torriero@unicatt.it

A.K. Naimzada et al. (eds.), *Networks, Topology and Dynamics*,
Lecture Notes in Economics and Mathematical Systems 613,

common in Europe nowadays (see the recent case Banco Santander-San Paolo) and networks of companies are created by connection through shareholding. Thus it is crucial to investigate the stability of such networks or, in other words, to ascertain how an external shock propagates through the network and which the effect on the market performance of the controlled and controlling companies can be. The Shareholding Network (SN) is the tool we use to analyze the shock transmission effects among companies which are connected by shareholding. We agree with [2] that although understanding the properties of a (national or supernational) system is an important strategic issue in economics, the topological structure of ownership and shareholding networks has not yet been thoroughly studied. Such networks seem sensitive to local bankruptcies and it is recognized that a failure on one or more companies related to each other by ownership ties can have unexpected effects on the market as a whole (see for instance the cases Enron and Parmalat). In the light of these anomalies one can wonder which is the *best* topological structure of a SN, i.e. whether a network which is well connected can stand disruption better than a network composed of many connected components. We will see, in an application to the Italian Market MIB, that the topology of the network can give indications on the market structure and its components.

Recent research has been carried out to model financial markets as a network. More precisely, in [3] and [4] a financial network is constructed representing the stock market data, which is obtained by calculating cross-correlations of stock price returns based on the opening prices over a certain period of time. Two vertices are connected if the correlation coefficient (for a given time span) exceeds a specified threshold. The structural properties of the market graph over time are also studied to obtain useful information on the dynamics of the stock market. In [7] a correlation-based network is constructed in the same way. A metric distance based on correlations, and structural properties, is introduced through minimum spanning trees at different time horizons, are deduced. In [13] a financial community is defined as a group of stocks selected by common opinion makers. By analyzing a large data base of web opinions, the authors relate connectedness to return correlations and find stocks that are hubs for information flows. Stocks with high centrality scores tend to have longer average covariance with other stocks than those with low scores.

In this paper we adopt a similar approach as in [1, 2], and [23], and model as a SN the universe of the quoted companies in the Italian Market MIB related to each other by shareholding ties.

In [17] some statistical properties of NYSE, NASDAQ and MIB are analyzed. In [23] the Japanese SN is studied with a special attention to the automotive industry and several properties of the sector are discussed.

Another important issue in SN refers to the detection of *dominant* subgraphs, i.e. a group (or groups) of companies that control the market through ownership. We believe that exploring topological measures to identify preeminent companies in the network is the right approach and in particular we refer to centrality. In the literature there are many contributions on centrality, both theoretical and empirical (see, for example [5, 6, 10] and [15]) and the concept is broad enough to allow us to choose which of the centrality measures fits our needs. A formal definition of betweenness and flow betweenness centrality was introduced first in [14] and [16] respectively; we refer to Borgatti [8] for other definitions and their implications. For a theoretical unifying approach see [18], where it is shown that all centrality measures, under certain conditions, are compatible with structural or automorphic equivalence. In [19] the structural properties related to betweenness are deeply investigated, in particular some cases in which betweenness attains its extremal values are examined; moreover, the new concept of total betweenness is defined and related to group betweenness, and a necessary and sufficient condition for the two measures to coincide is proved.

By introducing indegree, outdegree, betweenness and flow betweenness as centrality measures, we detect central companies under the different concepts: as far as the flow of information is concerned, the role of banks and insurance companies is clearly identified as the core of the Italian system. They play the role of hubs and we find evidence of a scale-free behaviour. Further, we relate asset volatility to centrality and quite reasonably find that central companies are less volatile. This means that external shocks tend to be *absorbed* by the central companies and the stability of the system is ensured by hubs. On the contrary, peripheral nodes are more volatile and then more sensible to external shocks. On the other hand, as soon as we recur to flow betweenness, i.e. we take into account the amounts of holding quotas, the scenario changes and industrial groups emerge, showing that large part of the MIB-quoted companies relate to seven companies, four of which lead to three major industrial groups.

Our model is not based on portfolio diversification (see, e.g. [2]): we claim, and this is confirmed by our results, that detection of ownership quotas can only in part be related to diversification purposes. In fact, as a side result of this research we confirm the intuition mentioned in [2]: the well known Capital Asset Pricing Model does not apply in our SN network. Companies diversify but not according to the market portfolio, i.e. investing in all companies in the market with a quota given by the respective capitalization relatively to the market capitalization, but instead tend to hold shares of companies of the same sector (this is a quite common practice in the production sector). Hubs, in our case essentially financial institutions, diversify. This result is supported by the distribution of holding quotas, showing a typical scale-free pattern.

The paper is structured as follows: in Sect. 2 we report preliminary definitions on graphs and scale-free networks; in Sect. 3 we describe all centrality measures used in this work; in Sect. 4 we introduce the model and the data set; the results are shown in Sect. 5. Conclusions and suggestions for further research are in the last section.

2 Preliminaries

A graph (denoted by $G(V,E)$) is a pair of sets (V,E), where V is the set of n nodes and E is the set of m unordered pairs of nodes of V; the pair $(i,j) \in E$ is called an edge of G and i and j are called adjacent $(i \sim j)$. A graph is simple if it is without loops and multiple edges. An undirected graph is a graph in which if $(i,j) \in E$, then $(j,i) \in E$. The degree $d(i)$ of a node i is the number of edges on which the node is incident.

A path is a sequence of adjacent nodes, where all nodes are distinct. The shortest path starting from a node i and ending in j is said geodesic. We will indicate as $i-j$ the geodesic starting from i and ending in j.

The distance $d(i,j)$ between two vertices i and j is the length of a shortest path from i to j.

A graph is connected if for each pair of nodes i and j, $(i,j = 1,2,\dots,n)$, there is a path beginning in i and ending in j. A connected component of $G(V,E)$ is the maximal connected subgraph of $G(V,E)$; obviously, a graph is connected if and only if it has only one connected component, otherwise G is disconnected.

A graph is complete if every pair of nodes is adjacent; we denote the complete graph on n nodes with K_n; the graph K_2 is called triangle.

The adjacency matrix $A(G)$ of an undirected graph is a matrix in which the ij-th entry is 1 if $i \sim j$, and 0 otherwise.

A weighted graph is a graph G where every edge (i,j) has a weight $w(i,j) = w_{ij}$ associated with it. We indicate a weighted graph with $N(V,E)$, (*network*) where W is the finite set of the m weights, each for every edge. If $N(V,E,W)$ is a network, $G(V,E)$ is the underlying graph.

A digraph $D(V,A)$ is a pair of sets (V,A), where A is the set of m ordered pairs of nodes of V; the pair $(i,j) \in A$ is an arc of D. A pair of arcs joining nodes i and j is said symmetric if both arcs (i,j), $(j,i) \in A$; an oriented graph is a digraph $D(V,A)$ having no symmetric pairs of arcs.

All the previous definitions about G can be naturally extended to D; some definitions, useful for our results, are recalled here.[1] If there is an arc starting from i and ending in j we say that i is adjacent to j and j is adjacent from i; the in-degree $d^{in}(j)$ of a node j is the number of adjacents to i, and the out-degree $d^{out}(j)$ of j is the number of adjacents from j.

The degree function $p(k)$ gives the number of nodes with exactly k edges. A random variable (r.v.) K is distributed as a power law if its probability distribution is

$$f(k) = ck^{-\alpha},$$

where α is a positive parameter and c is a positive scale factor (see for example [20]), so $f(k)$ gives the probability that a node in the network has degree k.

A r.v. has a lognormal distribution if $Y = \ln K$ is normally distributed. Hence the density function of a lognormal satisfies

$$f(k) = \frac{1}{\sqrt{2\pi}\sigma k} e^{-(\ln k - \mu)^2/2\sigma^2},$$

where μ and σ^2 are mean and variance of Y respectively.

The lognormal and power law distributions are similar in shape. In a log–log plot, while the power law can be easily fitted by a straight line, the lognormal will

---

[1] For more definitions and properties see for example [10].

generally appear to be nearly a straight line for a large portion of the body of the distribution. If in particular the variance of the degree distribution is large, the distribution may appear linear for a large range of values [11, 22].

In a scale-free network only few nodes have a high degree (hubs), while most nodes have a low degree. As a consequence, the degree distribution is characterized by a high variance or variation coefficient [20].

It has been recognized [11, 22] that generative processes leading to scale free networks (like the preferential attachment property) yield to power law or lognormal distributions. In many applications it has been proved that the degree distribution follows a power law where $2 < \alpha < 3$ [20]. In those cases $f(k)$ has infinite variance.

3 Centrality Measures

In this section we review some definitions about centrality measures which will be needed later. Among various centrality indices that have been introduced over the years, we will focus our attention on: indegree, outdegree, betweenness and flow betweenness centrality.

Actually, as we discuss later on, we consider them as the most appropriate measures for our analysis.

The degree centrality is the simplest centrality measure, related to degree, and indicates how many neighbours each node has: a node is degree central if it is adjacent to many other nodes. Formally:

Definition 1. Let $G(V,E)$ be a graph and let $i \in V$. The degree centrality $d(i)$ is the number of edges incident to i.

This definition can be generalized to a digraph $D(V,A)$ as follows:

Definition 2. The indegree centrality $d^{in}(i)$ and the outdegree $d^{out}(i)$ of a node i in a digraph are the number of arcs directed respectively to and from i.

The betweenness centrality, related to geodesics, identifies the intermediary role of a node in a graph: a node is betweenness central if it lies on many geodesics between other nodes. It gives the idea of traffic/information volume flowing between two nodes and passing through the intermediary. A node is central if it controls a large amount of the information flowing in the network. Formally:

Definition 3. Let $G(V,E)$ be a graph and let $k \in V$. The betweenness centrality $b(k)$ is defined as

$$b(k) = \sum_{i<j} \frac{g_{ij}(k)}{g_{ij}}, \qquad i,j \neq k,$$

where g_{ij} is the number of geodesics from i to j, and $g_{ij}(k)$ is the number of geodesics between i and j passing through k.

The normalized betweenness is

$$\frac{b(k)}{\binom{n-1}{2}} \,.$$

For valued graphs (networks), the flow betweenness centrality, related to the concept of maximum flow, generalizes the notion of betweenness. All paths between nodes, not only geodesics, are considered. Formally:

Definition 4. Let $N(V,E,W)$ be a network. For a node k in V, the flow betweenness centrality $fb(k)$ is

$$fb(k) = \sum_{i<j} m_{ij}(k)\,, \qquad i,j \neq k\,,$$

where $m_{ij}(k)$ is the maximum flow from node i to node j passing through k.

Thus the flow betweenness measures the contribution of a node to all possible maximum flows.

4 The Model

4.1 The Data Set

We considered the universe of companies ($n = 223$) quoted in MIB (Milan Stock Exchange) in the time span 1/1/2004, 31/12/2004. Shareholding data are taken from [9, 12, 21]. We computed the daily returns and the historical volatility as the daily return standard deviation σ_i $(i = 1,\ldots,n)$. Since the quantity held by investment funds (even in a short period) can be floating, to get a single value we referred to [21].

4.2 The Shareholding Network (SN)

We investigate the Italian market (MIB) and focus on some topological properties. The graph associated with the shareholders and quoted companies network is weighted and oriented. Companies are the nodes; an arc is drawn from the shareholder to the owned company, weighted by the proportion of shares held by the shareholder. Denoting by q_{ij} the number of shares of the j-th company held by the i-th company and by cap_j the market value of j, the weight is given by

$$\widetilde{w}_{ij} = \frac{q_{ij}}{cap_j} \qquad i,j = 1,\ldots,n$$

where n is the number of quoted companies. We considered as shareholders only quoted companies, i.e. we did not include shares held by companies not MIB-quoted. The weight w_{ij} can be interpreted as a measure of the capacity flowing through an arc: no more than 100% can flow towards the owned company and this is an obvious upper limit to shareholding. Another interpretation is based on reliability theory: the weight can be seen as the probability that an information can through flow freely the arc. We introduced a threshold w^0 of 0.01%, so our weights are

$$w_{ij} = \begin{cases} 0, & \text{if } \widetilde{w}_{ij} - w^0 \leq 0, \\ \widetilde{w}_{ij}, & \text{if } \widetilde{w}_{ij} - w^0 > 0. \end{cases}$$

(See, for instance, [13] or [3] under which the weights are set to zero.) The idea behind is that a low participation share has practically no influence on the owned company's strategy and consequently on its market performance.

A non-symmetric weight matrix $W = [w_{ij}]$, $i, j = 1, \ldots, n$ is generated.

Our first aim is to identify the SN central companies, taking in mind that centrality depends on the particular features we want to discover.

We computed some centrality measures: indegree, outdegree, betweenness[2] and flow betweenness. The first two are related to degree centrality and give an immediate idea of the direct connections of nodes: a high outdegree means high shareholding, a high indegree means a high number of company's shareholders. Note that while the indegree, possibly with the sum of weights, indicates precisely which fraction of the company is detained by other companies, the outdegree does not provide information on the amount of capital involved and indicates quite roughly the *diversification* attitude of the company.

On the other hand centrality measured by betweenness allows an interesting insight into the inner SN structure. Then we assess how an external shock propagates through the network and how companies are affected. In the light of the efficiency theory of capital markets, we assume here that the market is immediately receptive of external shocks so prices adjust instantaneously. Betweenness is based on the number of shortest paths passing through each node connecting other pairs of nodes and so relies on the idea that an information or a shock takes the shortest path to reach a node. Since it has to be computed on the undirected and not weighted graph, what we can deduce from a high betweenness is the fact that the company is a nodal point in the flow of information but it can be either a controlling or a controller.[3]

Viceversa a low betweenness shows that the company is not an information intermediary. On the other hand, flow betweenness, by definition, does not require the information flowing necessarily through the shortest path (thereby suggesting an unreasonable delay in transferring information) but, while maximizing the flow, considers all paths: this seems a more reasonable approach to describe control. An indirect control of a company can be achieved by reaching it via different paths, passing through more and more companies.

---

[2] Note that in order to calculate betweenness centrality we limited ourselves to the adjacency matrix $A = [a_{ij}]$ where 1 indicates the presence of shareholding. This implies that if $w_{ij} > 0$ then $a_{ij} = 1$. In this case the graph is neither weighted nor oriented.

[3] We can define roughly as controller a company whose capitals devoted to *diversification* are much higher than the own capitals held by others.

5 The Results

Taking out the isolated nodes[4] (102), the resulting graph is as in Fig. 1.

The SN is composed of five connected components: we will concentrate our analysis on the largest connected one (109 nodes).

5.1 Degrees, Betweenness and Information

The most outdegree and indegree central nodes are reported in Table 1.

Banca Intesa ($d^{out} = 88$) and S. Paolo IMI ($d^{out} = 74$) are the highest outdegree companies[5] and account for 29% of all outdegrees. Not surprisingly the first ten companies with highest outdegree are all financial institutions (banks or insurance companies) and account for 73% of all outdegrees. This confirms the key role of financial intermediaries in the Italian economy and qualify them as possible hubs. We checked the presence of the scale-free property in outdegree and indegree. See the outdegree logarithmic plot in Fig. 2.

The outdegree distribution shows a high volatility ($\mu = 5.18$, $\sigma = 13.73, \sigma/\mu = 2.65$) and the power law function ($\alpha = 1.39$) fits nicely the distribution ($R^2 = 0.94$).

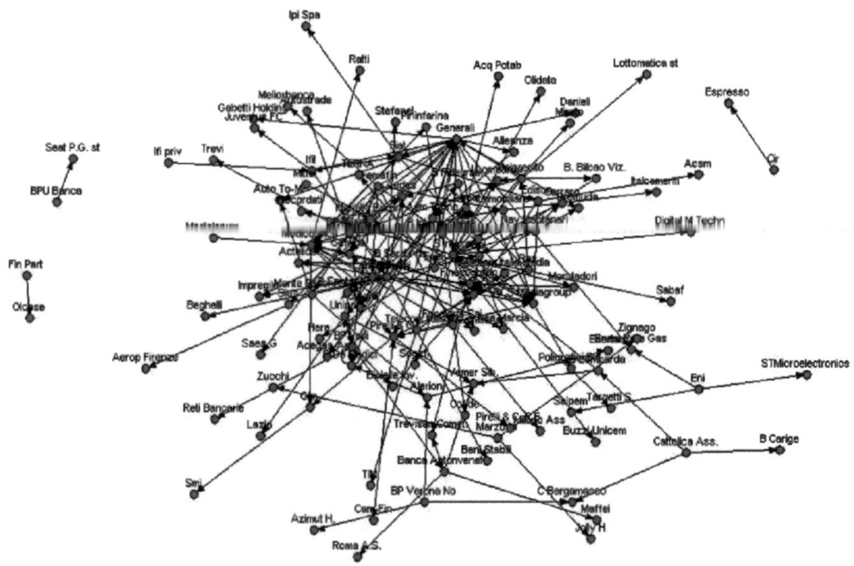

Fig. 1 SN without isolated nodes

[4] We do not consider control quotas held by non MIB-quoted companies.

[5] Banca Intesa and S. Paolo IMI merged at the end of 2006 creating the sixth largest financial institution in the world.

Table 1

Name	d^{out}	Name	d^{in}
B. Intesa	88	Autostrade	13
S. Paolo IMI	74	B. Intesa	12
RAS	60	RCS Mediagroup	11
B. Fideuram	46	Alleanza	10
B. Sardegna	37	Mediobanca	10
BP Lodi	27	BPU Banca	10
Generali	22	Autogrill	10
B. Intermobiliare	20	B. Antonveneta	9
Fondiaria-SAI	18	B.P. Verona No	9
BNL	18	Ass. Generali	9

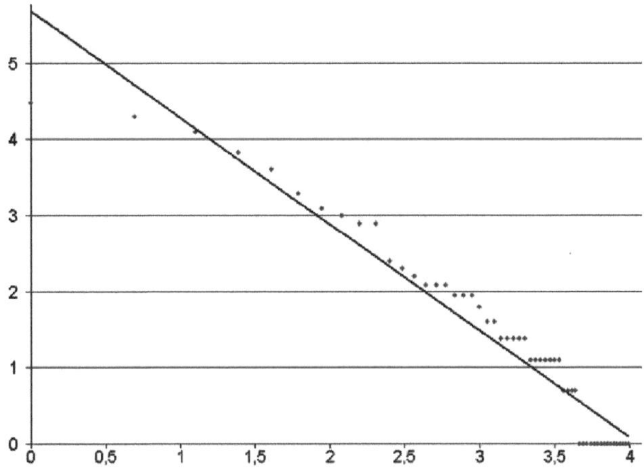

Fig. 2 Log outdegree – $R^2 = 0.94$, $\alpha = 1.39$

The indegree distribution shows a more regular pattern ($\mu = 4.15$, std. dev $\sigma = 2.79$, $\sigma/\mu = 0.672$) and a power law fitting of $\alpha = 0.62$ ($R^2 = 0.79$). Knowing that the scale-free property is a necessary condition for the presence of hubs (see [17, 20]), detection of scale-free in outdegree means the possible presence of hubs, while a less pronounced scale-free in the indegree sequence shows a kind of assortative mixing (see [23]). Thus there are very few companies that diversify (essentially banks and insurance companies), while most of the companies in the biggest connected component have scarce direct links or are controlled (56 out of 109 companies have zero outdegree).

This result is confirmed by betweenneess values.

Banca Intesa and S. Paolo-IMI are again the highest betweenness central companies and the 12 most central companies are all banks and insurance companies, but not necessarily the highest in outdegree, as can be seen in Fig. 3.

A scale free behaviour is confirmed by the fitting of betweenness centrality; a log-normal curve fits better the distribution than a power law ($R^2 = 0.95$ vs $R^2 = 0.82$,

Company	Betweenness (norm)	Out-degree	In-degree
B Intesa	0,7058162	88	12
S. Paolo - Imi	0,5335697	74	5
Ras	0,2986899	60	0
B Fideuram	0,2030733	46	8
Assicurazioni Generali	0,1234863	22	9
Bnl	0,1157525	18	8
BP Lodi	0,1043315	27	6
B Sard	0,0915177	37	1
Banca Antonveneta	0,0713815	5	9
B Intermobiliare	0,0691372	20	6
BP Verona No	0,0577655	3	9
Monte Paschi Si	0,0541684	7	4

Fig. 3 The 12 most betweenness central companies

$\alpha = 2.39$; the distribution: $\mu = 0.03$, $\sigma = 0.09$, $\sigma/\mu = 3.02$, see Fig. 4). This is explained by the magnitude of the variance. The distribution is not so volatile like in and out degrees and the second order interpolation gives a higher R^2 since the coefficient of the second order power is a function of the reciprocal of the variance [11, 22].

In this scenario it is interesting to understand how the information intermediaries react to external shocks and how the flow of information spreads along the paths. To this end we related historical stock volatility and betweenness. Our conjecture is that a highly betweenness central company, when it receives sudden information to be spread in the network, is actually more stable than less central nodes (we assume that stability means relatively low stock volatility). The central companies act as a *cushion* and partially absorb with their strength the incoming shocks. This can be seen in Fig. 5: the highest betweenness central companies show strong stability, while there are low betweenness companies that are highly volatile. This could also probably due to low capitalization. The four highest betweenness-lowest volatility are Banca Intesa, San Paolo-Imi, Ras, Fideuram; the three highest volatility-low betweenness are Geox, Bipielle Inv., Impregilo.

During our data collection two companies went bankrupt: Giacomelli and Cirio, the first a singleton, the second belonging to a triangle. Whether they should have survived had they belonged to the biggest component is a matter out of the scope of this work.

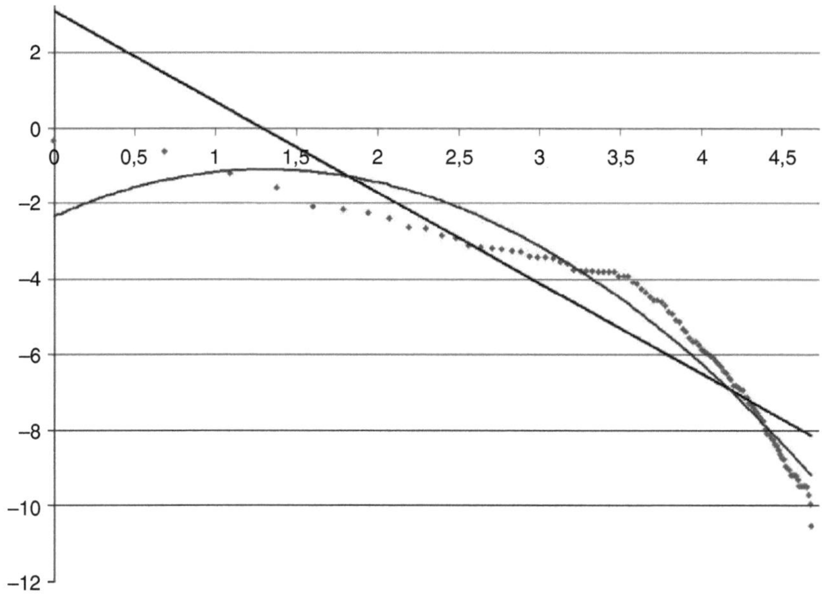

Fig. 4 Log Betweeness centrality Second order: $R^2 = 0.95$ (blue); first order: $R^2 = 0.82$ (black), $\alpha = 2.39$

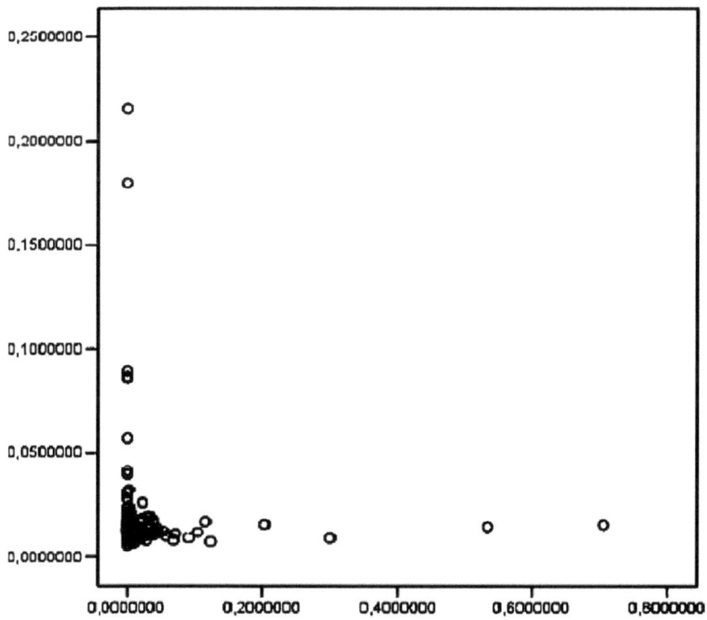

Fig. 5 Historical volatility and betweenness

5.2 Flow Betweenness and Control

Another interesting feature that can be studied in SN is related to control. In our opinion a possible way to detect the *real controllers* is by the use of flow betweenness. This centrality measure takes into account weights and considers all possible paths from a node to another, thereby revealing the possible ways a company can follow to control indirectly another company, while maximizing flow. By computing the flow betweenness of all nodes belonging to the largest connected component, we found out that 34 nodes out of 109 show a flow betweenness different than zero and seven companies emerge, as in the Table 2.

A bank and an insurance company (Mediobanca and Generali) are still at the top, even though they are not among the highest in betweenness and outdegree, but now the industrial groups show their importance: Pirelli, Fiat and Edison. Their central role is apparent. A particular case is RCS, an editorial company, owner of the famous newspaper Corriere della Sera. RCS was recently (after 2004) at the center of a struggle for control that ended up in judicial courts.

We checked again the presence of scale free in flow betweenness. As in betweenness, a better fitting is obtained by a second order interpolation, i.e. a lognormal instead of a power law ($R^2 = 0.97$ vs $R^2 = 0.86$ with $\alpha = 2.41$; the distribution: $\mu = 160.57$, $\sigma = 621.94$, $\sigma/\mu = 3.87$; see Fig. 6). Thus, the seven companies play the role of "control hubs".

6 Conclusions

This paper focuses on some topological properties of the financial Italian Market (MIB), from the point of view of the shareholding structure. More specifically, suitable centrality measures are used to perform our analysis. First we suggest that betweenness centrality can identify companies which are central in the information flow; then, recognizing that the control of a company can be exercised also in an *indirect* way, i.e. through a chain of shareholding not necessarily via the minimum path, we propose flow betweenness to detect the indirect control. The results

Table 2 Flow betweenness highest central companies

	Flow betw	Betw (norm.)	Outdeg	Indeg
Mediobanca	4,196.5	0.0375	9	10
Generali	3,565.28	0.1235	22	9
Pirelli	2,832.08	0.0098	5	7
RCS Mediagroup	2,463.08	0.0380	2	11
Fiat	1,858.42	0.0198	4	4
Edison	737.274	0.0232	4	1
Ifil	641.26	0.0000	3	5

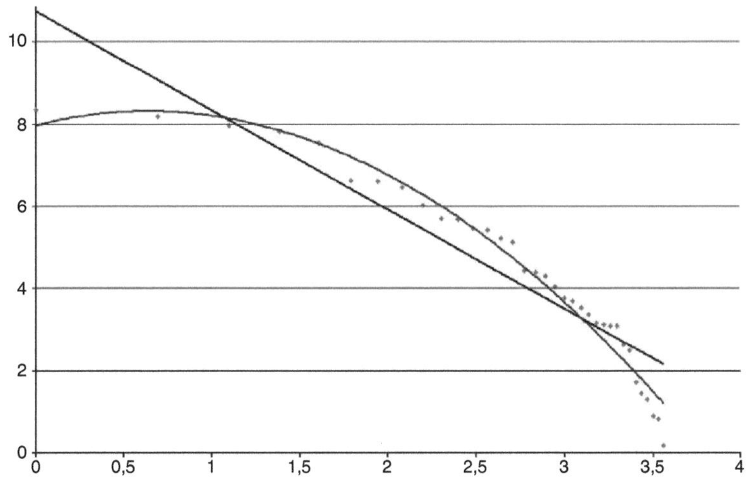

Fig. 6 Log Flow-betweenness centrality Second order (blue): $R^2 = 0.97$; first order (black): $R^2 = 0.86$, $\alpha = 2.41$

show, quite obviously indeed, that each centrality measure helps detecting revealing specific network properties and companies which are central in one sense are in fact playing a secondary role in another. An extension of this study to further years could help capturing the dynamic structure of the market and the possible changes in shareholding.

Acknowledgements We would like to thank Andrea Mangone for valuable computing assistance and the referees for their helpful comments. All errors are our own.

References

1. Battiston S (2004) Inner structure of capital control networks. Physica A 338:107–112
2. Battiston S, Garlaschelli D, Caldarelli G (2003) The hidden topology of market investment networks – draft. http://www.lps.ens.fr
3. Boginsky V, Butenko S, Pardalos P (2003) On structural properties of the market graph. In: Nagurney A (ed) Innovations in financial and economic networks. Edward Elgar, Cheltenham
4. Boginsky V, Butenko S, Pardalos P (2005) Statistical analysis of financial networks. Comput Stat Data Anal 48(2):431–443
5. Bonacich P (1987) Power and centrality: a family of measures. Am J Sociol 92(5):1170–1182
6. Bonacich P, Lloyd P (2001) Eigenvector-like measures of centrality for asymmetric relations. Soc Networks 23:191–201
7. Bonanno G, Caldarelli G, Lillo F, Micciché S, Vandewalle N, Mantegna RN (2004) Network of equities in financial markets. Eur Phys J B38:363–371
8. Borgatti SP (2002) Centrality and network flow. Soc Networks (2005) 27(1):55–71
9. Borsa Italiana. http://www.borsaitalia.it, April 2005
10. Buckley F, Harary F (1990) Distance in graphs. Addison–Wesley, Reading, MA

11. Caldarelli G (2007) Scale free networks: complex webs in nature and technology. Oxford University Press, Oxford
12. CONSOB, Commissione Nazionale per le Società e la Borsa. http://www.consob.it, April 2005
13. Das SR, Sisk J (2004) Financial communities. Working paper, Leavey School of Business, Santa Clara University and Anderson School of Management, University of California, Los Angeles, USA
14. Freeman LC (1977) A set of measures of centrality based on betweenness. Sociometry 40(1):35–41
15. Freeman LC (1979) Centrality in social networks: conceptual clarification. Soc Networks 1:215–239
16. Freeman LC, Borgatti S, White DR (1991) Centrality in valued graphs: a measure of betweenness based on network flow. Soc Networks 13:141–154
17. Garlaschelli D, Battiston S, Castri M, Servedio VDP, Caldarelli G (2003) The scale-free topology of market investments, e-print, arXiv:cond–mat/0310503
18. Grassi R, Stefani S, Torriero A (2006) Central vertices in networks: a unified approach. Working paper no. 105, Dept Quantitative Methods, University of Milano – Bicocca, Milan, Italy
19. Grassi R, Scapellato R, Stefani S, Torriero A (2008) Betweenness centrality: extremal values and structural properties. In: Naimzada A, Stefani S, Torriero A (eds) Lecture notes in economics and mathematical systems: networks, topology and dynamics: theory and applications to economic and social systems. Springer, Heidelberg
20. Li L, Alderson D, Doyle JC, Willinger W (2004) Towards a theory of scale-free graphs: definition, properties, and implications. Internet Math 2(4):431–523
21. Meridiana dell'investitore. http://www.bnl.it/pagina.asp
22. Mitzenmacher M (2003) A brief history of generative models for power law and lognormal distributions. Internet Math 1(2):226–251
23. Wataru Souma, Yoshi Fujiwara, Hideaki Aoyama (2005) Heterogeneous economic networks. arXiv:physics/0502005, v1

Further Reading

1. Barabàsi AL, Albert R (1999) Emergence of scaling in random networks. Science 286:509–512
2. Barrat A, Barthélemy M, Vespignani A (2005) The effects of spatial constraints on the evolution of weighted networks. J Stat Mech P05003
3. Barthélemy M (2004) Betweenness centrality in large complex networks. Eur Phys J B38:163–168
4. Barthélemy M, Barrat A, Pastor-Satorras A, Vespignani A (2005) Characterization and modeling of weighted networks. Physica A 346:34–43
5. Bollobás B (2001) Random graphs, 2nd edn. Cambridge University Press, Cambridge
6. Harary F (1969) Graph theory. Perseus, Cambridge, MA
7. Wasserman S, Faust K (1994) Social Network Analysis. Methods and Applications. Cambridge University Press, Cambridge
8. White DR, Borgatti SP (1994) Betwenness centrality measures for directed graphs. Soc Networks 16:335–346

Network Dynamics when Selecting Work Team Members

Arianna Dal Forno and Ugo Merlone

Abstract When selecting work team members several behavioral components concur. In this chapter we summarizes our line of research on this topic; here, we articulate our results and provide suggestions for extending our analysis in order to shed light on the selection of work team members. First, a computational model – together with a theoretical approach and the results of two human experiments where subjects interact in a similar game – allows us to identify some of the most important determinants. Our results suggest that the occurrence of two factors is crucial: the presence of leaders as aggregators of knowledge and the presence of agents able to expand and improve their higher profit projects. Second, we explicitly assume the presence of formal leaders. By analyzing the results of this modified model, we shed light on the conditions which allow the emergence of effective leaders.

1 Introduction

Human interaction and group formation in the workplace is an important aspect in terms of performance and satisfaction and has been studied by several authors (for instance, [16]). In [14] it is claimed that the literature about intra-organizational network has largely ignored the literature about formal teams. In social network analysis various methods are available. Among others, some important threads have included the development of mathematical tools (namely, graph theory) to characterize networks, and the development of statistical tools to analyze the interdependencies peculiar to networks.

A. Dal Forno and U. Merlone
Dipartimento di Statistica e Matematica Applicata, Università di Torino,
Piazza Arbarello 8 10122 Turin, Italy
dalforno@econ.unito.it, merlone@econ.unito.it

A.K. Naimzada et al. (eds.), *Networks, Topology and Dynamics*,
Lecture Notes in Economics and Mathematical Systems 613,

In [22] the Author presents a model of evolution of friendship network where the dynamics of the network structure is considered as the result of individual characteristics and of behavioral rules such as preferences for similar friends.

Previous work provides a description of the mathematical models for network evolution when ties are directed and the node set is fixed [1]. There, it is also shown that many of these models tend to an asymptotically connected network.

Some authors present an empirical study on group composition [10]. Their findings show that, when selecting group members, people are biased towards others of the same race, or towards others who have a reputation for being competent and hard working, or towards others with whom they have developed a strong working relationship.

The computational approach allows a sort of "What if" analysis, and simulation can be used to establish constructive sufficiency [21]. This may be helpful in complex models where analytical results may be difficult to obtain, and in which the consequences depend partly on random or pseudorandom processes. It may also be a source of other insights into the relationship between the assumptions and the consequences. Of course, the explanation has a greater impact when the relationship between the assumptions and the result is nonobvious, and is supported by empirical evidence.

This paper summarizes our line of research on this topic; here we articulate our results which were published in previous contributions and provide suggestions for extending our analysis in order to shed light on the selection of work team members. Our line of research is articulated in several stages. First, in [5] we proposed a theoretical model of social interaction for the study of network dynamics, then in [6] we analyze human subjects behavior when interacting in the proposed model. Finally, in [8], such behaviors were implemented to simulate the evolution of artificial agent populations in a similar context. From both the analysis and the comparison of these data two major issues emerged: the existence of leaders and their role in the interaction, which leads to a further stage of research, concerning the emergence of effective leaders. The results of this line of research were published in [7].

The structure of the paper is the following. In Sect. 2 the underlying theoretical model is presented and the unique Nash equilibrium and social equilibrium are provided. Section 3 displays the results obtained after the implementation of observed behavioral rules. Sections 4 and 5 are devoted to the proposal of our future lines of research on this topic, namely the role of communication among agents and its effectiveness, and a quantitative study of networks that takes into account a measure *ad hoc* of influence (i.e., leadership).

2 The Theoretical Model

The organization consists of n agents univocally identified by an index $i \in N = \{1,2,\dots,n\}$. Agents interact forming projects in which at most m members can participate. In the artificial simulations and the human subjects experiments we fixed $m = 7$ (for an empirical motivation of this choice the reader may refer to Chap. II in [15].

Each agent can choose its partners from a subset $M \subseteq N$ of known people. Acquaintance of agents in the organization is described using a sociomatrix **K**. Each element k_{ij} of the sociomatrix **K** indicates whether agent i knows agent j: zero indicates that i does not know j; by converse, value one indicates that i knows j. We assume that each agent knows itself; as a consequence all diagonal entries are set to one. **K** is not necessarily a symmetric $n \times n$ matrix.

Agents can participate in at most two projects; in each of them two decisions have to be taken:

1. They must specify the designated members of the project.
2. They must specify the effort they will exert in a (repeated) public goods game.

We consider exclusively the projects where all participating agents, agree on the team composition. The team composition, together with the participants' efforts, constitute an *implemented project*.

The relation "i works with j in an implemented project" defines a non-dichotomous symmetric matrix **W** where element $w_{ij} \in \{0,1,2\}$ is defined by the number of projects in which agents i and j work together. Matrix **W** defines the project network; when n agents work together on an n-member project we say they form a *size n clique* since in the graphical representation of matrix **W** they are depicted as a clique with n nodes.

Within each implemented project agents play a public goods game (for a survey on experimental research see [12]). The efforts of the participants are aggregated and used to produce a commodity with a production function f; the output is shared among the members of the team. We denote c_i agent i's cost of effort, and assume that greater effort means greater cost to the agent and increasing marginal cost. The payoff of agent i in project p can be formalized as follows:

$$\pi_{i,p} = \frac{f\left(\sum_{j \in T_p(i)} e_j\right)}{n} - c_i\left(e_i\right), \tag{1}$$

where e_i is agent i's effort and $T_p(i)$ is the set of partners of agent i in project p.

We assume that:

1. There is a unique level of effort maximizing agent's payoff.
2. There exists a unique Nash equilibrium e^N.
3. When all the agents exert the same effort, both the optimal effort e^N and the optimal payoff increase with the number of members participating to the project.

In order to keep the maths simple we considered in our experiments and simulations the following payoff formulation:

$$\pi_{i,p} = \frac{\left(\sum_{j \in T_p(i)} e_j\right)^2}{n} - e_i^3. \tag{2}$$

In this case it is easy to prove that $e^N = 2/3$, and that the socially optimal effort for a n-team is $e_n^S = 2n/3$. With this payoff formulation, when everybody exerts the socially optimal effort, the individual payoff increases with the number of agents in the team.

3 The Results of the Computational Model

The second stage of our research consisted in performing some experiments where human subjects interacted on the game we presented in the previous section. The results of these experiment were presented in [6]. When considering the simulation stage, the first step consists in the implementation of some of the behaviors we observed in the human subjects. Our purpose is not to replicate the observed human behavior in experiments, but rather, to use the empirical data to infer some of the implicit behaviors that generated them and model them in our artificial agents. The objective is to establish the constructive sufficiency of the model to produce behavior like that observed in human aggregation processes, such as the formation of partially connected cliques and leadership.

Human interaction and team formation is a complex phenomenon. To identify the different components we introduce a computational model of interaction and team formation among artificial agents (Fig.1). This way we are able to break down the agents' behavior in micro phases. We study the relative importance of each of these micro aspects of behavior when these lead towards the emergence of some macro behaviors in the artificial society we consider. Our agents are all utility maximizers but, at the same time, they are heterogeneous in terms of behavioral rules, described as follows. We study how heterogeneity (in our sense individual attributes at the micro level) affects, at the macro level, the network structure and its dynamics. Finally, the task our agents are asked to perform comprises both intragroup and intergroup levels of conflict and, for this reason, may be interpreted as a sort of generalized team game as studied in [3].

As mentioned, the computational model we consider is an *Agent Based Model* where the classes of the behavior of the agents have been obtained analyzing the results we acquired considering two human subject series of experiments. The first series consisted of 21 sessions each including about 93 individuals, while the second one consisted of 12 sessions with about 48 individuals (for a complete description of the experiment the reader may refer to [6], while for a discussion on the use of human subjects experiment the reader may refer to [4] and [7]).

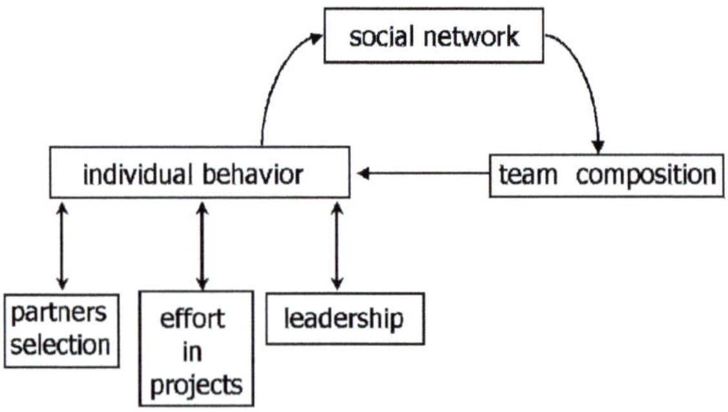

Fig. 1 Team composition as the result of individual behavior

While in the public goods game, agents are assumed to have the same role, in our human subject experiments some subjects took a different role. This is not surprising since, as it is well known, several aspects of human behavior cannot be completely explained in terms of rationality, see for instance, [11]. In particular some individuals took a leadership role within their group; this may be explained since in our interaction the agents were allowed to communicate and discuss the game before each repetition.

While several classes of behavior were implemented, those which resulted more interesting were the following:

- Agents considering the two best projects and expanding the second one either by adding one more subject or proposing a new project with at least one agent more than the second best project.
- Agents playing the socially optimal effort.
- Provided that the leader knows less than thirteen agents, when the first best project is not a size 7 clique or, the first best project is a size 7 clique but the second project is not, then it introduces all the agents it knows to one another.
- When the leader knows less than seven agents and the best project is not a size 7 clique or, the agent knows less than eight agents and the best project is a size 7 clique but the second best project is not, the agent expands the vector of its known agents in order to include all the agents that in the sociomatrix have a geodesic distance smaller than three; in a friendship relation this would simply mean that "the friends of my friends become my friends".

The results we obtain with our computational model are quite interesting. They are reported in [8] and for the sake of brevity cannot be reported here in full. Nevertheless, they can be articulated in diverse ways. A first important aspect is the role of communication and mutual acquaintance between potential group members. While our model was not intended to capture the individual communication between subjects, even in the much simpler model of project discussion that we considered, the importance of mutual acquaintance and agent coordination, in choosing the project to implement, is relevant. For example, our model explains both the difficulties in large groups with no leaders and the problems emerging when too many leaders are present. In fact, we compared the effectiveness of leaders in terms of number of links in the organization. According to our findings the number of leaders and their relative location is extremely important. In this sense the leadership role is necessary for a sort of implicit coordination of agents. In our model, leaders do not suggest projects, rather they act on the social network and may help the emergence of projects in the discussion phase. That is, social leaders are cardinal in stating the pace of a balanced expansion of the social matrix; essentially they control the combinatorial explosion while fostering mutual acquaintance among agents. Given the role of social leaders in group formation it is interesting to consider the model of group development by [19]. This model in its original form consists of four stages: forming, storming, norming and performing. While our model was not meant to replicate Tuckman's model, some considerations are in order. Firstly, we can interpret the role of social leader in sharing members' mutual acquaintance as a way

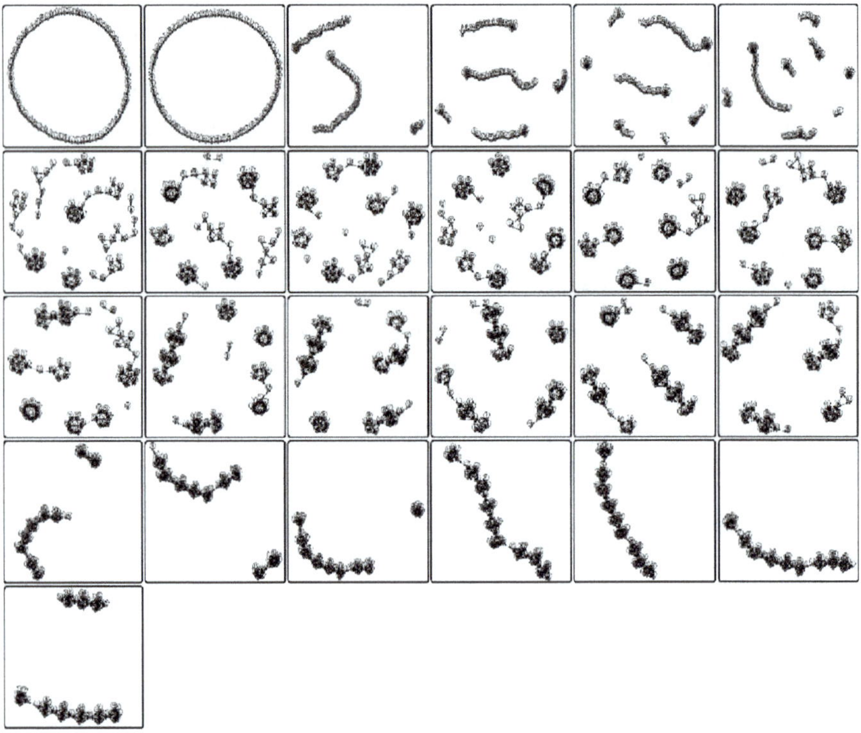

Fig. 2 Project network evolution in populations with 7-interval leaders and circle graph initial sociomatrix at turns 1–6, 10–60, 100–600, 1,000–6,000 and 10,000

to reduce uncertainty in the forming stage when importance is placed on making acquaintances, sharing information and testing each other. Secondly, we can also observe a sort of norming stage as the group members exert the effort which is optimal to the group and do not free ride. As a final point, it must be observed that, since in our model the leaders were those agents that incentivized acquaintance among the others, they were not the agents with more connections. By contrast, the agents in the "influence area" of two leaders were those with more connections. Another interesting aspect is the comparison of the human experiment and the computer simulation. In Figs. 2 and 3 we present respectively artificial population evolution at turns 1, 2, 3, 4, 5, 6, 10, 20, 30, 40, 50, 60, 100, 200, 300, 400, 500, 600, 1,000, 2,000, 3,000, 4,000, 5,000, 6,000 and 10,000 while human subject evolution is presented on the first 21 consecutive turns.

Finally, in Fig. 3, the project network evolution is reported for one of the human subject experiments we considered. In this case the reported turns are consecutive from the first to the last.

While with human subjects we found the same tendency to aggregation as in the artificial experiments, nevertheless two important differences must be observed. First, since the project selection process among humans is more interactive and effective than the simple model of communication we implemented, the network

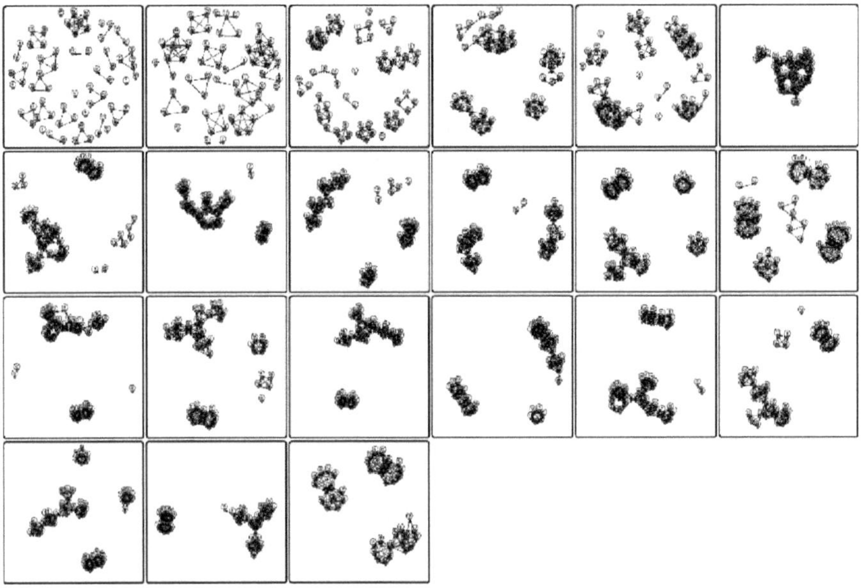

Fig. 3 Project network evolution for a human subjects experiment at turns 1–21

evolution is faster than in the artificial society; here size 6 weakly connected clusters are present on turn 2 while on turn 3 appear strongly connected clusters. Second, the human experiment took place on different dates and we had the no-turn up problem: not all of the subjects turned up at each session of the experiment. This may explain the project network disaggregation: subjects had to continuously adapt their projects according to the contingent situation. The analysis of our simulations and the comparison to the human experiment behavior indicates several directions where to extend our research.

4 Modeling the Leadership Role

In this case we focus on the emergence of leadership when some of the observed behaviors (see [7]) are incorporated in a network. In the first stage of our research we consider a similar theoretical model with different stress on the leadership role. While in the network dynamics research we studied a sort of implicit leadership, the approach we follow in this research explicitly models the leader role. Analyzing the different interactions, we study under what conditions individuals may emerge as effective leaders.

In the following stage of our research the leader role will be introduced more explicitly in the network interaction. In other words, agents will be able to decide whether they want to be followers of a particular leader, find a different leader or even stand as leaders themselves. The approach we follow is the one outlined in [4] where gathered data and observations, when performing classroom experiments with human subjects, are used to model bounded rationality agents. The theory we build is a "Grounded theory" in the sense of [18], i.e., a theory that is derived from data systematically gathered and analyzed through the research process.

The different results we were able to observe, both in the human experiments and the simulations, are the further confirmation of interaction complexity in organizations. The simulation proved to be extremely important in analyzing the system behavior when some variables were fixed; this would be impossible when considering only human subjects, as many uncontrollable factors determine their decisions. In our approach we were interested in discovering under which conditions we could observe different leaders *emerge*. At the beginning of the interactions, in the human subject experiments leadership roles seemed to be distributed to the agents. In fact, in the human subjects experiments the number and composition of initial groups and leaders seemed to be the result of subject physical positioning during the experiment, and this was someway replicated in the computer simulations. So the question was to understand on what basis only few of them could emerge and survive during the repeated interaction. These leaders emerged in the sense that they could "keep followers". This point is interesting because somehow it links some aspects of leadership described in popular books ([17], for instance) to empirical evidence. Also comparing the results obtained with the human subjects experiments to the simulations provides some interesting suggestions, since in both cases leaders rewarding followers according to distributive justice would emerge; this kind of leader was named "Fair".

The interaction we used with the human subjects represented many of the elements of complexity theory per se. The computational model we present retains some of these elements. For example, with some classes of leaders, we can observe different final outcomes with small variations in the parameters, namely, the responsiveness probability and the initial number of leaders. This was particularly evident when considering either free riders or leadership styles different from Fair. In particular, while in the absence of free riders the only perceived inequity is leaders' distributive fairness, in a population with free riders further inequity is perceived. Both kinds of inequity cause some turbulence which, under certain conditions, does not encourage the emergence of the most suitable leader. Furthermore, unlike the computer simulation, the human subjects experiments do not produce single final leaders, which may suggest that the human subject responsiveness is not immediate. In addition, if we assume that in the human subjects experiments emerging leaders were those with larger groups, we can observe that the computer simulations are consistent with the empirical evidence, as in both emerging leaders have the larger groups. As a final point, recall that the underlying theoretical model leads to a unique final group, because of the higher individual payoffs. We did not implement agents having the maximization payoff as an objective, rather, following evidence in human subjects experiments, we chose to model sensitiveness to inequity. Yet, inequity minimization led chiefly to the maximization of the aggregate payoff, when fewer Fair leaders were present. The results of both the human subjects experiment and the computer simulation, showed how leadership emergence is a complex phenomenon, where many aspects and variables interact.

5 Introducing Explicit Agent Communication

While in the original paper the communication phase is modeled as a sort of brainstorming, in this research we are incorporating a model of communication among agents. The agent will be able to communicate about future projects and will be endowed with a sort of social cognition about what happened in the past. This way issues like free riding, commitment and reputation will be considered. In other words, the communication phase we observed in the human subject experiment will be modeled more explicitly. Many important matters will be tackled in this research. For instance, how can agents identify free riders? How can a bargaining process on the team composition end in reasonable time with a project? Furthermore, how can teams sanction free-riders or expel them from the project? The answers to these questions are still under study and are part of our further research.

6 Social Network Analysis

As mentioned in the results of the computational model for the artificial population we found that social leaders were not the agents with more connections. Rather, the agents in the "influence area" of two leaders were those with more connections. It would be interesting to understand the different roles agents play during the interaction simply by analyzing the social network.

In the artificial population analysis we introduced some measures for quantifying the evolution. The density of a graph – which is a recommended measure of group cohesion (see [2]) – is proportional to the number of links; these statistics seem appropriate for our analysis. In the first stage, we had to adapt density measure accordingly to our situation which consists of multiple projects and non-dichotomous networks. We actually considered modified density together with the number of components, which seems to be appropriate for our analysis. This way the different situations could be described as follows:

- Several links and several components indicate the presence of isolated *size n cliques*.
- Several links and few components indicate the presence of several connected *size n cliques*.
- Few links and several components indicate the presence of several isolated agents.
- Few links and few components indicate the presence of simple structures such as the circle graph or chains of agents.

While on the hand it would be interesting to refine the measures we used, on the other hand it would be appealing to find some network measure able to identify the leaders from the social network data without any knowledge of their actions. We are analyzing some of the network measures introduced in [20] in order to be able to identify the social leader simply by analyzing the network evolution over time.

Another aspect we are considering is the modification of network representation algorithm such as [13] in order to have an efficient graphical representation of the social network evolution over time. The first step has been to analyze and reimplement [13] visualization procedure as described in [9].

References

1. Banks DL, Carley KM (1996) Models for network evolution. J Math Sociol 21(1–2):173–196
2. Blau PM (1977) Inequality and heterogeneity. Free Press, New York
3. Bornstein G (2003) Intergroup conflict; individual, group and collective interests. Personal Soc Psychol Rev 7(2):129–145
4. Dal Forno A, Merlone U (2004) From classroom experiments to computer code. J Artif Soc Soc Simul 7(3)
5. Dal Forno A, Merlone U (2004) A model of social interaction for the study of network dynamics. In: Proceedings of Wild@Ace 2004, Workshop on industry and labor dynamics, the agent-based computational economics approach, Laboratorio R. Revelli, Centre for Employment Studies, University of Torino
6. Dal Forno A, Merlone U (2005) The evolution of coworker networks. A comparison between experimental and computational results. In: Troitzsch KG (ed) Representing social reality, Proceedings of the third conference of the European social simulation association, Fölbach, Koblenz, pp 223–227
7. Dal Forno A, Merlone U (2006) The emergence of effective leaders: an experimental and computational approach. Emergence 8(4):36–51
8. Dal Forno A, Merlone U (2007) The evolution of coworker networks. An experimental and computational approach. In: Edmonds B, Hernándes C, Troitzsch KG (eds) Social simulation technologies: advances and new discoveries. InformationScience Reference, Hershey, NY, pp 280–293
9. Frison A (2006) Analisi di reti sociali, algoritmi di visualizzazione. Undergraduate dissertation, Università degli Studi di Torino
10. Hinds PJ, Carley KM, Krackhardt D, Wholey D (2000) Choosing work group members: balancing similarity, competence, and familiarity. Organ Behav Hum Decis Process 81(2):226–251
11. Hirschhorn L (1985) The psychodynamics of taking the role. In: Colman AD, Geller MH (eds) Group relations reader, vol 2. AK Rice Institute Series, AK Rice Institute, Springfield, VA, pp 335–351
12. Kagel JH, Roth AE (1995) The handbook of experimental economics. Princeton University Press, Princeton, NJ
13. Kamada T, Kaway S (1989) An algorithm for drawing general undirected graphs. Inf Process Lett 31(1):7–15
14. Lazer D, Katz N (2003) Building effective intra-organizational networks: the role of teams. Research paper. Centre for Public Leadership, J.F. Kennedy School of Government, Harvard University
15. Miller EJ, Rice AK (1967) Systems of organization: the control of task and sentient boundaries. Tavistock, London, UK
16. Moreland RL, Levine JM (1992) The composition of small groups. In: Lawler E, Markovsky B, Ridgeway C, Walker H (eds) Advances in group processes, vol 9. JAI, Greenwich, CT, pp 237–280
17. O' Toole J (1999) Leadership A to Z. A guide for the appropriately ambitious. Jossey-Bass, San Francisco, CA

18. Strauss AL, Corbin JM (1998) Basics of qualitative research: techniques and procedures for developing ground theory, 2nd edn. Sage, Thousand Oaks, CA
19. Tuckman BW (1965) Developmental sequences in small groups. Psychol Bull 63:384–399
20. Wasserman S, Faust K (1999) Social network analysis. Cambridge University Press, Cambridge, MA
21. Young J (1998) Using computer models to study the complexities of human society. Chron High Educ 4(46A):17–19
22. Zeggelink E (1995) Evolving friendship networks: an individual-oriented approach implementing similarity. Soc Networks 17:83–110

Empirical Analysis of the Architecture of the Interbank Market and Credit Market Using Network Theory

Giulia De Masi

Abstract The credit relationships in an economic system are essentially three: the interbank market, the lending market between banks and firms, the commercial credit among firms. Here the focus is on the first two kinds of credit, using network tools. The graph theory, which is at the basis of network analysis, is used as the natural mathematical environment to investigate the architecture (*topology*) of these markets. The interbank market is represented as a network where the nodes are banks and the links are the reciprocal exposures. The lending relationships between banks and firms is represented by a bipartite graph where the nodes are of two kinds: banks and firms. From the bipartite graph, the network of cofinancing banks is extracted. We observe the leading role of large Italian banks which form a strong core in the network both in the interbank and in the lending market. The small banks act as lenders and the large as borrowers in the interbank market. Both of them finance firms on the lending market, the large ones financing the most of the large firms, while the small ones the small local firms.

1 Introduction

In the recent years the study of Complex Networks has acquired increasing importance. Several real systems have been represented as networks. This kind of approach allows to get insights into the architecture of complex systems composed by many objects, linked to each other in a non trivial way [1, 2]. Most of them show peculiar scaling properties. In particular many networks are scale-free, that means that the degree distribution is power-law tailed. Applications of network theory in Economics can be useful in order to consider explicitly the relations among economic agents [3]. Many empirical analysis of economic systems have been done using

G. De Masi
Dipartimento di Economia, Università Politecnica delle Marche,
P.le Martelli 8 60121 Ancona, Italy
giuliademasi@gmail.com

A.K. Naimzada et al. (eds.), *Networks, Topology and Dynamics*, Lecture Notes in Economics and Mathematical Systems 613,

network tools: the main examples are the applications to world trade web [4, 5], interbank market [6, 7], stock market [8], e-commerce [9].

In this paper we focus on the problem of credit relationships, defining the concept of *credit network* [10]. A complete understanding of the architecture of credit relationships in economic systems is of primary importance. The problem of the economic stability in fact is strongly related to the underlying structure of credit/debt relationships among its components: this structure plays a crucial role in the exposure to risk of avalanche failures and domino effects in systems characterized by credit relationships [11]. In general three kinds of credit relationships can be identified: financial credit from banks to firms, inter-bank credit and commercial credit between firms.

In the following we focus on the first and the second type of credit relationships. The presence of credit/debt relationship allows us to define a network of credit where nodes are the economic agents (banks and/or firms) and the links are credit relationships.

The interbank market emerges as a consequence of the need of banks to manage their liquidity. The loans are originated by the fact that every bank needs liquidity in order to satisfy the demands of customers. To buffer liquidity shocks the European Central Bank requires that on average 2% of all deposits and debts owned by banks are stored in the National Central Banks. Given this constraint, the banks can exchange excess reserves on the interbank market with the objective to satisfy the reserve requirement and in order to minimize the reserve implicit costs [12–14].

In the lending market, the firms behave very differently from each other. Some firms obtain credit just from one bank, other ones have loans from many banks (multiple relationships). Given the relevance of the problem, a recent literature focuses on this topic [15–17].

In this paper the network approach in applied to two real cases: the Italian interbank data and the Italian credit relationships between banks and firms. This paper differs from the previous studies not merely for the method used, but because the network analysis allows to obtain information on the structure of relationships and in particular in the architecture of second order or higher. While standard econometric analysis are useful to detect correlations among different features of banks and firms, they do not allow to study the topology of the underlying architecture of bank-credit relationships. Nevertheless the structure of relationships plays a crucial role in bankruptcy diffusion. We organize the paper in the following way:

- Section 2: network approach
- Section 3: the application to the interbank market in Italy
- Section 4: the description of multilending in Italy
- Section 5: discussion and conclusions

2 The Network Representation

We represent this system as a network, in order to use an approach based on the graph theory to analyze the complex structure of credit relationships in the Italian economic system.

A network is defined as a set of nodes and links and it is mathematically represented as a graph.

In our particular case the nodes are banks and firms [10]. The links represent the credit relationships among them. These kinds of networks, composed by two kinds of nodes, are called "bipartite networks".

Moreover, we can extract two networks from the overall network, each one composed by just one kind of node: this method is called one-mode reduction and the two networks *projected networks*, in the sense that they are obtained as a projection of the initial graph in the subspace composed by only one kind of node.

A network is represented from a mathematical point of view by an adjacency matrix. The element of the adjacency matrix a_{ij} indicates that a link exists between nodes i and j, that is $a_{ij} = 1$ if the bank i provides a loan to the firm j; otherwise $a_{ij} = 0$.

The *degree* of a node i is the number of links outgoing from it and is calculated by

$$k_i = \sum_j a_{i,j} \,. \tag{1}$$

The *assortativity* is a measure of similarity among nodes and it is defined as

$$k_{nn}(i) = \frac{1}{k_i} \sum_{j \in v(i)} k_j . \tag{2}$$

A graph is said assortative if the nodes with high degree are connected to nodes with high degree, on the contrary it is said disassortative, if they are connected to nodes with low degree.

In a graph, the distance between two vertices is defined as the length of the shortest path joining them.

The *distance* d_{ij} between two vertices i, j is *the shortest* number of edges to go from i to j. Therefore the neighbors of a vertex i are all the vertices j which are connected to that vertex by a single edge ($d_{ij} = 1$). Using the adjacency matrix this can be written as

$$d_{ij} = \min \left\{ \sum_{k,l \in \mathcal{P}_{ij}} a_{kl} \right\} , \tag{3}$$

where $\mathcal{P}_{ij}$ is a path connecting vertex i and vertex j.

The *diameter* of a graph is given by the maximum of all distances between all the pairs of nodes.

An important concept is the "importance" of the node, or *centrality*. The concept of importance is related to the system under study. An usual definition of "centrality" is given from dynamical properties of the graph. A sensible measure is given by the number of times that one vertex k is crossed going from one vertex i to another j following the path of minimal length (distance $d(i, j)$). This quantity is called *site betweenness* $b(i)$ and it is usually defined by

$$b(i) = \sum_{\substack{j,l=1,n \\ i \neq j \neq l}} \frac{\mathcal{D}_{jl}(i)}{\mathcal{D}_{jl}} , \tag{4}$$

where $\mathcal{D}_{jl}$ is the total number of different shortest paths (distances) going from j to l and $\mathcal{D}_{jl}(i)$ is the subset of those distances passing through i. The sum runs over all pairs with $i \neq j \neq l$.

The *clustering coefficient* $cc1(i)$ is a measure of the density of connections around a vertex and is defined as

$$cc1(i) = \frac{2}{k_i(k_i-1)} \sum_{j,h} a_{ij} a_{ih} a_{jh}. \tag{5}$$

Hence, the clustering coefficient allows to calculate the proportions of the nearest neighbors of a node that are linked to each other. The average clustering coefficient,

$$< cc1 >= \frac{1}{N} \sum_i cc1(i)$$

indicates the statistical level of cohesiveness measuring the global density of interconnected vertex triplets in the network.

Many tentative of studies have been done in the field of bipartite graphs [18–20]. While we can apply the above measures in networks where all nodes are of the same kind, in bipartite networks some of these measures can not be applied, because of the different nature of the nodes. Therefore the main statistical quantities under study are degree distribution of each of the two kinds of nodes and correlations among the degree of the two kinds of nodes. A measure of connectedness in bipartite networks is given by the density of cycles of size 4 surrounding a node [21]. Explicitly this clustering coefficient reads

$$cc2(i) = \frac{\sum_{m=1}^{k_i} \sum_{n=m+1}^{k_i} q_i(m,n)}{\sum_{m=1}^{k_i} \sum_{n=m+1}^{k_i} [a_i(m,n) + q_i(m,n)]}, \tag{6}$$

where m and n label neighbors of node i, $q_i(m,n)$ are the number of common neighbors between m and n and $a_i(m,n) = (k_m - \eta_i(m,n))(k_n - \eta_i(m,n))$ with $\eta_i(m,n) = 1 + q_i(m,n) + \theta_{mn}$ and $\theta_{mn} = 1$ if neighbors m and n are connected with each other and 0 otherwise.

In the case of projected networks, the quantities under study are degree distribution, clustering coefficient, assortativity, betweenness centrality, diameter [18].

3 Application to the Italian Interbank Market

The interbank market emerges in order to allow the banks to manage their liquidity. Liquidity management in the banking system is essential for a smooth operate of payment systems. For example, in the Euro area the European Central Bank (ECB) normally aims to satisfy the liquidity needs of the banking system via its open market operations (main and long-term re-financing operations, fine-tuning and structural interventions) the most relevant of which are the weekly auctions.

Banks submit the amount of money they want to deal and interest rate they are ready to pay for it. The ECB collects bids and executes the auctions. The allocations are settled on the bank's account to the National Central Bank (NCB) on Wednesdays.

Credit institutions in the Euro area are required to hold minimum reserve balances with NCBs (set at 2% of all deposits and debts issued with a maturity of less than two years, excluding repos and interbank liabilities, but with a minimum threshold applied). Reserves provide a buffer against unexpected liquidity shocks, mitigating the related fluctuations of market rates. They have to be fulfilled only on average over a *one-month maintenance period* that runs from the 24th of a month to the 23rd of the following month (when this is not a holiday in which case is anticipated to the previous working day). Given this constraint, banks can exchange excess reserves on the interbank market with the objective to satisfy the reserve requirement and in order to minimize the reserve implicit costs [12–14].

The interbank markets are basically managed by each European country [12]. These markets are in almost all cases phone-based, that means that each bank has some brokers doing their transactions by phone. The only exception is the Italian market, which is totally screen-based, implying that each banks operator can see real time quotes of all other banks and do its transaction.

This paper analyses on the network analysis of the Italian interbank market. This market is unique in the Euro area in being screen based and fully electronic: outside Italy interbank trades are largely bilateral or undertaken via voice brokers. While banks can still choose with whom to trade, the information about the rates and the trades are public. The Italian electronic broker market e-MID (electronic Market for Interbank Deposits)[1] covers the entire existing domestic overnight deposit market in Italy. Both Italian banks and foreign banks can exchange funds on the e-MID. The participating banks were 215 in 1999, 196 in 2000, 183 in 2001 and 177 in 2002.

Our data set is composed by banks operating on the Italian market for which we have the complete record of all transactions. The data set used to construct the interbank network includes only overnight transactions and consists of encoded name of first bank, encoded name of second bank, amount of transaction, rate applied to transactions, date and time. To de-codify the name of banks, a list of banks has been provided: the first number is the label of the bank, the second one is the group of classification of that bank provided by Italian National Bank (1 = foreign banks, 2 = big Italian banks, 3 = medium Italian banks, 4 = small Italian banks, 5 = cooperative credit banks).

For every day of trading, we compute the network of debts/loans. Our dataset also allows us to build the matrices associated to directed graphs. We can make directional links by allowing them to follow the flow of money, so that a link is incoming to the buyer and outgoing from the seller. A directed graph may be more relevant in assessing the risk of contagion[2] and systemic default in the system. Hence we define six more matrices A^b,A^l, C^b,C^l and W^b,W^l. The elements a^b_{ij} (a^l_{ij}) indicate

[1] e-MID is run by e-MID S.p.A. Società Interbancaria per l'Automazione (SIA), Milan. The central system is located in the office of the SIA and the peripherals on the premises of the member participants.

[2] A shortage of liquidity can propagate from a bank to the ones that have a relation of credit to it.

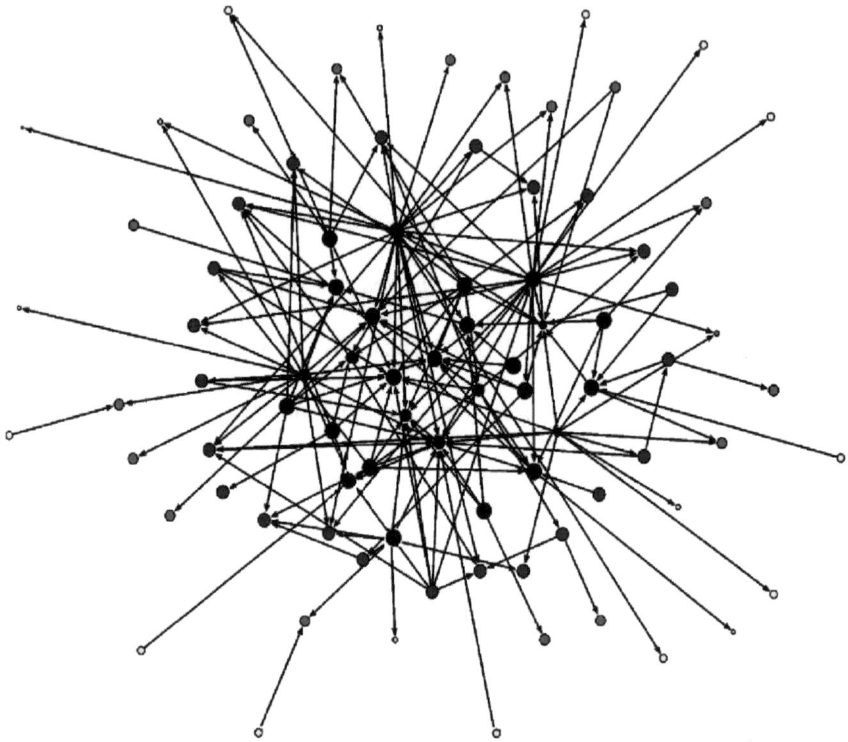

Fig. 1 A plot of the inter bank network. The color codes for the various groups are the following: 1 = *yellow*, 2 = *red*, 3 = *blue*, 4 = *black*. Note that the black vertices (bank of group 4) form the core of the system [6]

if at least one transaction has occurred on a given day between bank i and bank j, with bank i as the borrowing (lending) bank. The elements of the connectivity matrix c^b_{ij} (c^l_{ij}) denote the number of transactions on a given day between bank i and bank j, with bank i as the borrowing (lending) bank. The elements of the weighted connectivity matrix w^b_{ij} (w^l_{ij}) denote the overall volume exchanged on a given day between bank i and bank j, with bank i as the borrowing (lending) bank. Obviously $w^l_{ij} = w^b_{ji}$. We define the flow between two banks as $f_{ij} = w^l_{ij} - w^b_{ij}$. The flow is positive if the bank is a net lender. With this convention we define a weighted graph, whose plot is in Fig. 1

The distribution of banks degree and of weights of the links indicates that banks have an highly heterogeneous behavior, since the number of their partners varies very widely. In Fig. 2 the distribution of degree for a particular day is plotted: we observe high skewness, proving that some banks have many partners, but most of the banks have few partners. Moreover in same figure the assortativity and clustering distributions show a peculiar architecture of the network. The system shows disassortative mixing, i.e. banks with higher degree are more likely connected to banks with lower degree.

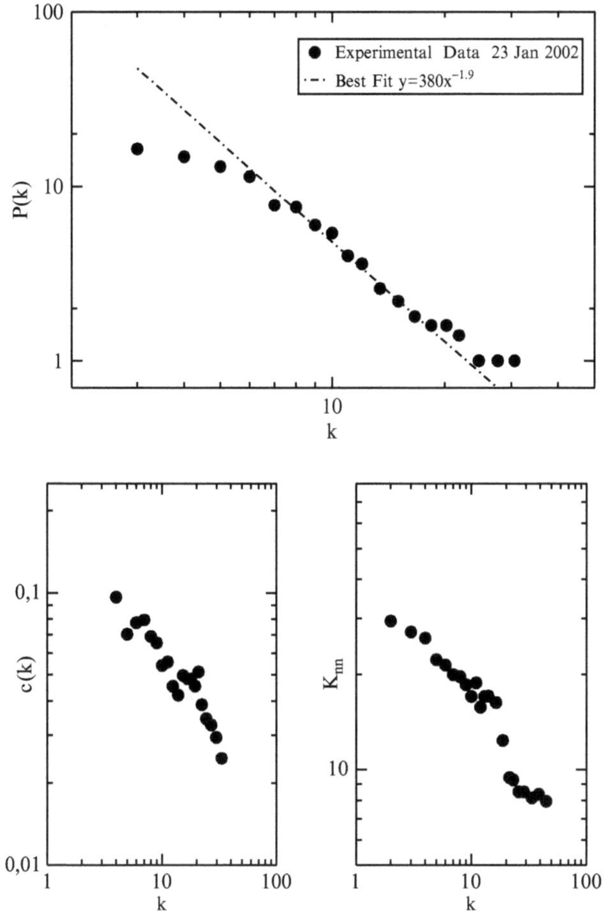

Fig. 2 Cumulative distribution of banks degree (*left*); distribution of clustering coefficient c and assortativity $k_{nn}(k)$ versus k [6]

In Fig. 3 (top panel) we show that the banking system is highly heterogeneous with fat tailed cumulative distribution of banks weights (volume of single contract).

The total volume transacted by each bank is even not equally distributed among its links as we observe from the plot of the participation ratio $Y_2^c(i)$ (Fig. 3, bottom panel): real participation ratio (dots) differ from a random case of equally distributed weights (straight line); banks with an high number of connections have heterogeneous volumes of contracts with their partners [7].

Interestingly, regardless the change in volumes, all the above topological measurements remain similar when computed in different days of month.

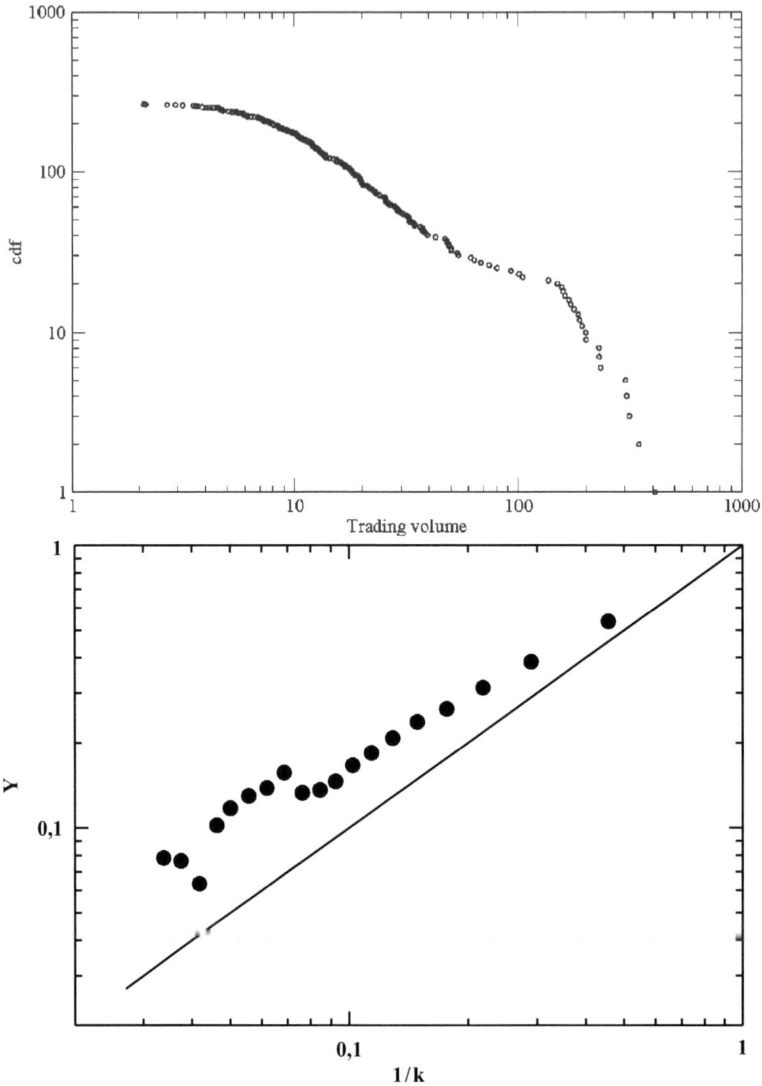

Fig. 3 Distribution of weights w_{ij}, that represent the contract volumes (*top panel*) and participation ratio $Y_c(k)$ (*bottom panel*)

4 Application to the Italian Multilending

Among European countries, Italy has the biggest average number of banks relationships per firm. The multiple bank relationships became more relevant in Italy in the 50s [22]. In fact, after the second world war Italian firms were able to finance themselves: therefore the interaction among banks and firms was very small. In the

70s the worsening of financial conditions of Italian firms induced entrepreneurs to ask for credit to the banks. In these years the multiple bank relationships emerged. On the one hand, this implied that a firm received credit from more than one bank, on the other hand implied that a good knowledge of the real economic conditions of firm was not easily available to banks. In many cases credit was provided also for personal trust reasons, without a deep investigation on the actual financial condition of the firm [22].

Detragiache et al. studied data of medium-small size Italian manufacturing firms (firms with less than 500 employees). In 1994, 89% of Italian firms have multiple links; the median number of relationship is 5 and the 75th percentile is 8 [17]. The phenomenon of multiple linking is more striking also respect to U.S..

While it is quite clear why big firms have multiple links, it is not clear why small firm have multiple links [23].

In the following we analyze a subset of the AIDA database [24]. We study data regarding Italy. This set contains information on the largest 170,000 Italian societies from 1992. We have detailed information on the characteristics of each firm, as the total net worth, the total asset, the solvency ratio, the number of workers, the added value. Moreover we know the identity of banks financing each firm. This allows us to study how the number of relationships varies for firms of different size. We merge this dataset with another one derived from Bank of Italy classification of the whole set of Italian banks in five groups [25]: large banks, medium banks, small banks, cooperative credit banks and rural banks. In this way we can identify the attributes of the banks lending to each individual firm and to detect different behavior of banks of different sizes. The Bank of Italy provides a classification of Italian banks in five categories:

- Larger banks have funds of more than 45 billion Euros.
- Large banks between 20 and 45.
- Medium between 7 and 20.
- Small between 1 and 7.
- The remaining banks are minor banks.

Our sample is composed by 11 larger banks, 11 large banks, 34 medium banks, 125 small banks and 307 smaller banks. In this paper we focus on credit relationships of Italian firms with Italian banks in the year 2003. The set is composed by 55,005 contracts among 488 Italian banks and 33,468 Italian firms. The study of the evolution of the structure of the credit network will appear in a further paper.

We distinguish firms following the ISTAT classification [26]: in 2003 the Italian firms were 4.1 million: the 95% micro, 4.5% small, 0.5% medium. In our sample 29% are micro, 40% small, 23% medium, 7% large [27]. In this sense our sample privileges the analysis of big firms and does not consider small firms and small banks.

Unfortunately we do not have access to the amount, the rate and the duration of credit, but this dataset is more extended than other ones studied in the literature.

4.1 The Bipartite Network of Banks and Firms

We denote firms with the index $_f$ and banks with the index $_b$.

By definition each bank is linked just with firms and each firm is linked just with banks.

We have recently done a detailed analysis of the network of banks and firms [28, 29]. It emerges that small banks have a few credit links and large banks a great number of contracts. Moreover, big firms tend to receive loans from big banks but also from some small banks; on the contrary, small firms receive loans from small (local) banks. In general big firms prefer multiple relationships, while small firms have mostly single relationship. According to the literature this is just a consequence of the exposure to risk: in fact firms with a single linkage are in general the smallest, with low solvency ratio and therefore less solvent.

The largest k_b is 6,699, the mean 149, the median 3 and the 75th percentile 30; whereas the largest k_f is 15, the mean 1.8, the median 2 and the 75th percentile is 2. The behavior of banks in our sample is more heterogeneous than that of firms: in fact the degree ranges in three orders of magnitude from 1 to 6,699. There is also a fat-tailed degree distribution of the degree of the banks; some banks finance a large number of firms: they are the hubs of the network [29].

It can be emphasized that large firms receive credit from many large banks, while small firms are financed by small local banks and often by one bank alone. It is not uncommon, that large firms with multiple linkages have relationships with banks of very different size. Firms with multiple linkages are big, often regarded as financially safe: because of it, small banks finance them, even without doing a rating investigation. Moreover large firms are usually able to provide hard information. On the contrary, small firms often find financing problems: therefore they are usually financed by small local banks, often better able to collect soft information than large banks [29].

4.2 The Projected Network of Co-Financing Banks

In the study of bipartite graph a very widely used approach is to study separately two networks that can be defined from the original network. If we call the two kinds of nodes as nodes A and B, we can study the network G_{A+B} which has the total set of nodes $(A+B)$ or the networks G_A and G_B which have only nodes of kind A or B respectively [31, 32].

In our case we can define the network of banks and the network of firms. The first is the network of cofinancing banks: two banks are linked if they finance the same firm. The second is the network of co-financed firms, that is two firms are connected if they are financed by the same bank. The network of firms is highly disconnected; therefore we focus just on the bank network. In Fig. 4 is provided an example of projection. Let consider a set of firms $F = 1,2,3,4$ (squares) and a set of banks

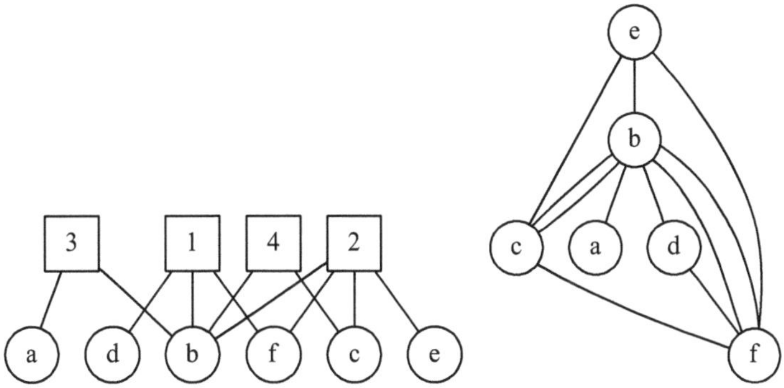

Fig. 4 Bipartite graph and projected graph (one mode reduction on banks space). Firms are *square* and banks *circles*

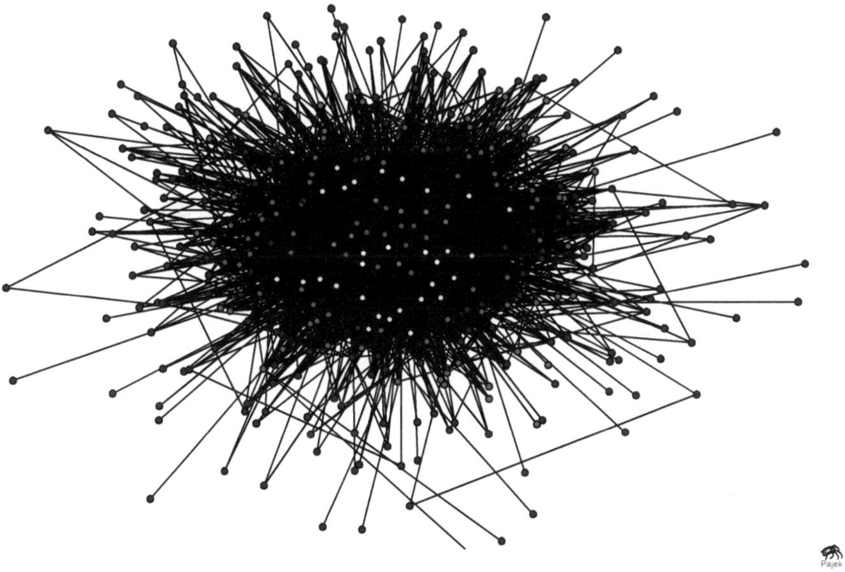

Fig. 5 Graph representation of the one-mode reduction network on the subspace of banks. The convention for the use of colors is the same given above (the *white dots* are the foreign banks)

$B = a,b,c,d,e,f$ (circles). The corresponding graph is given in the top panel. The corresponding projected network is in the bottom panel.

This kind of approach allows to identify common characteristics (if they exist) among banks linked with the same firm.

The bank network in Fig. 5 is composed by 507 Italian banks. Banks are colored, using the Bank of Italy classification: red nodes are larger banks, orange large banks, yellow medium banks, green small banks, blue smaller banks.

In the projection process, we lose some information related to how many firms finance in common two banks. In fact two banks have a link both if they have only one firm in common and more than one. To maintain this information we define a weighted network: the weight associated to the link between two banks is the number of common firms they finance.

In the bank network the average degree is 5.7, while the average clustering coefficient is 0.3. The distributions of the degree and of the weights are fat-tailed distributions, reproducing the heterogeneous behavior of banks [29].

In Fig. 6 the two measures of clustering coefficients are plotted: $cc1$ (density of cycles of size 3) and $cc2$ (density of cycles of size 4).

In Fig. 7 (left panel) the distribution of distances is plotted. The graph is connected and the maximum distance is 6. This is an evidence of the fact that the bank graph shows small-world properties. In the right panel the betweenness b is plotted as a function of k. The nodes with high degree have also high betweenness. This is a sign of the underlying structure characterized by a strong core of hubs and not by small subgraphs weakly linked each other (in this latter case we should observe in

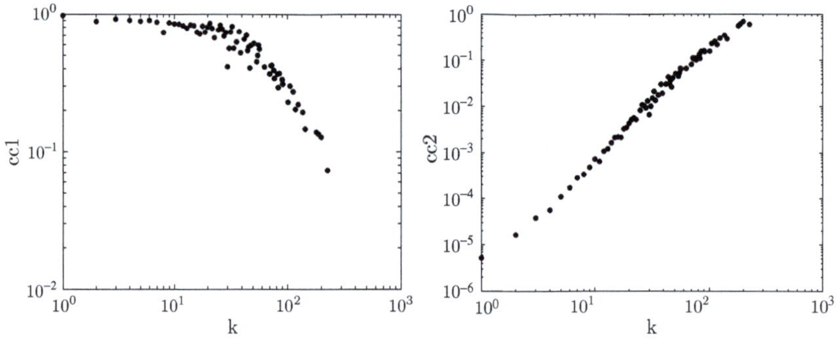

Fig. 6 Clustering coefficients $cc1$ and $cc2$

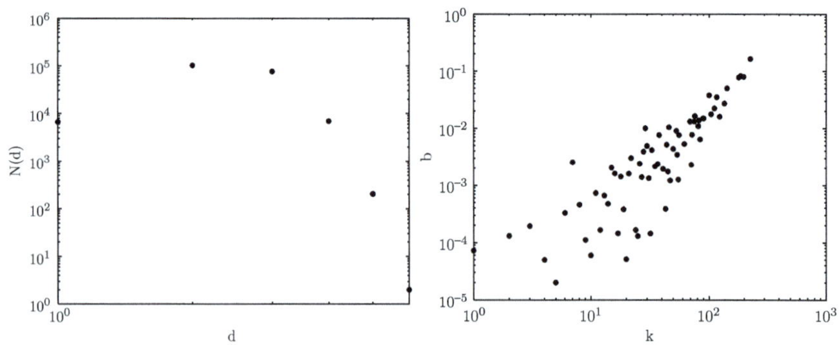

Fig. 7 Distributions of distances among all the pairs in the graph (*left*) and betweenness b vs degree k (*right*)

fact high betweenness for low values of k). The measure of betweenness is not only important to study the architectural properties but also the dynamical properties.

To investigate the role of relationships among large banks (hubs), we measure the rich-club properties. The rich-club phenomenon is one of the main structural properties of complex networks: this property refers to the tendency of high degree nodes, the hubs of the network, to be very connected to each other; nodes with high degree are much more likely to form tight and well interconnected subgraphs (clubs) than low degree nodes [33]

$$\phi(k) = \frac{2E_{>k}}{N_{>k}(N_{>k}-1)}, \tag{7}$$

where $N_{>k}$ is the number of nodes with degree higher than k, $E_{>k}$ is the number of links between nodes having degree higher than k and $N_{>k}(N_{>k}-1)/2$ is the maximum possible number of links among the $N_{>k}$ nodes. The rich club phenomenon is not trivially related to the properties of assortative–disassortative mixing in networks. Many different networks (both assortative and disassortative) display rich club phenomenon, that is $\phi(k)$ increasing with increasing k, indicating the presence of oligarchies in financial networks.

In Fig. 8 we observe that the system of Italian banks is only slightly disassortative, that is the hubs (large banks) are linked partially with banks with high degree, partially with banks with low degree. The rich club properties are present only for values of degree $k < 30$.

We want to progressively isolate the most important links. Therefore we clear progressively links with values lower than 100, 300, 500 (Fig. 9). That means that we progressive focus on banks which co-finance in common at least 100, 300, 500 firms. As we see from the figure (where the cases with 100 and 500 common firms are plotted), progressively the surviving banks are the core of the Italian large banks.

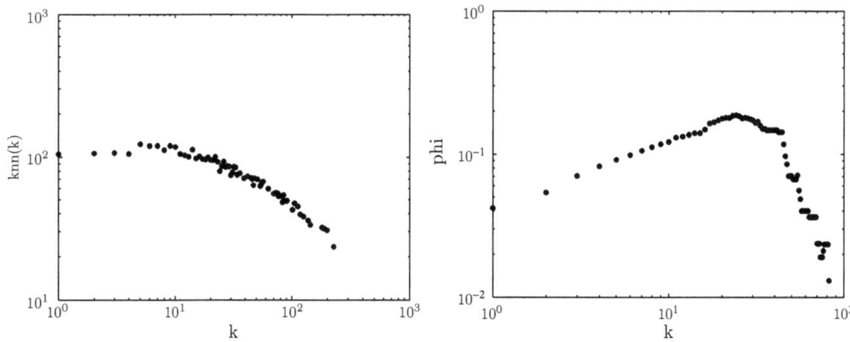

Fig. 8 Assortativity $k_{nn}(k)$ and rich club coefficient $\phi(k)$

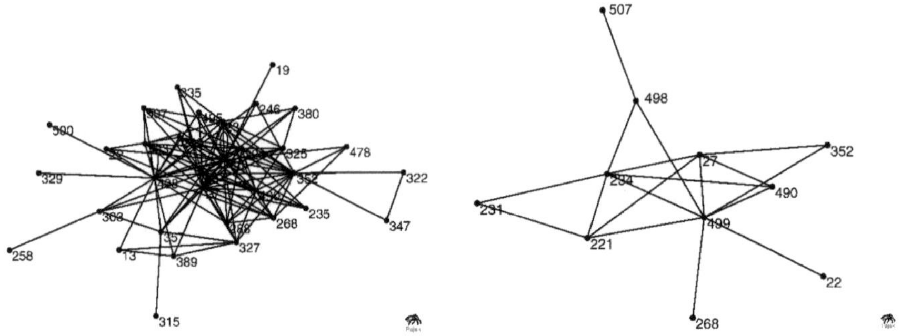

Fig. 9 The networks of banks cofinancing at least 100 (*left*) and 500 (*right*) firms in common

5 Discussion and Conclusions

In this paper we have analyzed the Italian interbank market, credit market and the network of Italian banks in Italy focusing on their cofinancing relationships.

The interbank network is fairly random, preferential lending is limited and cash flows directly from the lender to the borrower without intermediaries. Banks also do not seem able to exploit short term profit opportunities by borrowing from some and lending to others on the same day. All these observations suggest that the interbank market is relatively efficient [7].

This also due to the fact that the banking system is highly heterogeneous and is arranged in a configuration with large banks borrowing from a large number of small creditors.

Moreover, the multilending in Italy is very widespread and it may be a channel of diffusion of financial distresses. The topology of the underlying credit network plays a crucial role in bankruptcy propagation.

The large banks are the hubs of the network and they form the core of the network. This is an evidence of the fact that the Central Bank may guarantee the stability of the whole banking system controlling the financial status of large banks. The small banks on the contrary are the leaves of the network. In a recent paper we have shown that the structure detected by the MST method points out that they are linked in regional branches, where the central node is the largest regional bank surrounded by small local banks [29]. These ones are largely exposed to failures in the case of bankruptcy of the large bank. In this case, if the large bank is able to absorb the shock, this prevents the diffusion of the distress on the whole network.

Acknowledgements The author acknowledges the coauthors M. Gallegati, G. Iori and G. Caldarelli of the cited papers on the Italian credit market [28, 29] and the Italian interbank [6, 7] market and participants to the conference Net2007 at Urbino. Part of this work has been financed by the Eurace project (EU contract No.035086).

References

1. Dorogovtsev SN, Mendes JFF (2003) Evolution of networks from biological nets to the internet and the WWW. Oxford University Press, Oxford
2. Caldarelli G (2006) Scale-free networks. Complex webs in nature and technology. Oxford University Press
3. Caldarelli G Kirman A (1999) Beyond the representative agent. Edward Elgar, Cheltenham, UK
4. Serrano MA, Boguñá M (2003) Topology of the world trade web. Phys Rev E 68:015101
5. Garlaschelli D, Loffredo MI (2004) Fitness-dependent topological properties of the world trade web. Phys Rev Lett 93:188701
6. De Masi G, Iori G, Caldarelli G, (2006) A fitness model of the Italian interbank market. Phys Rev E 74:066112
7. Iori G, De Masi G, Precup O, Gabbi G and Caldarelli G (2007) A network analysis of the Italian overnight money market. J Econ Dynam and Control 32(1):259–278
8. Garlaschelli D, Battiston S, Castri M, Servedio VDP, Caldarelli G (2005) Th scale free topology of market investments. Physica A 350:491–499
9. Reichardt J, Bornholdt S (2005) eBay users form stable groups of common interest. arXiv:physics/0503138
10. Souma W, Fujiwara Y, Aoyama H (2003) Complex networks and economics. Physica A 324(1–2):396–401
11. Delli Gatti D, Gallegati M, Greenwald B, Russo A, Stiglitz JE (2006) Business fluctuations in a credit-network economy. Physica A 370:68–74
12. Hartmann P, Manna M, Manzanares A (2001) The microstructure of the euro money market. J Int Money Finance 20:895–948
13. Barucci E, Impenna C, Reno R (2004) Res Banking Finance 4:321–362
14. Boss M, Elsinger H, Summer M et al. (2004) Network topology of the interbank market. Quantitative Finance 4:677–684
15. Ogawa K, Sterken E, Tokutsu I (2007) Why do Japanese firms prefer multiple bank relationship? Some evidence from firm-level data. Econ Syst 31:49–70
16. Ongena S, Smith DC (2000) What determines the number of bank relationships? Cross-country evidence. J Financial Intermediation 9(1):26–56
17. Detragiache E, Garella P, Guiso L (2000) Multiple vs single banking relationships: theory and evidence. J Finance LV(3)
18. Peltomaki M, Alava M (2006) Correlations in bipartite collaboration networks. J Stat Mech P01010
19. Sneppen K, Rosvall M, Trusina A (2004) A simple model for self organization of bipartite networks. Europhys Lett 67(3):349–354
20. Guillaume JL, Latapy M (2004) Bipartite structure of all complex networks. Inf Process Lett 90(5)
21. Lind PG, Hermann HJ (2007) New approaches to model and study social networks. arXiv:physics/0701107
22. Stanca L, Arosio S, Bonanno V, Cavalieri C (2005) La struttura finanziaria in Italia: evoluzione del rapporto banche e imprese. Politica Economica
23. Forestieri G, Tirri V (2002) Rapporto banca-impresa. Struttura del mercato e politiche del prezzo, Ente Luigi Einaudi. Quaderni di Ricerche 31
24. www.bvdep.com
25. Annual report on vigilance activites of Bank of Italy, May 2005
26. The Italian Institute for Statistics provides a classification of Italian firms in micro-firms, small-size, medium-size and large. The classification depends on the numeber of workers: (1) less than 10 workers, micro; (2) between 10 and 50 small; (3) between 50 and 250 medium; (4) more than 250 large
27. Salza E (2004) Il sistema produttivo e il rapporto banca-impresa. Impresa e Stato 67

28. De Masi G, Gallegati M (2007) Debt-credit economic networks of banks and firms: the Italian case. Springer, Berlin Heidelberg New York. ISBN: 978-88-470-0664-5
29. De Masi G, Gallegati M (2007) The Bank-Firm topology in Italy (submitted)
30. Pastor-Satorras R, and Vespignani A (2001) Epidemic spreading in scale-free networks. Phys Rev Lett 86:3200–3203
31. Strogatz SH (2001) Exploring complex networks. Nature 410:268
32. Newman MEJ (2001) Scientific collaboration networks. II. Shortest paths, weighted networks, and centrality. Phys Rev E 64(1):026118
33. Colizza V, Flammini A, Serrano MA, Vespignani A (2006) Nat Phys 2:110
34. Mantegna RN (1999) Hierarchical structure in financial markets. Eur Phys J B 11:193–197

Network Measures in Civil Air Transport: A Case Study of Lufthansa

Aura Reggiani, Sara Signoretti, Peter Nijkamp, and Alessandro Cento

Abstract Air transport networks have exhibited a trend towards complex dynamics in recent years. Using Lufthansa's networks as an example, this paper aims to illustrate the relevance of various network indicators – such as connectivity and concentration – for the empirical analysis of airline network configurations. The results highlight the actual strategic choices made by Lufthansa for its own network, as well in combination with its partners in Star Alliance.

1 Introduction

Network analysis has already a long history in operations research and quantitative social science research. In the past, much attention has been paid to shortest-route algorithms (for example, the travelling salesman problem), where the spatial configuration of networks was put in the centre of empirical investigation. Integer programming, linear and nonlinear programming turned out to offer a proper analytical toolbox. In recent years, we have seen several new trends, in particular, the rise of hub-and-spoke systems in liberalized networks, the emergence of dynamic adjustments to new competitive conditions and the increase in complexity in international networks.

A. Reggiani and S. Signoretti
Department of Economics, Faculty of Statistics, University of Bologna, Bologna, Italy
aura.reggiani@unibo.it, sara98@supereva.it

P. Nijkamp
Department of Spatial Economics, Faculty of Economics and Business Administration VU University Amsterdam, Amsterdam, The Netherlands
pnijkamp@feweb.vu.nl

A. Cento
KLM Royal Dutch Airlines, Milan, Italy
Alessandro.Cento@klm.com

A.K. Naimzada et al. (eds.), *Networks, Topology and Dynamics*,
Lecture Notes in Economics and Mathematical Systems 613,

Furthermore, it appears that in the past decades many social, spatial and economic systems show an organized pattern characterized by network features, such as transportation, telecommunication, information or energy systems. As a consequence, much attention has recently been paid to the study of network properties emerging in many social, spatial and economic fields, as witnessed by the vast amount of literature published in the past years [8, 23, 24, 35, 36, 42, 43]. Air transport is a prominent example of modern network constellations and will be addressed in this paper from a connectivity perspective.

Air transport shows indeed clear network features, which impact on the way single airline carriers operate [15]. The abundant scientific literature on airline networks has addressed this topic in terms of theoretical modelling and empirical measurements on different typologies of airline network configurations. This strand of recent research aimed to measure the network structure in relation to the effects of: (a) the market deregulation in United States in 1978 and in the European Union in the 1990s, (b) new trends in recent airline business strategies denoted as "low cost" principles. Low cost carriers developed rather fast after the deregulation policy, by acquiring a competitive network advantage on traditional airlines, which consequently seemed to reorganise rapidly their airline network to respond to the new market dynamics.

In this context, interesting research has emerged that mainly addressed the issue of describing and classifying networks by means of geographical concentration indices of traffic or flight frequency [13, 16, 17, 28, 30, 38–40, 44]. These measures, such as the Gini concentration index or the Theil index, provide a proper measure of frequency or traffic concentration of the main airports in a simple, well-organized network. However, if a real-world network structure is complex, including multi-hub or mixed point-to-point and hub-spokes connections, the concentration indices may record high values for all types of structure, but fail to clearly discriminate between different network shapes [3]. There is a need for a more appropriate measurement of connectivity structures in complex networks.

Starting from the above considerations and research challenges, the present paper aims to investigate the scientific potential and applicability of a series of network connectivity/concentration indices, in order to properly typify and map out complex airline network configurations. Specifically, these various network indicators will be adopted and tested to describe the main properties – in terms of the network connectivity and configuration – of Lufthansa's airline system. The aims of the present paper are then: (a) to detect the extent to which the real network configuration is close to typical network models that evolved over time; (b) to examine how concentration measures can point to the different network topologies; and (c) to study the way nodes are connected, that is, to analyse their distribution function.

The present article is organized as follows: Sect. 2 will provide a brief description of the main models of network connectivity that have been developed in the framework of (spatial or social) network analysis. In this section, the focus will mainly be on the concept of vertex degree distribution in a network and on the main indicators used for the analysis of air transport. Next, Sect. 3 presents a novel empirical analysis of Lufthansa's network; the methods provided by network analysis are applied in order to understand more thoroughly the real network's topology. Finally, Sect. 4 will offer some conclusions from the present paper, as well as some further research challenges.

2 Network Models and Measures in Air Transport Systems

2.1 Preface

Many economic activities are currently characterized by network characteristics with a high degree of complexity, since their processes and outcomes depend not only on the choices of the single agents but also on the dynamic – often nonlinear – interactions between them in a continuous dynamic interplay [36].[1] A clear example of a complex spatial-economic network is the geographical network of the air transport industry: understanding its peculiarities and responding to these features can bring about substantial advantages for both consumers and producers [15]. Airline network analysis has gained much popularity in recent years.

Modelling complex networks is also a great challenge: on the one side, the topology of the network is governing the complex connectivity dynamics (see, for instance, [6]); on the other side, the functional-economic relationships in such networks might also depend on the type of connectivity structure. The understanding of these two interlinked network aspects may be instrumental for capturing and analysing airline network patterns. Starting from the above considerations, we will review, in the next subsection, the main connectivity models and measures which have recently gained a great deal of attention in the scientific literature, with a particular view to air transport networks.

2.2 Network Models

In the last decades network theory has gained scientific interest and sophisticated network models have been used in different fields, including economics and geography [46]. This trend faced also quite some difficulty, because existing models were not able to clearly describe the network properties of many real-world systems, whose complexity could not fully be understood [4].

Spatial-economics systems – including air transport networks – are complex, because agents interact, obtaining significant benefits by means of a joint activity [11]. This interacting process may become a permanent feature thus leading to a new meso- or macro structure, for example, to the creation of clusters.

[1] These authors point out that the main feature of complexity is that the outcome (of the activity of a complex system) "should not be obvious from the single building blocks" [12]. Consequently, the term complexity indicates that the final result cannot be foreseen even when the single components of a system are known and studied.

Air transport systems have over the past years been experiencing such clustering processes. An example is provided by airlines' alliances.[2] The main reason why airline carriers cooperate of aggregate stems from cost reductions they can thus obtain. Being a member of an alliance impacts on the carriers' strategy for a long time and also influences the network configuration they adopt. It is worth noteworthy that alliances play also an important role in determining market dynamics; in 2005, the three main alliances in air transport accounted for 80% of the total capacity offer.[3] Therefore, we need to develop airline network models that can adequately take into account clustering and merger processes.

A further important trend many real networks show is the so-called "Small-World (SW) effect". This term indicates that the diameter[4] of a network is so short that it takes only a few movements along links in order to move between any two nodes of a network [37]. In air transport systems, we can point out the SW effect by taking into consideration and comparing the network configuration of single carriers or of alliances; such systems exhibit a clear SW effect when it takes only a small number of flights to link the two most distant airports in the network.

Alongside the SW effect, the SW network model has been developed in order to take into account both the SW effect and the related clustering processes [47]. The main features of this model are a short diameter and a high clustering coefficient.

A further elaboration of the SW model is the so called Scale-Free (SF) network introduced by Barabási and Albert [4] in order to incorporate two mechanisms upon which many real networks have proven to be based: *growth* and *preferential attachment*. The former points to the dynamic character of networks, which grow by the addition of new nodes and new vertices; the latter explains how new nodes enter the network, namely by connecting themselves to the nodes having the highest number of links.

An important feature of SF networks is represented by their vertex degree distribution[5] $P(k)$ which is proportional to $k^{-\gamma}$ (with k being the number of links), that is, to a power law. The value of the degree exponent γ depends on the attributes of the single systems and is crucial to detect the exact network topology, in particular the existence of the hubs (highly connected nodes). As Barabási and Oltvai [6] highlight, a SF network embeds the proper hub-and-spoke model only when $\gamma = 2$, while for $2 < \gamma \leq 3$ a hierarchy of hubs emerge. For $\gamma > 3$, the hub features are absent and the SF network behaves like a random one.

In air transport systems, we can point out SW networks by considering full-service carriers. Without national or political impediments in a free market, these carriers typically organize their network into a hub-and-spoke system, where one or a few central airports called "hubs" have a high number of links to the other airports called "spokes". Passengers travelling from a place of origin to a place of destination

---

[2] The processes underlying the creation of an alliance can be clearly depicted by considering the integration of Lufthansa and Swiss, described in the Lufthansa Annual Report (2005); available on the website http://konzern.lufthansa.com/en/html/ueber_uns/swiss/index.html).

[3] See http://www.tourismfuturesintl.com/special%20reports/alliances.html.

[4] The concept of a diameter will be defined in Sect. 2.3.

[5] $P(k)$ is the probability that a chosen node has exactly k links [6]. See also (1).

have to stop typically in one or a few hubs to change aircraft. Hubs are organised in order to allow flight connectivity by coordinating the scheduled timetable of the arriving and departing flights. Investigating the airline strategy in designing hub connectivity and timetable coordination has been the aim of several empirical network studies. Some examples of theoretical and empirical investigation of hub connectivity can be found in the works of Bootsma [10], Dennis [18], Rietveld and Brons [41], Veldhuis and Kroes [45], and Burghouwt and de Wit [14]. As a consequence, the hub has to manage normally a high volume of traffic at the same time, due to their central connecting role in the network.

In contrast to SF networks, we have to highlight also random networks [19], which display homogeneous, sparse patterns, without cluster characters. Their vertex degree distribution follows a Poisson distribution.[6]

In air transport, random networks are useful to map point-to-point connections, as it is the case for low-cost airlines [17]. In the ideal point-to point network all airports are connected to each other, so that passengers can fly from one airport to any other directly without stopping in any hub to change aircrafts. These networks have a low diameter, as a consequence of the high number of direct links between airports. Reggiani and Vinciguerra [[37], p. 148] point out that a random network can be seen as "*a homogeneous system which gives accessibility to the majority of the nodes in the same way*". Furthermore, as it is evident by looking at the plot of the exponential function, the probability to find highly connected nodes is equal to 0. Therefore, no clear hubs exist, and the network configuration appears to be random because no single airport displays a dominant role in a connected network.

In Sect. 2.3, we will address two main degree (connectivity) distributions that have often been observed in empirical experiments, vis-à-vis exponential and power-law.

2.3 Network Degree Distributions

The vertex degree distribution is one of the key tools we may use to point out the network configuration [37], since this function determines the way nodes are connected. It can be defined as the probability $P(k)$ of finding nodes with k links.

In general, we can state that

$$P(k) = N(k)/N, \tag{1}$$

where $N(k)$ is the number of nodes with k links and N is the number of nodes of the network.

With regard to the network topologies developed in the framework of graph theory, complex systems tend to show two main degree distributions: the *Poisson* distribution [19] and the *power-law* function [5].

[6] For a review of random models, SW models and SF models, see Albert and Barabási [2] and Jeong [27].

The former is defined as

$$P(k) \sim e^{<k>} \frac{<k>^k}{k!}, \tag{2}$$

and describes networks – so-called random networks – where the majority of nodes have approximately the same number of links, close to the average degree $<k>$ [4]. Equation (2) is a distinctive feature of point-to-point networks, such as those adopted by low-cost airlines; this network topology is typical of equilibrated economic-geographical areas, where a high number of direct links can be profitably operated.

The power-law function is defined as

$$P(k) \sim k^{-\gamma}, \tag{3}$$

and characterizes networks having a small number of nodes with a very high degree while the majority of nodes have a few links. Equation (3) has important economic implications: it characterizes SF networks, where the term SF refers to the fact that "*the power-law distribution does not change its form no matter what scale is used to observe it*" [[37], p. 150], and that, in these networks, distances are irrelevant. Therefore, we expect to find SF networks in "global networks", such as the Internet and air transport, and in general in those networks where relevant economic aggregation clusters (preferential attachments) attract flows from distant nodes.

It interesting to note that from the above distribution functions (2) and (3) we can extrapolate the related cost/utility/impedance functions [37]. However, when the identification of the two functions is ambiguous, we need to obtain additional information from network theory (for example, centrality indices, dominance indices). A multidimensional approach is needed in this respect, where not only the way airports are connected is relevant (spatial network components), but also the geometrical architecture of the network, as well as its degree of network homogeneity (physical network components).

In Sects. 2.4 and 2.5 we will now introduce some indicators and measures we can adopt to study the network configuration of carriers: the computation of the above indices is crucial to understand the tendency to agglomeration of concentration patterns, and hence the possibility of hierarchical network relations among nodes.

It is moreover important to identify a SF network because of its strong features in terms of robustness and vulnerability. In the case of a random attack (or disturbance) on nodes, the SF network will strongly persist, because a random attack will probably damage nodes that have only a few connections, which are the majority. Nevertheless in case of an attack against the main hubs, the network will easily be fragmented. Consequently, we might also talk of "vulnerability/permeability" of the SF network: if a strategic input, for example, a virus, is dispersed in the hubs, it is certainly diffused all over the network. On the other hand, random networks are weak against a random attack which will cause the split of the network.

Consequently, it is important to identify hubs in the network in order to prevent targeted attacks and to preserve the system [23]. The identification of such characteristics is certainly useful to the understanding of the dynamics of air network configurations, also from the perspective of policy/planning interventions.

2.4 Network Topology Indices

Airline networks may exhibit simple or complex topologies. Networks have been given several definitions in the framework of graph theory, as for instance by Harary [25]: "*a network is a graph, or directed graph, together with a function which assigns a positive real number to each edge*". In this context it is useful to outline the geometrical indicators most frequently used to represent the network shape; they are illustrated in Table 1.

Table 1 Network's topology indices

Index or measurement	Description	Formulation	Variables	Source
Degree	The degree of a node is given by the number of its links	$k(v)$	$k(v)$ is the number of links of node v	Barabási and Oltvai [6]
Closeness	It indicates a node's proximity to the other nodes	$C(v) = \frac{\sum_{t \in V} d_{vt}}{n-1}$	d_{vt} is the shortest path (geodesic distance) between nodes v and t; n is the number of nodes in the network	Newman [31]
Betweenness	It indicates a node's ability to stand between the others, and therefore, to control the flows among them	$B(v) = \sum_{s \neq t \neq v \in V} \frac{\sigma_{st}(v)}{\sigma_{st}}$	$\sigma_{st}(v)$ and σ_{st} are, respectively, the number of geodesic distances between s and t that pass through node v, and the overall number of geodesic distances between nodes s and t	Freeman [20]
Diameter	It measures the maximum value of the geodesic distances between all nodes	$D = \max_{s,t \in V, s \neq t} d_{st}$	d_{st} is the geodesic distance between nodes s and t	Boccaletti et al. [9]
Clustering coefficient	It measures the cliquishness of a node	$Cl(v) = \frac{l_v}{_{\max} l_v}$	l_v and $_{\max} l_v$ are, respectively, the number of existing and maximum possible links between the nodes directly connected to node v (its neighbours)	Watts and Strogatz [47]

It should be mentioned that the first three indices measure the centrality of a vertex in a graph, while the last two can be used to investigate the networks' topological properties [37]. It is necessary to underline that the "geodesic distance", used to compute closeness, betweenness and diameter, represents the shortest of all distances between two nodes [21].

In the context of our empirical experiments, we will apply the above indicators to explore Lufthansa's network structure and configuration, since all complex systems characterized by a network structure share properties exclusively depending on network's configuration (see also [46]). Before starting our empirical analysis, Sect. 2.5 will illustrate additional indices that we may use to investigate the networks' concentration.

2.5 Network Concentration Indices

If we want to detect the networks' configuration (random versus SF) we also need to understand to what extent these networks are concentrated, because the existence of hubs implies a high degree of concentration [40]. To this purpose we will use: (a) the Gini concentration index; (b) the Freeman centrality index;[7] and (c) the entropy index. These three indices are illustrated in Table 2.

The first index G measures the inequality existing in a distribution, and ranges between 0 and 1; the higher its value, the more uneven is the distribution [21]. The second index F takes into account the structure of the system, and measures the network shape as the degree of inequality in a network with respect to a perfect star network [21].

The third is the entropy function E, which shows the degree of variety existing in an economic or spatial network [22]. In particular, entropy can be employed as a tool for studying spatial differentiation, that is, heterogeneity in a system: "*for instance, by investigating whether certain spatial configurations are completely arbitrary and disordered or whether these configurations show a certain degree of spatial organization or regularity*" [[32], pp. 18–19]. Therefore the entropy function indicates how organized a system is: the higher is the value of E, the more diversified the network [22].

Next, in Sect. 3 we will carry out an empirical study on four networks – based on Lufthansa's airline network – by means of the analytical tools previously described.

---

[7] The concept of concentration aims at discerning whether or not the activity we are studying is located homogeneously over a geographical area, without considering the form of corresponding system. In the framework of our experiments, networks are concentrated to the extent that some nodes have a share of flights which is higher than the area they occupy [21].

The concepts of centrality – referring to single nodes – and centralization – referring to a whole network – are closely related: a network is centralized when a node, or a group of nodes, can control the flows the network represents and are consequently given higher centrality values [20]. We can, therefore, state that centralized networks are always concentrated as well, while the opposite does not always holds.

Table 2 Network's concentration indices

Indicator	Formula	Use	Variables used	Sources
Gini concentration index	$G = \frac{\sum_{i=1}^{n}\sum_{j=1}^{n}\lvert x_i - x_j\rvert}{2n^2\mu}$	It is a measure of geographical concentration	x_i,x_j are the number of weekly flights from airports i and j, ranked in increasing order; n is the number of airports in the network; μ is $\sum_i x_i/n$	Cento [17]
Freeman centrality index	$F_B = \frac{\sum_i [F_B(x^*) - F_B(x_i)]}{n^3 - 4n^2 + 5n - 2}$	It is a measure of similarity to a perfect star network	$F_B(x_i) = \sum\sum b_{jk}(x_i)$ is the $j < k$ $j < k$ betweenness centrality of node x_i; $F_B(x^*)$ is the highest betweenness centrality value of the distribution	Cento [17]
Entropy function	$E = -\sum_{ij} p_{ij} \ln p_{ij}$	It measures the degree of spatial organisation and variety in a system	p_{ij} is the probability of a link between nodes i and j	Nijkamp and Reggiani [32]; Frenken [22]

3 An Empirical Application to Lufthansa's Airline Network

3.1 The Data Base

This section will focus on the geographical analysis of Lufthansa's aviation network in the year 2006. The airline network measurement is essential for exploring the airline behaviour and its implications for the supply, the traffic demand, the airports' infrastructure and aviation planning. The airline network can be subdivided into domestic, international or intercontinental configurations depending on whether the airports connected are located within a country, a continent or in different continents. Furthermore, an airline network can be interconnected or interlined to partner's networks within the alliance concerned. This classification is based on geographical, air transport-political and economic characteristics, such as airlines' degree of freedom from the Chicago Convention (see [17]) market liberalization, or costs and traffic demand. Therefore, the overall network configuration is the result of the integrated optimisation of the domestic, international, and intercontinental parts of the total network. These sub-network configurations may range from fully-connected or point-to-point to hub-and-spokes configurations to alliances (fully-contracted) or to a mix of these configurations. Within this conceptual framework, we will position our analysis of four sub-networks of Lufthansa. As summarized in Table 3, we coin networks A1 and A2, referring respectively to the flights operated

Table 3 Lufthansa's network constellation (2006)

Network	Area under consideration	Carrier or alliance operating the flight	Nodes	Total number of links
A1	Europe	Lufthansa	111	522
A2	World	Lufthansa	188	692
B1	Europe	Star Alliance	111	3,230
B2	World	Star Alliance	188	6,084

by Lufthansa in Europe and in the whole world, while networks B1 and B2 take into consideration – respectively at a European and at a global level – the flights operated by all the carriers which are members of Star Alliance (to which Lufthansa belongs).[8]

The variable under analysis is represented by the number of direct connections of each airport in the summer season of the year 2006, measured on a weekly basis. The networks are represented in Figs. 3–6 in Appendix 2.

In all four cases we only consider those airports where Lufthansa operates with its fleet and not by partner's airlines. When we consider A1 and A2 networks, we clearly see that the majority of Lufthansa's flights are operated at a continental level. On the contrary, nearly half of Star Alliance's flights are operated outside Europe. This finding is not surprising, if we consider that the carriers making up Star Alliance are mainly from non-European countries.

Sections 3.2, 3.3 and 3.4 will now illustrate the empirical results of our experiments, aiming at analysing the connectivity and concentration patterns in the above mentioned networks.

3.2 Lufthansa's Network Geometry

On the basis of the indicators illustrated in Table 1, we will now show the results emerging from the related applications to the four Lufthansa's network domains A1, A2, B1 and B2. In particular, since all the indicators displayed in Table 1 characterise the nodes in a network, we will investigate by means of these indicators – in our four networks – the single nodes' features as well as the relations among nodes.

More specifically, in order to examine the nodes' location, we have computed the three centrality measures (degree, closeness and betweenness) described in Table 1. Concerning the investigation of the nodes' relations, we have examined the diameter and the clustering coefficient of the network (see again Table 1).

[8] The Star Alliance member carriers are currently: Air Canada; Air New Zealand; ANA; Asiana Airlines; Austrian; bmi; LOT Polish Airlines; Lufthansa; Scandinavian Airlines; Singapore Airlines; South African Airlines; Spanair; Swiss; TAP Portugal; THAI; United Airlines; US Airways; VARIG (the list was retrieved from www.staralliance.com).

The degree of a node (Table 1) can be seen as a measure of centrality if we assume – in the framework of our analysis – that the best connected airports have a greater power over the whole network, as they can control a considerable amount of all flights. In all networks we find that the airports of Frankfurt and Munich have always the highest degree (see Table 10 in Appendix 1).

A further analysis of nodes' centrality focuses on their "ease-of-access" to the other nodes.[9] In order to investigate this concept we have computed the closeness centrality[10] (Table 1). The values of this index for the networks under consideration (listed in Table 11 in the Appendix 1) show that the highest values usually correspond to the best connected nodes; therefore, closeness centrality is able to map out – in the framework of our study – the most important airports in terms of connectivity. A similar trend can be observed by considering betweenness centrality (Table 1; the values for networks A1, A2, B1 and B2 are listed in Table 12 in Appendix 1). This finding is not surprising, since hubs – in the framework of the hub-and-spoke model – are chosen from those airports falling among the highest possible number of pairs of other airports [15, 33].

The networks' topology can also be explored by examining how the various nodes relate and link, since this last attribute impacts the configuration of the whole structure. For this purpose we have computed the clustering coefficient (defined in Table 1; the ten highest values for the nodes of the four networks of our experiments are listed in Table 13 in the Appendix 1). The values indicate a significant difference between the networks A1 and A2 and the networks B1 and B2; in the former case the airports of Frankfurt and Munich dominate the chart; in the latter case, other airports appear to emerge, thus showing that flights are spread more equally on the whole network.

In addition, we will also consider the diameter of the above networks in order to investigate how the links' patterns influence the ability to move inside the network. Both A1 and A2 have a diameter of 4, while B1 and B2 have a diameter of 2. This can be justified only if there is no significant difference in the geographical configuration between A1 and A2, approximately a hub-and spoke, while B1 and B2 can be a mixture of hub-and-spoke and point-to-point networks. In other words, the integration of Lufthansa network in the Star Alliance reduces the travel distance, as the passengers can benefit from more connections and thus shorter paths to travel between the origin and the destination. This has important implications in the context of our study, because it entails that Lufthansa's networks shrink, when we consider the flights of all Star Alliance members.

Having examined now Lufthansa nodes' characteristics, we will explore Lufthansa's network features, in particular its network concentration and connectivity. The related results will be offered in the following Sects. 3.3 and 3.4.

---

[9] It can be assumed that access to the network is easier when nodes are closer [21].

[10] We compute the closeness centrality, as well as the subsequent betweenness centrality, using the Pajek software (http://vlado.fmf.uni-lj.si/pub/networks/pajek/).

3.3 Lufthansa's Network Concentration

The study of the networks' degree of concentration – which is carried out in the present subsection – is crucial in order to detect the exact network topology, because the hub-and-spoke model is highly concentrated, while point-to-point networks do not show this feature.

First, Table 4 presents the normalized Gini index (see Table 1) for the four networks under consideration. Both Star Alliance networks are less concentrated than the Lufthansa counterparts, meaning that when we enlarge the measurement to a broader network including intercontinental destinations and partners' networks, the configuration will probably evolve into a mix of multi hub-and-spoke and point-to-point structures. In particular, network A2 appears to be the most concentrated.

The information provided by the Gini index refers to the degree of concentration existing in a network, without any evidence on how this concentration impacts on the network topology. For this last purpose the Freeman centrality index (Table 1) has been computed. Its normalized values are represented in Table 5. This index assumes the value 1 for a hub-and-spoke network, and the value 0 for a point-to-point network [17].

According to the Freeman index, again networks A1 and A2 turn out to be the most concentrated ones. In particular, A2 network seems to be again the closest to the hub-and-spoke model; we may suppose that this network is characterized by a strong hierarchy among nodes.

Finally, concerning the last concentration index, that is, entropy (Table 1), Table 6 shows the related values for the networks A1, A2, B1 and B2. The results from

Table 4 Normalized Gini index

Network	Gini index
A1	0.762
A2	0.813
B1	0.524
B2	0.600

Table 5 Normalized Freeman index

Network	Freeman index
A1	0.504
A2	0.757
B1	0.059
B2	0.056

Table 6 Entropy values

Network	Entropy
A1	5.954
A2	6.194
B1	7.790
B2	8.389

Table 6 show that the entropy values are higher when we consider those flights operated by Lufthansa's partners (networks B1 and B2). A likely explanation for this increase is given by the process of construction of these networks, obtained by the addition of flights to the nodes of A1 and A2, respectively. Both B1 and B2 are therefore the "sum" of the networks implemented by the different carriers that are members of Star Alliance, and hence they are not the result of a specific strategy, as is the case for A1 and A2. Clearly, the above values indicate that A1 and A2 networks are more concentrated and less dispersed than the B1 and B2 networks; more specifically, A1 appears to be the most concentrated network.

In conclusion, from the above three indicators, networks A1 and A2 appear to be the most concentrated. However, among these two networks, A2 seems the most concentrated with respect to two indicators (Gini and Freeman), while A1 seems the most concentrated with respect to the entropy index.

In order to formally detect hub-and-spoke models, our next step will be the analysis of the vertex connectivity distribution functions of the four networks A1, A2, B1 and B2, in the light of their performance indicators (see also Sect. 2.3).

3.4 Lufthansa's Network Configuration

In Sect. 2.3 we have already stressed the importance of the vertex degree distribution function, in order to detect the most plausible network configuration. In this section, we will explore whether the variable "number of weekly connections" is rank-distributed – over A1, A2, B1 and B2 – according to either an exponential or a power function.

The R^2 values and the b coefficients of the two interpolating functions (exponential and power) concerning the four ranked distributions (in log terms) are listed in Table 7. The plots of both functions for the four networks under consideration are displayed in Figs. 1 and 2.

Both Table 7 and Figs. 1 and 2 highlight that our data sets better fit a power function, as the higher R^2 values indicate. It is worth noting that the b coefficient of the power function for A1, A2, B1 and B2 is respectively equal to 0.99, 0.82, 0.67

Table 7 Exponential and power fitting of rank distributions

Network	A1		A2		B1		B2	
Network parameters	R^2	b	R^2	b	R^2	b	R^2	b
Distribution function								
Power	0.95	0.99	0.93	0.82	0.75	0.67	0.70	0.65
Exponential	0.75	0.03	0.67	0.01	0.66	0.02	0.48	0.01

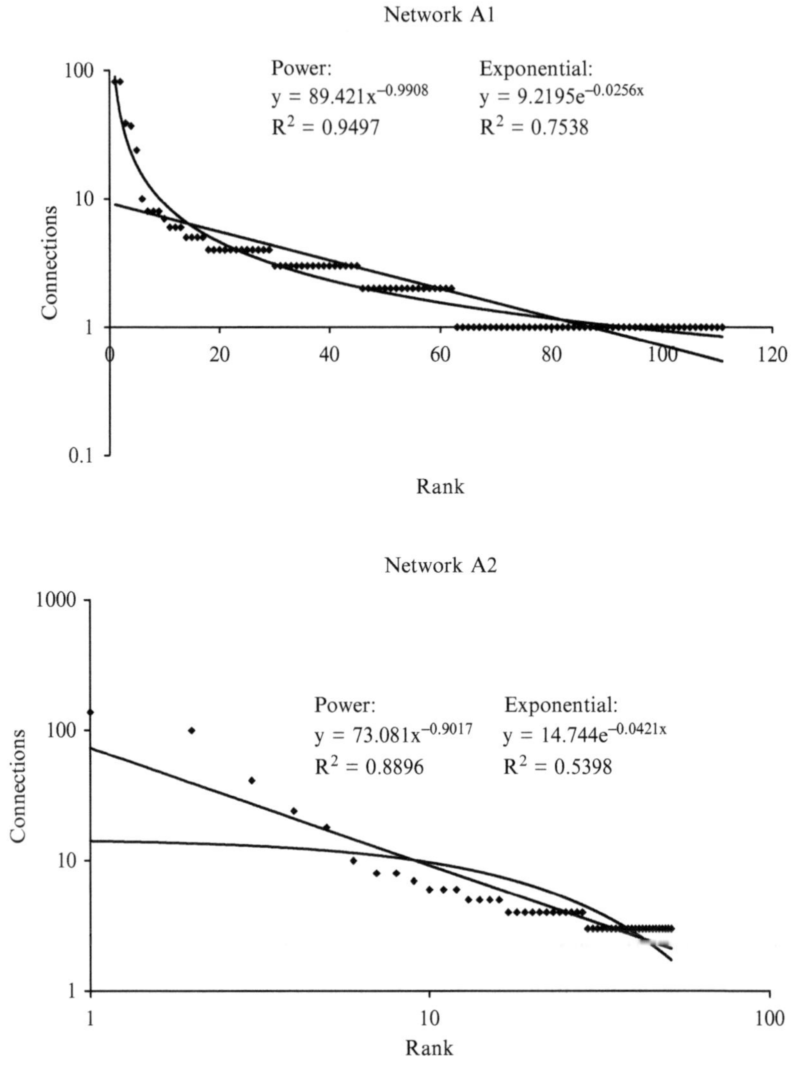

Fig. 1 Rank distribution fitting for networks A1 and A2

and 0.65. If we carry out a transformation[11] of these coefficients, we observe that the A1 network displays a power-law exponent equal to 2, thus indicating a stronger tendency to a hub-and-spoke system according to Barabási and Oltvai [6], while the other three networks A2, B1 and B2 display power-law exponent between 2 and 3, thus indicating a tendency to a hierarchy of hub/agglomeration patterns.

---

[11] Adamic [1] shows that the power-law exponent γ (emerging from the nodes' probability distribution (3)) is related to the power function coefficient b (emerging from the distribution relating the degree of the nodes to their rank (rank size rule); see Figs. 1 and 2) as follows: $\gamma = 1 + (1/b)$.

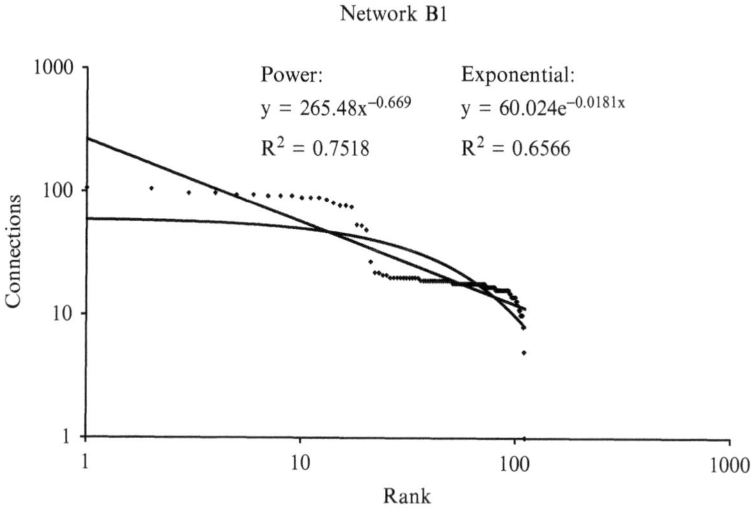

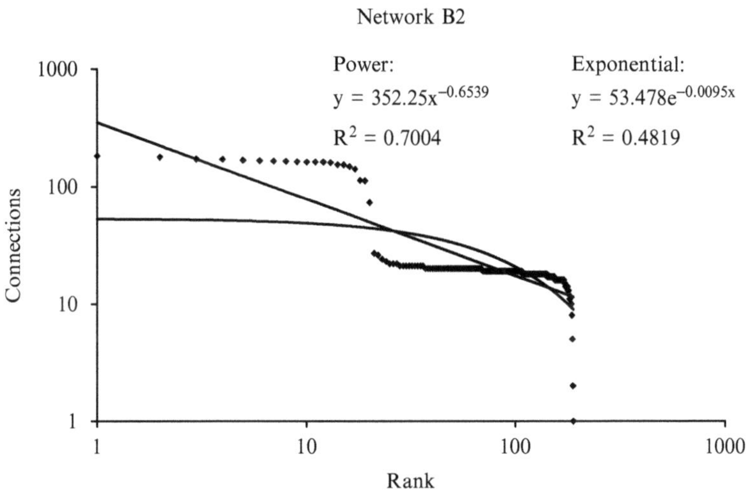

Fig. 2 Distribution fitting for networks B1 and B2

A further issue concerns the fitting of the exponential function. Also in this case we obtain high R^2values, although inferior to the ones emerging in the power case; however, the coefficient of the exponential function is always very low, ranging from 0.01 to 0.03 (Table 7).

Therefore, if we look at the R^2indicators, all networks under consideration appear to be in a "border-line" situation (that is, an ambiguity between a power and exponential fitting). Nevertheless, if we look at the coefficient values, the four networks seem to show a tendency toward an agglomeration structure of SF type, expressed by a clear power-law vertex degree distribution, with the degree exponent γ equal to 2 (network A1), or varying between 2 and 3 (networks A2, B1, B2).

A further consideration concerns the plots of networks B1 and B2 (Fig. 2). We can clearly see that both identify a power function with a cut-off. Thus, if we eliminate – in both networks B1 and B2 – those nodes which have less than 10 links, we slightly improve the fitting of their power function, obtaining for networks B1 and B2 respectively R^2 values of 0.84 and 0.75, but still lower than the R^2 values regarding A1 and A2.

In conclusion, from the estimation results displayed in Table 3, the networks A1, A2 appear to show the strongest characteristics of concentration and preferential attachment. In particular, network A1 appears to be the closest to the hub-and-spoke model, from the perspective of Barabási and Oltvai's approach. Given these preliminary results, it is worth to examine these configurations, by exploring further indicators of the network concentration, such as those defined in Sects. 2.4 and 2.5. Consequently, a multidimensional method, such as Multi Criteria Analysis (MCA), taking into account – by means of an integrative approach – all adopted indicators and related results, was next carried out and applied.[12] The alternatives are the four networks A1, A2, B1, B2 under consideration, while the criteria have been grouped according to three macro-criteria: network concentration, topology and connectivity (Table 8). It should be noted that, concerning the topology criteria, we have considered the diameter and the clustering coefficient, since these two indices provide the network geometry's features (see Sect. 3.2). In particular, concerning the latter, the average clustering coefficient has been adopted [6].

The first group of macro-criteria is related to the networks' concentration. It should be noted that in our MCA procedure, the entropy indicator needs to be transformed positively because the real values of the entropy function increase

Table 8 Alternatives and criteria

Alternatives	A1 (Lufthansa, Europe) A2 (Lufthansa, World) B1 (Star Alliance, Europe) B2 (Star Alliance, World)
"Concentration" criteria	Gini index Freeman index Entropy
"Topology" criteria	Diameter Average clustering coefficient
"Connectivity" criteria	R^2 of the fitted power function (ranked degree distribution) Coefficient of the power function R^2 of the fitted exponential function (ranked degree distribution) Coefficient of the exponential function

[12] In particular, the Regime method and software has been used [26].

Table 9 Findings of multi-criteria analyses

Criteria considered	All criteria combined	Concentration criteria	Topology criteria	Connectivity criteria	Concentration and topology criteria
Hierarchy of the alternatives	A1	A2	B1	A1	A1
	A2	A1	B2	B1	B1
	B2	B2	A1	A2	A2
	B1	B1	A2	B2	B2

when networks are more heterogeneous, that is, less concentrated.[13] The second group of macro-criteria refers to the networks' physical measurement. Here, the diameter needs to be converted in utility, because its value is higher when networks are less centralized. The third group of macro-criteria is related to connectivity. This property is investigated through the interpolation of the ranked degree distributions, where – in the power function – the highest exponent of 0.99 implies a value of the exponent degree[14] – in the associated power-law distribution – close to 2 (perfect hub-and-spoke). The $R^2$and the coefficient of the exponential function need to be converted to utility, since both values indicate random and homogeneous patterns.

We have carried out five scenarios by considering: (a) all the criteria mentioned above; (b) each macro-criteria separately; (c) concentration and topology criteria together. In each scenario an equal weight, that is, unknown priority, has been given to the single criteria. The results are listed in Table 9.

These findings point out that network A1 prevails, however with two exceptions. The former is represented by network A2, which is the top-scorer when we consider the criteria related to the networks' concentration/geography: this finding comes from the higher centralization and concentration degree of network A2, as demonstrated by the Freeman and Gini indices. The latter exception is represented by network B1, which prevails when we consider the criteria related to the physical measurement of networks.

It turns out that the Lufthansa network A1 is the most concentrated one; we can conjecture that A1 is close to a hub-and-spoke system, according to the values expressed by its exponent degree in the power-law distribution (see Table 7). This result confirms the dual-hubs network strategy advocated by the German carrier [29]. Frankfurt and Munich act as central hubs, where all intercontinental flights depart and arrive in conjunction with the European and domestic flights. This timetable coordination is designed to allow passengers to transfer from one flight to another for different national and international destinations. The general conclusions of the present article are included in the Sect. 4.

[13] The relation between concentration and centralization is described in Footnote 6 in Sect. 2.5.

[14] See Footnote 12.

4 Conclusions

Airline networks are fascinating examples of emerging complex and interacting structures, which may evolve in a competitive environment under liberalized market conditions. They may exhibit different configurations, especially if a given carrier has developed a flanking network framework together with partner airlines.

The present paper has investigated the network structure of four networks of Lufthansa by considering several indicators concerning the concentration, topology and connectivity (degree distribution) functions characteristics of this carrier. An integrated multidimensional approach, in particular multicriteria analysis has been adopted, in order to take into account all information obtained by the above indicators.

The related results point out that all the four Lufthansa networks can be properly mapped into the SF model of the Barabási type. In particular, network A1 can be formally identified as a hub-and-spoke structure. In general, we can conjecture a "tendency" towards a hubs' hierarchy or hub-and-spoke configuration in Lufthansa's European network, as also witnessed by the emergence of various nodes (Frankfurt, Munich and Dusseldorf) which are organized as hubs in the framework of Lufthansa's activities. All in all the four networks exhibit a hierarchical structure mainly dominated by German airports.

The results obtained thus far highlight various characteristic features of complex aviation networks, but need to be complemented with additional investigations, in particular, on the structure and driving forces of the demand side (types of costumers, in particular). Furthermore, the market is decisive in a liberalized airline system, and hence also price responses of customers as well as competitive responses of main competitors would need to be studied in the future.

From a methodological viewpoint a refined weighted network analysis – taking into account the strength of each connecting link – might offer better insights into the topological structure of the airline network at hand (see, for example, [7]).

Acknowledgements The authors wish to thank Roberto Patuelli (University of Bologna at Rimini) for his comments on the present chapter, as well as for his cooperation in the editing process.

Appendix 1

In this Appendix, we will present the top ten scores of the airports – according to the main topological indices illustrated in Table 1– belonging to the four airline networks A1, A2, B1 and B2. These networks are visualized in the subsequent Appendix 2.

Table 10 Top-ten scores of airports according to the degree index (corresponding values in brackets)

A1	A2	B1	B2
MUC (82)	FRA (138)	FRA (106)	FRA (183)
FRA (81)	MUC (100)	MUC (105)	MUC (179)
DUS (39)	DUS (41)	BRE (97)	HAM (172)
HAM (24)	HAM (24)	HAM (97)	DUS (171)
STR (18)	STR (18)	BSL (94)	STR (168)
TXL (10)	TXL (10)	DUS (94)	LEJ (166)
CDG (8)	CDG (8)	LEJ (92)	ZRH (165)
NUE (8)	NUE (8)	NUE (92)	TXL (164)
BRU (7)	BRU (7)	STR (92)	NUE (163)
LHR (6)	MXP (6)	CGN (89)	BRE (162)

Table 11 Top-ten scores of airports according to the closeness index (corresponding values in brackets)

A1	A2	B1	B2
MUC (0.78)	FRA (0.79)	FRA (0.96)	BRE (1)
FRA (0.76)	MUC (0.64)	MUC (0.95)	DUS (1)
DUS (0.60)	DUS (0.53)	HAM (0.89)	ZRH (1)
HAM (0.55)	HAM (0.51)	DUS (0.87)	FRA (0.98)
STR (0.54)	STR (0.50)	NUE (0.86)	MUC (0.95)
TXL (0.51)	CDG (0.49)	STR (0.86)	HAM (0.93)
CDG (0.51)	NUE (0.49)	LEJ (0.85)	STR (0.91)
NUE (0.51)	BRU (0.48)	CGN (0.84)	LEJ (0.89)
LHR (0.51)	LHR (0.48)	TXL (0.84)	NUE (0.89)
MXP (0.51)	MXP (0.48)	ZRH (0.84)	FMO (0.85)
	VIE (0.48)		

Table 12 Top-ten scores of airports according to the betweenness index (corresponding values in brackets)

A1	A2	B1	B2
MUC (0.51)	FRA (0.76)	MUC (0.06)	MUC (0.06)
FRA (0.50)	MUC (0.03)	FRA (0.06)	FRA (0.06)
DUS (0.06)	DUS (0.03)	DUS (0.05)	DUS (0.06)
KUF (0.05)	BKK (0.02)	HAM (0.05)	BRE (0.05)
HAM (0.03)	KUF (0.02)	STR (0.05)	CGN (0.05)
GOJ (0.02)	HAM (0.01)	BRE (0.04)	HAM (0.05)
STR (0.01)	CAI (0.01)	HAJ (0.04)	NUE (0.05)
CDG ($4.5e^{-4}$)	CAN (0.01)	NUE (0.04)	STR (0.05)
CGN ($9.5e^{-5}$)	GOJ (0.01)	TXL (0.04)	ZRH (0.05)
BRU ($1.9e^{-5}$)	GRU (0.01)	CGN (0.04)	CGN (0.05)
	JED (0.01)		DRS (0.05)
	KRT (0.01)		LEJ (0.05)
	LOS (0.01)		
	PHC (0.01)		

Table 13 Top-ten scores of airports according to the clustering coefficient (corresponding values in brackets)

A1	A2	B1	B2
MUC (0.82)	FRA (0.75)	FRA (0.96)	BRE (1)
FRA (0.80)	MUC (0.48)	MUC (0.89)	DUS (1)
DUS (0.24)	DUS (0.11)	LEJ (0.77)	ZRH (1)
HAM (0.10)	HAM (0.04)	ZRH (0.67)	FRA (0.96)
STR (0.06)	STR (0.02)	BSL (0.66)	MUC (0.88)
CDG (0.01)	TXL (6e-3)	STR (0.57)	LEJ (0.84)
TXL (0.01)	CDG (5e-6)	DUS (0.55)	BSL (0.81)
NUE (9e-3)	NUE (4e-3)	HAM (0.55)	GVA (0.67)
BRU (6e-3)	BRU (2e-3)	GVA (0.48)	HAM (0.63)
MXP (4e-4)	ZRH (2e-3)	TXL (0.47)	STR (0.60)
VIE (4e-4)			

Table 14 Nomenclature of airports under study

BKK	Bangkok
BRE	Bremen
BRU	Bruxelles
BSL	Basel
CDG	Paris Charles de Gaulle
CGN	Koln
DRS	Dresden
DUS	Dusseldorf
FMO	Munster
FRA	Frankfurt
GOJ	Novgorod
GRU	Sao Paulo
GVA	Geneva
HAM	Hamburg
JED	Jedda
KRT	Khartoum
KUF	Samara
LEJ	Leipzig
LHR	London-Heathrow
LOS	Laos
MUC	Munich
MXP	Milano-Malpensa
NUE	Nuremberg
PHC	Port Harcour
STR	Stuttgart
TXL	Berlin-Tegel
VIE	Wien
ZRH	Zurich

Appendix 2

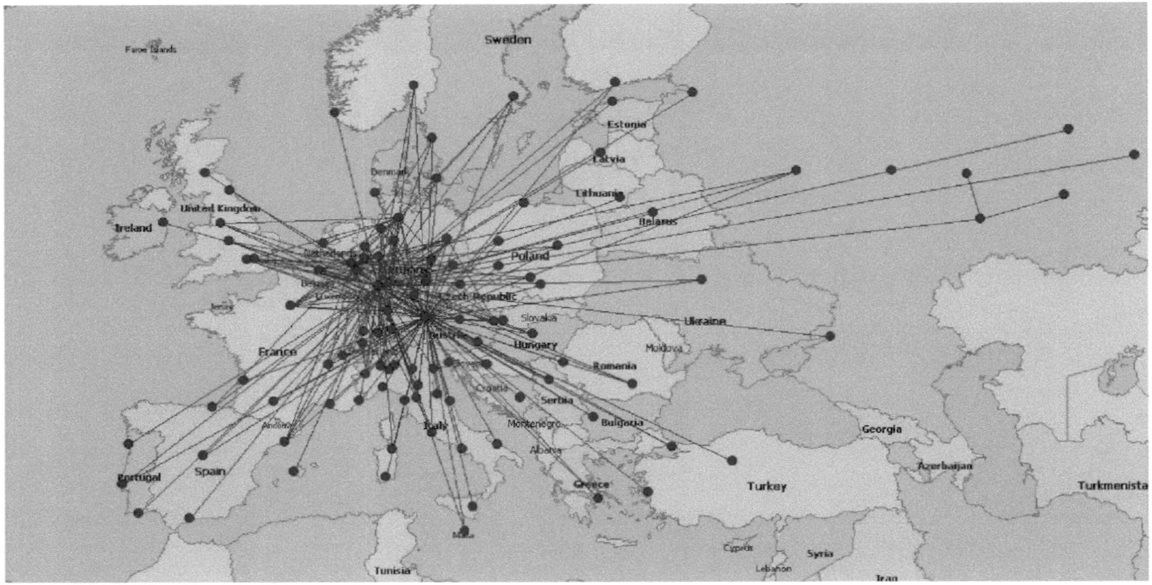

Source: OAG [34].

Fig. 3 Lufthansa medium haul network in Europe (Network A1)

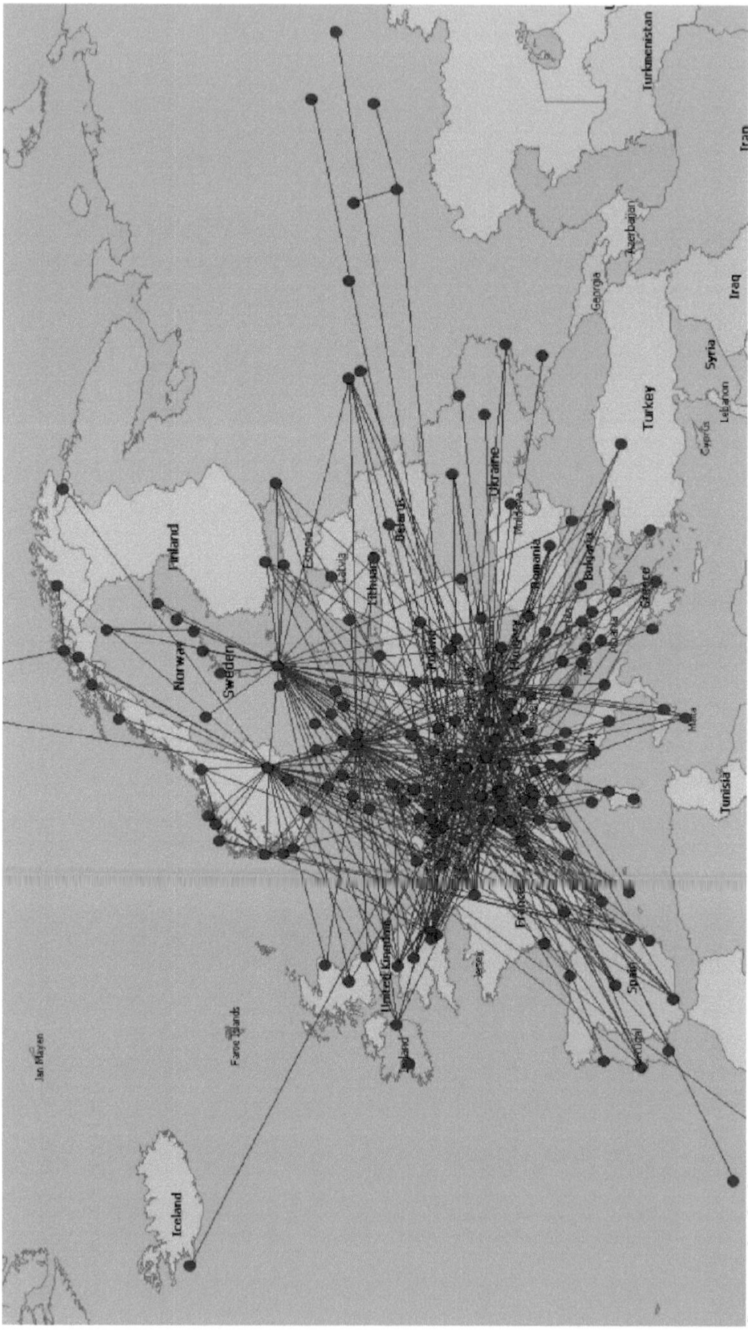

Source: OAG [34].

Fig. 4 STAR Alliance medium haul network in Europe (Network B1)

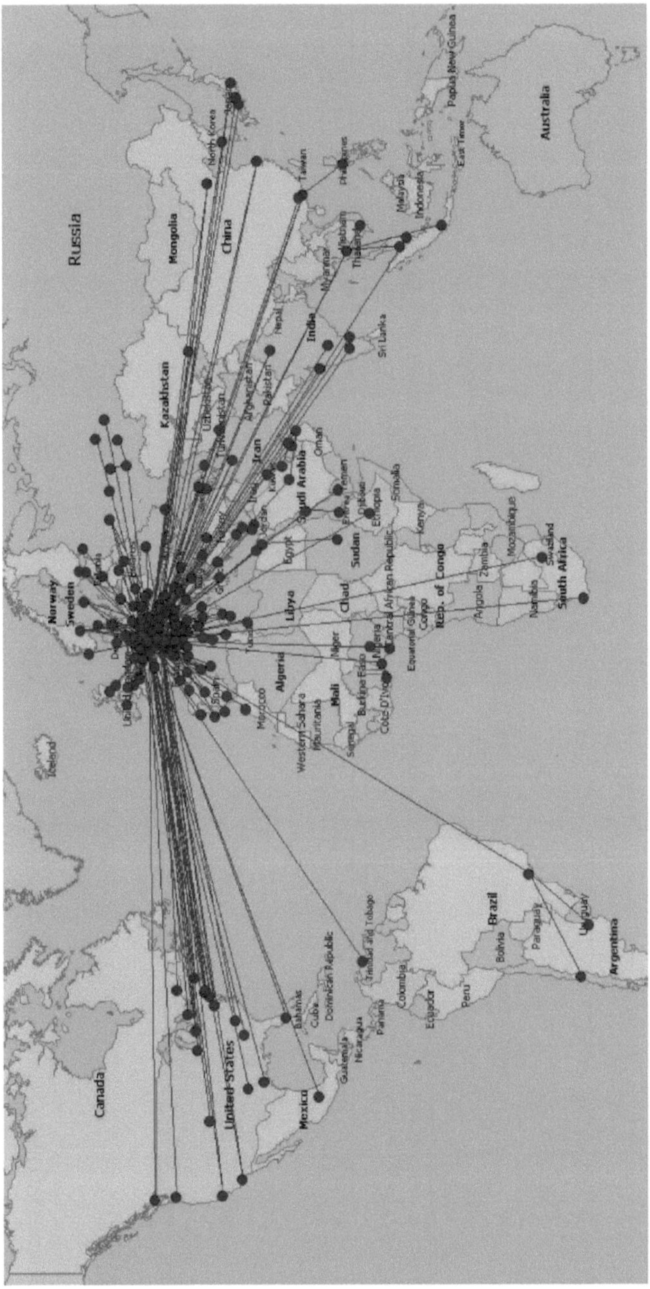

Source: OAG [34].

Fig. 5 Lufthansa global network (Network A2)

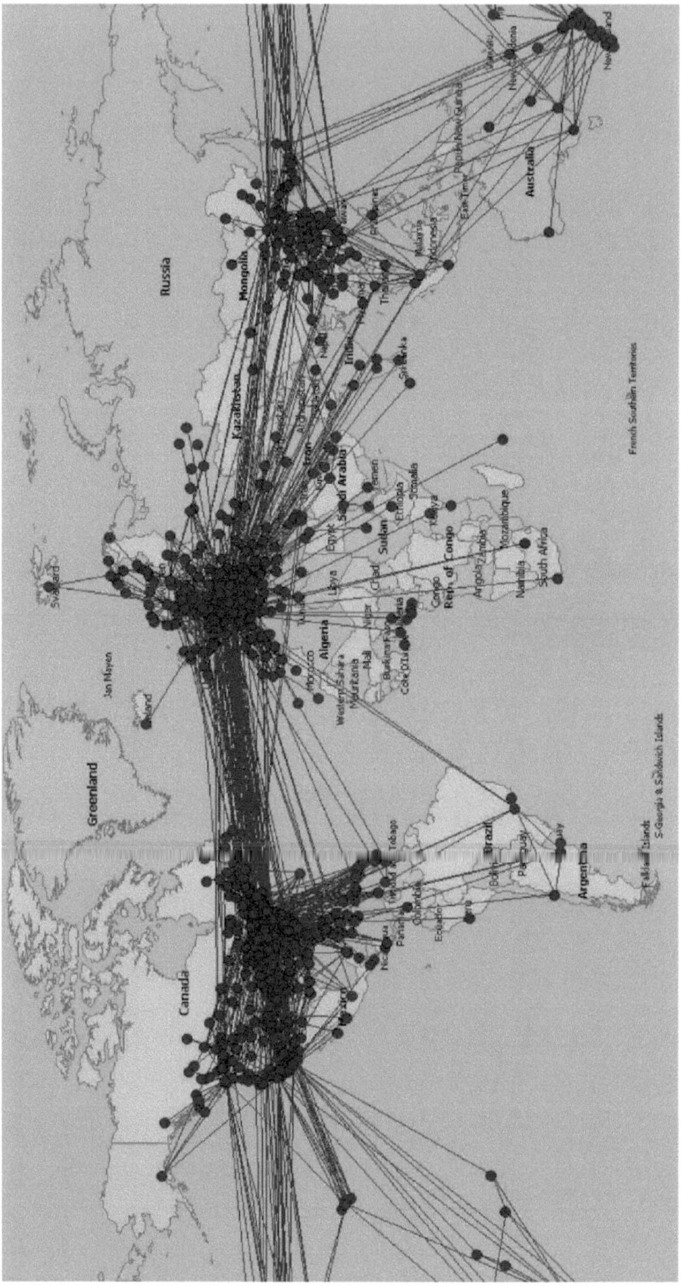

Source: OAG [34].

Fig. 6 STAR Alliance global network (Network B2)

References

1. Adamic LA (2000) Zipf, power-laws, and Pareto – ranking tutorial. http://www.hpl.hp.com. Accessed 16 Apr 2007
2. Albert R, Barabási A-L (2002) Statistical mechanics of complex networks. Rev Mod Phys 74:47–97
3. Alderighi M, Cento A, Nijkamp P, Rietveld P (2007) Assessment of new hub-and-spoke and point-to-point airline network configurations. Transp Rev 27:529–549
4. Barabási LA, Albert R (1999) Emerging of scaling in random networks. Science 286:509–512
5. Barabási A-L and Bonabeau E (2003) Scale-Free Networks, *Scientific American* 288(5), pp. 50–59
6. Barabási LA, Oltvai ZN (2004) Network's biology: understanding the cell's functional organization. Nat Rev Genet 5:101–113
7. Barrat A, Barthélemy M, Pastor-Satorras R, Vespignani A (2004) The architecture of complex weighted networks. Proc Natl Acad Sci USA 101(11):3747–3752 http://www.pnas.org/cgi/doi//10.1073/pnas.0400087101
8. Barthélemy M (2003) Crossover from scale-free to spatial networks. Europhys Lett 63:915–921
9. Boccaletti S, Latora V, Moreno Y, Chavez M, Hwang D-U (2006) Complex networks: structure and dynamics. Phys Rep 424:175–308
10. Bootsma PD (1997) Airline flight schedule development; analysis and design tools for European hinterland hubs. University of Twente, Utrecht
11. Boschma RA (2005) Proximity and innovation. a critical assessment. Reg Stud 39:61–74
12. Bossomaier TRJ and Green DG (2000) Introduction. In TRJ Bossomaier and DG Green (eds), *Complex Systems* (pp. 2–9). Cambridge (UK): Cambridge University Press
13. Bowen J (2002) Network change, deregulation, and access in the global airline industry. Econ Geogr 78:425–439
14. Burghouwt G, de Wit J (2003) The temporal configuration of European airline networks. University of Montreal, Publication AJD-74
15. Button K, Stough R (2000) Air transport networks: theory and policy implications. Edward Elgar, Cheltenham
16. Caves DW, Christensen LR, Tretheway MW (1984) Economics of density versus economies of scale: why trunks and local service airline costs differ. Rand J Econ 15:471–489
17. Cento A (2006) Challenge to the airline industry: emergence of carriers and low cost carriers. PhD Thesis, VU University, Amsterdam
18. Dennis NPS (1998) Competition between hub airports in Europe and a methodology for forecasting connecting traffic. In: 8th World Conference on Transport Research, Antwerp
19. Erdös P, Rényi A (1959) On random graphs I. Publ Math 6:290–297
20. Freeman LC (1977) A set of measures of centrality based on betweenness. Sociometry 40:35–41
21. Freeman LC (1979) Centrality in social networks: a conceptual clarification. Soc Netw 1:215–239
22. Frenken K (2006) Innovation, Evolution and Complexity Theory. Edward Elgar, Cheltenham
23. Gorman SP (2005) Networks, security and complexity. Edward Elgar, Cheltenham
24. Gorman SP, Kulkarni R (2004) Spatial small worlds. Environ Plan B 31:273–296
25. Harary F (1969) *Graph Theory*. Addison-Wesley, Reading, MA
26. Hinloopen E, Nijkamp P (1990) Qualitative multiple criteria choice analysis: the dominant regime method. Qual Quant 24:37–56
27. Jeong H (2003) Complex scale-free networks. Phys A 321:226–237
28. Lijesen MG (2004) Adjusting the Herfindahl index for close substitutes: an application to pricing in civil aviation. Transp Res E 40:123–134
29. Lufthansa (2005) Annual report. http://konzern.lufthansa.com/en/html/ueber_uns/swiss/index.html

30. McShan WS (1986) An economic analysis of hub-and-spoke routing strategy in the airline industry. PhD Thesis, Northwestern University
31. Newman MEJ (2003) A measure of betweenness centrality based on random walks. arXiv:cond-mat/0309045
32. Nijkamp P, Reggiani A (1992) Interaction, evolution and chaos in space. Springer, Berlin
33. O' Kelly EM, Miller HJ (1994) The hub network design problem: an overview and synthesis. J Transp Geogr 2:31–40
34. OAG (2006) Official Airline Guides. Worldwide flight guide. Reed Publications. http://www.oag.com
35. Patuelli R (2007) Regional labour markets in Germany: statistical analysis of spatio-temporal disparities and network structures. PhD Thesis, VU University Amsterdam
36. Reggiani A, Nijkamp P (eds) (2006) Spatial dynamics, networks and modelling. Edward Elgar, Cheltenham
37. Reggiani A, Vinciguerra S (2007) Network connectivity models: an overview and empirical applications. In: Friesz T (ed) Network science, nonlinear science and infrastructure systems. Springer, New York
38. Reynolds-Feighan AJ (1994) EC and US air freight markets: network organisation in a deregulated environment. Transp Rev 14:193–217
39. Reynolds-Feighan AJ (1998) The impact of US airline deregulation on airport traffic patterns. Geogr Anal 30:234–253
40. Reynolds-Feighan AJ (2001) Traffic distribution in low-cost and full service carrier network in the US air transport market. J Air Transp Manag 7:265–275
41. Rietveld P, Brons M (2001) Quality of hub-and-spoke networks; the effects of timetable coordination on waiting time and rescheduling time. J Air Transp Manag 7:241–249
42. Schintler LA, Gorman SP, Reggiani A, Patuelli R, Nijkamp P (2005a) Small-world phenomena in communication networks: a cross-Atlantic comparison. In: Reggiani A, Schintler LA (eds) Methods and models in transport and telecommunications: cross atlantic perspectives. Springer, Berlin
43. Schintler LA, Gorman SP, Reggiani A, Patuelli R, Gillespie A, Nijkamp P, Rutherford (2005b) J Complex network phenomena in telecommunication systems. Netw Spat Econ 5:351–370
44. Toh RS, Higgins RG (1985) The impact of hub-and-spoke network centralization and route monopoly on domestic airline profitability. Transp J 24:16–27
45. Veldhuis J, Kroes E (2002) Dynamics in relative network performance of the main European hub airports. In: European Transport Conference, Cambridge
46. Waters N (2006) Network and nodal indices. Measures of complexity and redundancy: a review. In: Reggiani A, Nijkamp P (eds) Spatial dynamics, networks and modelling. Edward Elgar, Cheltenham
47. Watts DJ, Strogatz SH (1998) Collective dynamics of small world networks. Nature 393:440–442

On Certain Graph Theory Applications

Klavdija Kutnar and Dragan Marušič

Abstract Recent developments with regards to certain open problems in fullerenes and torusenes together with the methods used to solve these problems in a graph-theoretic context are presented. The stability of fullerenes and torusenes via the number of perfect matching and Hamiltonian cycles is considered in more detail.

1 Introductory Remarks

Graph theory has come into its own through many important contributions to a wide range of fields (such as economy, statistics, interconnection networks, biology, chemistry, etc.) and is now one of the fastest-growing areas in discrete mathematics and computer science. It turns a real-world problem into a mathematical one.

For instance, designing computer interconnection networks based on graph theory gives more efficient networks (see [2]). To model computer interconnection networks with graphs topologies, the following correspondences are used: graph vertices model computer processors; and graph edges model connections between individual processors. Some such networks are two-dimensional meshes and tori [16, 32], hypercubes [15, 21], cubic symmetric graphs [3] and others [8].

In this paper we restrict ourselves to graph theory applications to chemistry. In particular, we survey some recent developments with regards to certain open problems in fullerenes and torusenes together with the methods used to solve these problems in a graph-theoretic context.

K. Kutnar
University of Primorska, FAMNIT, Glagoljaška 8, 6000 Koper, Slovenia
klavdija.kutnar@upr.si

D. Marušič
University of Primorska, FAMNIT, Glagoljaška 8, 6000 Koper,
Slovenia and University of Ljubljana, IMFM, Jadranska 19, 1000 Ljubljana, Slovenia
dragan.marusic@upr.si

A.K. Naimzada et al. (eds.), *Networks, Topology and Dynamics*,
Lecture Notes in Economics and Mathematical Systems 613,

Throughout this paper graphs are finite, undirected and connected, unless specified otherwise. For adjacent vertices x and y in X, we write $x \sim y$ and denote the corresponding edge by xy. Given a graph X we let $V(X)$ and $E(X)$ be the vertex set and the edge set of X, respectively. If $S \subseteq V(X)$, then $S^c = V(X) \setminus S$ denotes the complement of S and the graph induced on S is denoted by $X[S]$. The line graph $L(X)$ of X is a graph with vertex set $E(X)$ and any two vertices of $L(X)$ are adjacent if and only if their corresponding edges share a common endvertex in X. A graph is 3-edge-connected if three edges are needed to be removed in order to disconnect the graph. A graph X is said to be *vertex-transitive*, *edge-transitive* and *arc-transitive* if its automorphism group Aut$X$ acts transitively on $V(X)$, $E(X)$ and $A(X)$, respectively. A *Hamiltonian cycle* in a graph is a cycle passing through all vertices. A *Hamiltonian path* in a graph is a path passing through all vertices. A *perfect matching* in a graph is a set of disjoint edges covering every vertex of the graph.

2 Graph Theory Applications to Chemistry

With the revolution in carbon chemistry and physics that stemmed from the discovery of fullerenes, graph theory has come to new prominence in chemistry, as a means of obtaining systematic qualitative information about isomerism, shape, stability, electronic structure and reactivity of these previously unsuspected forms of the most well-studied element in the Periodic Table. An exciting spin-off from this frantic activity (over 10,000 papers in chemistry and physics journals in the past decade) has been the upsurge of mathematical interest in the graphs and polyhedra related to the fullerenes. This in turn has suggested new directions for chemical application, for example, the use of independence numbers as indicators of stable addition patterns in fullerene compounds.

A main challenge in this multidisciplinary area is the relation between graph invariants and real-world applications. Many classical graph invariants have a role in chemistry. For example, qualitative theories of stability invoke the Kekulé number or the number of perfect matchings, and in the fullerene area counting of pentagon–pentagon adjacencies gives a first filter to separate stable from unstable isomers.

In Sect. 2.1 some open problems with regards to fullerenes and torusenes from a graph-theoretic viewpoint are presented. Section 2.2 deals with existence of Hamiltonian cycles, and existence of Hamiltonian paths in fullerenes.

2.1 *Fullerenes and Torusenes*

Fullerenes, discovered in 1985, are all-carbon "sphere"-shaped molecules with trivalent polyhedral skeletons, having pentagonal faces and all other hexagonal faces. This important class of molecules is a basis of thousands of patents for a broad range of pharmaceutical, electronic and other commercial applications [4, 28]. The most stable fullerene is Buckminsterfullerene that consists of 60 carbon atoms. It was obtained for the first time in the graphite vaporization experiment [19] and spectroscopic evidences of his structure are given in [20]. R. F. Curl, H. Kroto and R. E. Smalley received the Nobel Prize for this discovery. From a mathematical point of view, *fullerene graphs*, in short *fullerenes*, correspond to cubic (trivalent) and 3-edge-connected planar graphs which have, in view of the well known Euler formula for connected planar graphs, 12 pentagons and the remaining faces are hexagons. (Recall that by Euler's formula the number of faces of a connected planar graph X in its planar embedding is equal to $|E(X)|+|V(X)|-2$.) It is therefore not surprising that many questions about the chemistry of fullerenes together with the methods used to answer these questions find their natural environment in a graph-theoretic context.

In a classical paper by Grünbaum and Motzkin [13] the existence of fullerenes on n vertices was established for all even $n \geq 20$ except for $n = 22$. The smallest fullerene is the dodecahedron. For a systematic introduction on fullerene graphs we refer the reader to the monograph [11].

One of many open problems with regards to fullerenes cast in a graph-theoretic language, concerns the number of perfect matchings in a fullerene. The concept of perfect matchings corresponds to the notion of Kekulé structure in chemistry [31] and plays a very important role in analysis of benzenoid systems, fullerenes and other carbon molecules [14, 29]. The existence of a perfect matching in every fullerene follows from a classical result of Petersen that every connected cubic graph with no more than two cut-edges has a perfect matching. However, in [35] it was shown that a fullerene of order n has at least $\lceil 3(n+2)/4 \rceil$ perfect matchings. Recently it was shown that for all sufficiently large n there is a fullerene of order n that has exponentially many perfect matchings in terms of the number of vertices (see [10]).

The *leapfrog-fullerene* $\mathrm{Le}(F)$ is obtained from a fullerene F by performing the so called tripling (leapfrog transformation) which consists in the truncation of the dual of F. Hence $\mathrm{Le}(F) = \mathrm{Trun}(\mathrm{Du}(F))$ (see [7]).

In a particular situation a perfect matching in a cubic graph arises as a complement of a Hamiltonian cycle. Therefore, knowing that a fullerene contains a Hamiltonian cycle, and hence possibly lots of such cycles, is a useful information when analyzing stability properties of fullerenes. This brings us to the next open problem on fullerene, concerning existence of a Hamiltonian cycle in an arbitrary fullerene, which will be dealt with in more detail in Sect. 2.2.

The synthesis of fullerenes prompted a natural question as to whether similar carbon structures exist on other closed surfaces. Apart from the sphere only three other surfaces are possible: the torus, the Klein bottle and the projective plane [5]. Of these only torus-shaped graphite-like carbon structures, known as *torusenes*, have received enough research attention [6, 7, 12, 17, 22] in view of their direct experimental relevance, [23] but also since the other two types are meaningless from a chemical point of view as they do not admit a realization in the Euclidean 3-space.

Only two fullerenes, notably the dodecahedron fullerene C_{20} and the buckminsterfullerene C_{60} (that is, the leapfrog-fullerene of C_{20}), are vertex-transitive admitting a transitive group of automorphisms and thus a high degree of symmetry. The situation with torusenes is completely different since all torusenes are vertex-transitive.

From a graph-theoretic viewpoint, a torusene is a cubic (trivalent) graph, embeddable onto the torus in such a way that each face is a hexagon. The description uses three parameters p, q and t, explained hereafter. A torusene $H(p,q,t)$ is obtained from pq hexagons stacked in an $p \times q$ parallelogram where the two sides are glued together in order to form a tube, and then the top boundary of the tube is glued to the bottom boundary of the tube in order to form a torus. At this last stage the top part is rotated by t hexagons before the actual gluing takes place. In Fig. 1 the $H(8,1,2)$ torusene realization of the Moebius–Kantor graph is shown. Note that this graph admits also an $H(4,2,1)$ torusene realization [25].

Since every torusene is vertex-transitive the existence of a Hamiltonian cycle in an arbitrary torusene follows from results proved by Thomassen [30]. Perfect matchings in torusenes have also be investigated (see [17]). However, by Voorhoeve's result [34], every bipartite torusene of order n has at least $(4/3)^{n/2}$ different perfect matchings.

In chemistry, a molecule is said to be *chiral* if it is not superimposable on its mirror image regardless of how it is contorted. (A graph X embedded onto the torus is *chiral* if Aut$X$ contains no automorphism which "flips" an edge of a face while preserving this face.) Since biomolecules, such as proteins and enzymes, in living organisms are chiral, therapeutics possessing this property significantly enhance the potential impact of a drug product. Consequently, research involving the concept of chirality is very important.

In the case of an arc-transitive torusene the automorphism group of the underlying cubic graph must contain, by the classical theorem of Tutte [33], a subgroup acting regularly on the corresponding set of 1-arcs. In other words, vertex stabilizers are isomorphic to $\mathbb{Z}_3$ and the graphs are therefore without reflections. Such a graph is referred to as *1-regular* provided this subgroup coincides with the full automorphism group. In [22] the following natural question with regards to chirality of torusenes $H(p,q,t)$ was posed.

Problem 1. [22, Problem 1.1] For which parameters p, q and t there exists a chiral torusene $H(p,q,t)$?

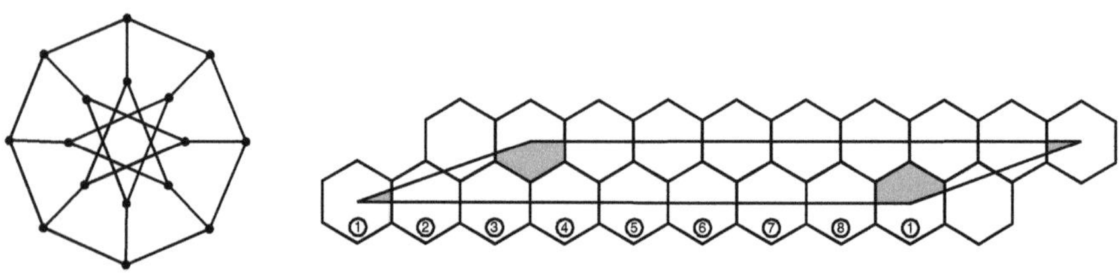

Fig. 1 The $H(8,1,2)$ torusene of the Moebius–Kantor graph

For example, the $H(3,1,1)$ torusene of the complete bipartite graph $K_{3,3}$ on six vertices is achiral since it possesses reflective symmetry. On the other hand, the $H(13,1,3)$ torusene corresponding to the unique 1-regular graph on 26 vertices possesses no reflective symmetry, and is thus chiral. In particular, Problem 1 has definitely a positive answer for a triple (p,q,t) whenever there exists a cubic 1-regular graph having an $H(p,q,t)$ torusene realization. A partial answer to Problem 1 follows from a complete classification of cubic arc-transitive Cayley graphs of dihedral groups [25]. Furthermore, in [22] from a chiral torusene $H(n,1,r)$ a chiral torusene $H(pn,p,pr)$, where p is a prime coprime with n, was constructed.

The chirality has also been studied in the framework of fullerenes. For detailed information on this topic we refer the reader to [18].

2.2 Hamiltonian Cycles and Paths in Fullerenes

A long standing conjecture suggests that an arbitrary fullerene contains a Hamiltonian cycle (see [26]). Not much progress has been made with regards to this conjecture. It has been verified for graphs on at most 176 vertices (see [1]). To the best of our knowledge, the following proposition is the most general result in this topic.

Proposition 1. [24, Theorem 1.1] *Let F be a fullerene. Then the leapfrog-fullerene* $\mathrm{Le}(F)$ *has a Hamiltonian cycle if $|F|$ (and thus also $|\mathrm{Le}(F)|$) is congruent to* 2 *modulo* 4, *and contains a long cycle missing out only two adjacent vertices (and thus contains a Hamiltonian path) if $|F|$ (and thus also $|\mathrm{Le}(F)|$) is divisible by* 4.

The method used in the proof of Proposition 1 to construct Hamiltonian cycles and paths in an arbitrary leapfrog-fullerene consists in identifying, in a given fullerene F, a subset of vertices inducing a tree, the complement of which has as few edges as possible. Since by [9, Theorem 2] fullerenes are cyclically 5-edge-connected graphs (a graph is *cyclically k-edge-connected* if at least k edges need to be removed in order to disconnect the graph into two components each containing a cycle), [27, Théorème 5] implies that the complement has either no edge when $|F| \equiv 2 \pmod 4$ or a single edge when $|F| \equiv 0 \pmod 4$. Hence, in the corresponding leapfrog-fullerene $\mathrm{Le}(F)$ this tree gives rise to a tree of faces the boundary of which is either a full Hamiltonian cycle when $|F| \equiv 2 \pmod 4$ and a long cycle missing only two (adjacent) vertices when $|F| \equiv 0 \pmod 4$.

Here we give examples of two leapfrog-fullerenes, arising respectively from fullerenes on 24 vertices and 26 vertices. The above mentioned method gives in the respective leapfrog-fullerenes a long cycle missing out two adjacent vertices (and therefore a Hamiltonian path) in C_{72} and a Hamiltonian cycle in C_{78}.

Example 1. In the left picture of Fig. 2 we show a fullerene on 24 vertices with an identified maximal induced tree with 17 vertices whose complement is a set of 7 vertices inducing a single edge. In the right picture the corresponding tree of faces whose boundary gives rise to a long cycle missing only two adjacent vertices in its leapfrog-fullerene C_{72} is shown.

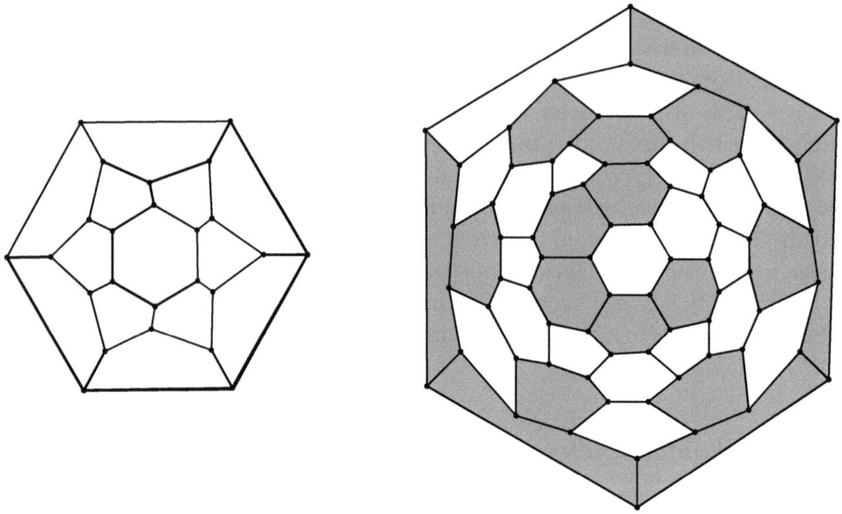

Fig. 2 A fullerene on 24 vertices and its leapfrog-fullerene with a Hamiltonian path

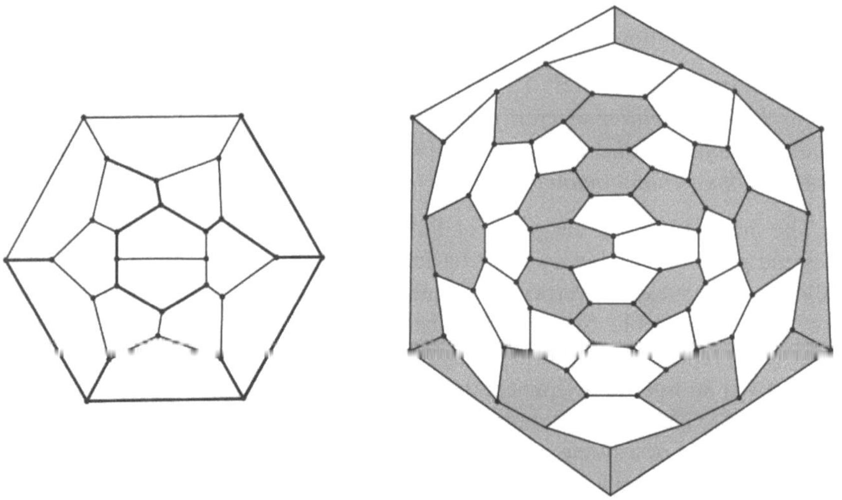

Fig. 3 A fullerene on 26 vertices and its leapfrog-fullerene with a Hamiltonian cycle

Example 2. In the left picture of Fig. 3 we show a fullerene on 26 vertices with an identified maximal induced tree with 19 vertices whose complement is an independent set of 7 vertices. In the right picture we show the corresponding tree of faces whose boundary gives rise to a Hamiltonian cycle in the corresponding leapfrog-fullerene on 78 vertices.

A natural question is whether this approach could be further explored, say in the context of other classical fullerenes transformations, such as chamfering (quadru-

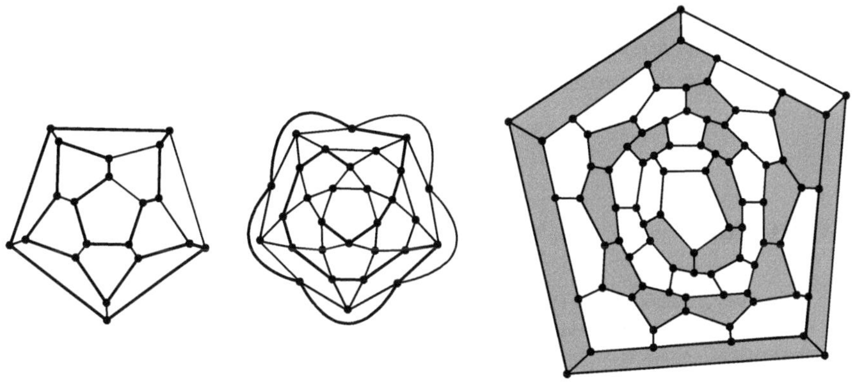

Fig. 4 The dodecahedron, its line graph and its chamfering-fullerene with a cycle of length 78

pling) and capra (septupling) transformations [18]. The theorem below shows the existence of a Hamiltonian path in certain chamfering-fullerenes. The chamfering-fullerene $\mathrm{Q}(F)$ is obtained from a fullerene F by performing the so called quadrupling transformation which consists in the truncation of the dual of the line graph $L(F)$ of F. Hence, $\mathrm{Q}(F) = \mathrm{Trun}(\mathrm{Du}(L(F)))$ (in other words, $\mathrm{Q}(F)$ is obtained from $L(F)$ by performing the leapfrog transformation).

Theorem 1. *Let F be a fullerene of order n that contains a Hamiltonian cycle. Then $\mathrm{Q}(F)$ contains a long cycle missing out only two nonadjacent vertices lying on a common hexagon. Moreover, $\mathrm{Q}(F)$ contains a Hamiltonian path.*

Proof. Let F be a fullerene of order n that contains a Hamiltonian cycle C. Note that the chamfering-fullerene $\mathrm{Q}(F)$ is of order $4n$.

Clearly, C gives rise to a cycle C' of length n in the line graph $L(F)$ whose complement is an independent set of $n/2$ vertices (recall that $L(F)$ is of order $3n/2$). Now let $v \in V(L(F)) \cap C'$. Then $L(F)[C' \setminus \{v\}]$ is a path whose complement $(C' \setminus \{v\})^c$ induces a graph with two incident edges (therefore two vertices of valency 1 and a vertex v of valency 2 exist in $L(F)[(C' \setminus \{v\})^c]$). Moreover, since every edge in $L(F)$ lies on a triangle the set $C' \setminus \{v\}$ is a maximal cyclically stable subset of $V(L(F))$ and $|C' \setminus \{v\}| = n-1$.

Now the path $L(F)[C' \setminus \{v\}]$ gives rise to a path of faces in the chamfering-fullerene $\mathrm{Q}(F)$ whose boundary is a cycle $\mathcal{C}$ of length $4n-2$ in $\mathrm{Q}(F)$. This may be seen, for example, by counting the number of vertices on the boundary of this path of faces. First, the boundary is clearly a cycle missing out two nonadjacent vertices, say v' and v'', lying on a hexagon corresponding to the vertex v in $L(F)$. And second, since all faces in this path are hexagonal the length of this cycle is $6(n-1) - 2(n-2) = 4n-2$. Finally, since v' and v'' are adjacent to neighboring vertices of $\mathcal{C}$ it follows that $\mathrm{Q}(F)$ contains a Hamiltonian path (see also Example 3).

Example 3. In the left picture of Fig. 4 we show the dodecahedron fullerene with a Hamiltonian cycle. In the middle picture we show the line graph of the dodecahedron fullerene, with an identified maximal induced tree with 19 vertices whose complement is a set of 11 vertices inducing a graph with two incident edges. In the right picture the corresponding tree of faces whose boundary gives rise to a long cycle missing only two vertices in the chamfering-fullerene C_{80} of the dodecahedron is shown.

References

1. Aldred REL, Bau S, Holton DA, McKay BD (2000) Nonhamiltonian 3-connected cubic planar graphs. SIAM J Discrete Math 13:25–32
2. Bermond JC, Delorme C, Quisquater JJ (1985) Strategies for interconnection networks: some methods from graph theory. J Parallel Distrib Comput 3:433–449
3. Boršnik U, Kutnar K, Marušič D, Janežič D (2008) Interconnection networks for parallel molecular dynamics based on hamiltonian cubic symmetric topology. J Math Chem, doi: 10.1007/s10910-008-9412-5 (in press)
4. Da Ross T, Prato M (1999) Medicinal chemistry with fullerenes and fullerene derivatives. Chem Commun 8:663–669
5. Deza M, Fowler PW, Rossat A, Rogers KM (2000) Fullerenes as tilings of surfaces. J Chem Inf Comput Sci 9:550–558
6. Diudea MV (2002) Topology of naphthylenic tori. Phys Chem Chem Phys 19:4740–4746
7. Diudea MV, John PE, Graovac A, Primorac M, Pisanski T (2003) Leapfrog and related operations on toroidal fullerenes. Croat Chem Acta 76:153–159
8. Dobravec T, Robič B, Žerovnik J (2003) Permutation routing in double-loop networks: design and empirical evaluation. J Syst Arch 48:387–402
9. Došlić T (2003) Cyclical connectvity of fullerene graphs and $(k,6)$-cages. J Math Chem 33:103–112
10. Došlić T (2007) Fullerene graphs with exponentially many perfect matchings. J Math Chem 41:183–192
11. Fowler PW, Manolopoulos DE (1995) An atlas of fullerenes. Clarendon, Oxford
12. Graovac A, Plavšić D, Kaufman M, Kirby EC, Pisanski T (2000) Application of the adjacency matrix eigenvectors method to geometry determination of toroidal carbon molecules. J Chem Phys 113:1925–1931
13. Grünbaum B, Motzkin TS (1963) The number of hexagons and the simplicity of geodesics on certain polyhedra. Can J Math 15:744–751
14. Gutman I, Polansky O (1986) Mathematical concepts in organic chemistry. Springer, Berlin
15. Hodoščeк M, Borštnik U, Janežič D (2002) CROW for large scale macromolecular simulations. Cell Mol Biol Lett 7:118–119
16. Horiguchi S, Ooki T (2000) Hierarchical 3D-torus interconnection network. In: Proc. IEEE Intl Symp. on Parallel Architectures, Algorithms and Networks (ISPAN2000). IEEE CS Press: Richardson, TX, USA
17. John PE (1998) Kekulé count in toroidal hexagonal carbon cages. Croat Chem Acta 71:435–447
18. King RB, Diudea MV (2006) The chirality of icosahedral fullerenes: a comparison of the tripling (leapfrog), quadrupling (chamfering), and septupling (capra) trasformations. J Math Chem 93:567–604
19. Kratschmer W, Lamb L, Fostiropoulos K, Huffman D (1990) Solid C_{60}: a new form of carbon Nature 347:354–358
20. Kroto HW, Heath JR, O'Brien SC, Curl RF, Smalley RE (1985) C_{60}: Buckminsterfullerene. Nature 318:162–163
21. Kutalek V, Dvorak V (2005) On complexity of collective communications on a fat cube topology. J Univ Comp Sci 11:944–961

22. Kutnar K, Malnič A, Marušič D (2005) Chirality of toroidal molecular graphs. J Chem Inf Model 45:1527–1535
23. Liu J, Dai HJ, Hafner JH, Colbert DT, Smalley DTRE, Tans ST, Dekker C (1997) Fullerene crop circles. Nature 385:780-781
24. Marušič D (2007) Hamilton cycles and paths in fullerenes. J Chem Inf Model 47:732–736
25. Marušič D, Pisanski T (2000) Symmetries of hexagonal molecular graphs on the torus. Croat Chem Acta 73:969–981
26. Myrvold WJ (2006) Investigating conjectures about fullerenes. In: "Applications of graph theory to chemistry" minisymposium. 2006 SIAM Discrete Mathematics, June 25–28, 2006. University of Victoria, Canada
27. Payan C, Sakarovitch M (1975) Ensembles cycliquement stables et graphes cubiques. Cahiers du Centre D'etudes de Recherche Operationelle 17:319–343
28. Planeix JM, Coustel N, Coq B, Brotons V, Kumbhar PS, Dutartre R, Geneste P, Bernier P, Ajayan PM (1994) Application of carbon nanotubes as supports in heterogenous catalysis. J Am Chem Soc 116:7935–7936
29. Randić M (2003) Aromaticity of polycyclic conjugated hydrocarbons. Chem Rev 103: 3449–3606
30. Thomassen C (1991) Tilinigs of the torus and the Klein bottle and vertex-transitive graphs on a fixed surface. Trans Am Math Soc 323:605–635
31. Trinajstić N (1992) Chemical graph theory. CRC, Boca Raton
32. Trobec R (2000) Two-dimensional regular d-meshes. Parallel Comput 26:1945–1953
33. Tutte WT (1947) A family of cubical graphs. Proc Camb Phil Soc 43:459–474
34. Voorhoeve M (1979) A lower bound for the permanents of certain $(0;1)$-matrices. Indagationes Mathematicae 41:83–86
35. Zhang H, Zhang F (2001) New lower bound on the number of perfect matchings in fullerene graphs. J Math Chem 30:343–347

MIX
Papier aus verantwortungsvollen Quellen
Paper from responsible sources
FSC® C105338

If you have any concerns about our products,
you can contact us on
ProductSafety@springernature.com

In case Publisher is established outside the EU,
the EU authorized representative is:
Springer Nature Customer Service Center GmbH
Europaplatz 3, 69115 Heidelberg, Germany

Printed by Libri Plureos GmbH
in Hamburg, Germany